Modelling with the Master Equation

Günter Haag

Modelling with the Master Equation

Solution Methods and Applications in Social and Natural Sciences

Springer

Günter Haag
University of Stuttgart
Institute of Theoretical Physics II
Germany

ISBN 978-3-319-60299-8 ISBN 978-3-319-60300-1 (eBook)
DOI 10.1007/978-3-319-60300-1

Library of Congress Control Number: 2017945902

Printed on acid-free paper

This Springer imprint is published by Springer Nature
The registered company is Springer International Publishing AG
The registered company address is: Gewerbestrasse 11, 6330 Cham, Switzerland

This book is dedicated to Eva, Manuel and Stephan and all the beautiful outcomes based on chaotic decisions.

Foreword

All of us have to make decisions on our actions, based on our expectations of the future.

This holds true for individuals as well as for institutions such as companies, city councils, state governments, and international agencies. But, as we all know, we live in an ever increasingly complex world with all its uncertainties. Can we find any guidelines how to plan and act? May be even enviously, we look at physics with its high predictive power and technical realizations which allow us to send a rocket to distant planets with high precision. Thus, it is certainly tempting to transfer insights gained in physics to the treatment of social, socio-economic or economic processes. The great predictive power of physics is due to its use of mathematics. In my view, mathematics and its way to think will become more and more important also in areas I just have mentioned.

An important step in transferring concepts from physics to sociology has been done in 1972 by my late friend and colleague Wolfgang Weidlich who could draw a close analogy between the formation of public opinion and order–disorder transitions in magnetism. The mathematical vehicle he used is the master equation that I will discuss below. The author of this book, Günter Haag, was one of Weidlich's prominent students and co-workers. Weidlich and Haag published a monograph on *"Concepts and models of a quantitative sociology: The dynamics of interacting populations"* (1983). As I know from Weidlich, Haag contributed considerably because of both his mathematical skills and his openness for practical applications. This has led, in particular, to the Weidlich–Haag model and to important further contributions e.g. to decision theory and regional planning by very concrete studies, which are also part of this book. But why use the master equation? It provides us with an excellent means to deal with uncertainties. In fact, it allows us to calculate probabilities of future states and thus to study various scenarios of developments. These developments result from purely deterministic relationships—so to speak, stringent laws, and chance events, on which we can make only guesses. The master equation combines these effects. In his book, Haag introduces the basic concepts of probability theory in an easy to understand way, derives the master equations and presents numerous explicit examples of its application in physics, chemistry and especially socio-economics. This book is an important contribution to Synergetics that deals quite generally with the self-organized formation of structures in Nature and Society. Haag's book addresses many problems of great public interest, such as migration and regional planning. I am sure that it will become an important reading for students, professors and practitioners.

Stuttgart
March 2017

Hermann Haken

Preface

Since the last decades, the modelling potential of the Master equation has been widely ignored, especially in the social sciences. Possibly, because the mathematics looks so complicated and a general introduction to the framework of the Master equation and its application was not available. Of course, one has to learn about the solution formalisms and the theory of the Master equation to apply the framework. The best way to do this is often to consider examples and applications of the Master equation in different fields and to learn to apply by doing.

The Master equation provides a general framework for model building in different disciplines like physics, chemistry and biology as examples in the natural sciences and economy, sociology, psychology and geography in the social sciences.

During the last decades, a rather big set of mathematical solution methods for the Master equation have been developed. It depends on the system under consideration which solution method seems to be most appropriate to apply. Therefore, it is one aim of this book not only to present different mathematical solution methods but also to show their potential in case of practical examples.

The book is based on courses of mine in the field of interdisciplinary research held at the University of Stuttgart during the last two decades. And, in fact, some examples of the book are related to those lectures and courses. But some applications and research issues are based on consultancy work of STASA (Steinbeis Applied Systems Analysis GmbH) which I founded in 1995.

To make the book easier to read, it is subdivided into three Parts. Part I comprises Chaps. 1–4 and is dealing with some statistical fundamentals, the derivation of the Master equation, the Fokker–Planck equation and other relevant statistical issues. In addition, solution methods of the Master equation including some rather new solution tools for a group of special problems are presented. Part I is rather technical and can be used as a toolbox.

However, benchmarking of different solution methods is important to learn about the advantages of the Master equation framework compared with other modelling approaches.

Therefore, in Part II and Part III of the book, we do not only apply the Master equation framework to different case studies but also compare it with other solution methods. A set of examples out of the field of physics, chemistry, population dynamics, dynamic decision theory, opinion formation and urban and regional dynamics are treated. However, the main focus of the examples is related to the social sciences. The examples underline the interdisciplinary modelling potential of the Master equation approach.

Of course, it is not necessary to follow in Part II and Part III one chapter after the other. Each chapter can be understood independently of the others. But sometimes it is helpful to compare applications out of different fields, Especially, since different methods of solution are applied and compared.

Since the book is written for graduate students, researchers and professionals, it was my aim to perform all mathematical steps which are relevant to come step by step to the final solution. Furthermore, it was my intention to introduce the reader by additional information to the different fields of application. The examples are selected to explain how the Master equation framework works, but also to introduce into different important interdisciplinary research topics of our scientific community.

The target audience therefore consists of interdisciplinary interested scientists, namely economists, physicists, biologists, geographers, sociologists, computer scientists, mathematicians and psychologists who are interested in modelling, simulations and mathematical methods and real-world applications.

Friendly relations with a number of colleagues from many universities all over the world have influenced the different applications and, therefore, the structure of the book.

A Nato Advanced Study Institute held in July 1982 in San Miniato, Italy, on evolving geographical structures focused my interest on interdisciplinary research of socio-economic space–time processes and patterns as well as real-world planning problems. Many international cooperation and resulting research projects were mainly initiated and supported by conferences and workshops organized and financed by *Deutsche Physikalische Gesellschaft* (DPG), the *International Institute for Applied Systems Analysis* (IIASA), the *Istituto Ricerche Economico-Sociali Del Piemonte* (IRES), the *Institut National D'Etudes Démographiques* (INED) and the *Centre for Regional Science Research Umea* (CERUM) to mention a few. It was not self-evident to find such a friendly acceptance and willingness to cooperate among economists, geographers, sociologists and regional and transport scientists with me as a physicist. This shows, however, that the field of interdisciplinary research is open for new ideas. I wish to thank all of them.

My special thanks go to my friend and mentor Wolfgang Weidlich, who unfortunately passed away far too early. His encouragement and many intensive and fruitful discussions and common work that have taken place over many years made the book possible.

Last but not least, special thanks go to the Springer-Verlag, especially to Barbara Feß for perfect managing of the publication task. My thanks also go to two unknown referees for important and helpful remarks and valuable advice.

Stuttgart
May 2017

Günter Haag

Contents

List of Figures

List of Tables

Chapter 1
Introduction

Contents

The understanding of the evolution and the internal structure of our world is one fundamental stimulus of human research. We strive to find out "Was die Welt im Innersten zusammenhält" (Goethe, Faust I) and what will happen in our future or, as Douglas Adams (1979) formulated in his famous book "The Ultimate Hitchhiker's Guide to the Galaxy" in a simple but realistic way, ...where do we come from, where do we go and where do we get the best Wiener Schnitzel?

But all attempts to understand more about issues such as the interactions of particles and molecules, the complexity of our biological world of plants, cooperation and competition of species, biological cells and the formation of organs in the fields of physics, chemistry and biology or in the social sciences the dynamics of economic and social conflicts, decision making, opinion formation and group dynamics, the building of networks, of urban and regional systems, the traffic dynamics and collective phenomena in the election of political parties—all these issues are based on models.

Models can be formulated and built in different languages. In agriculture, there exists a long tradition of farmers, applying models based on hundreds of years of experience, condensed in country sayings and weather proverbs. Meteorology formulates models for weather forecasting in the language of physics and mathematics. Models are based on rules. Rules are based on experience and experiments. The aim of modelling is always to develop a mathematical model as an image or picture of reality, formulated in logical symbols instead of words and rules representing the interactions of the symbols. This means we share the viewpoint of John L. Casti (1992a, b) "The study of natural systems begins and ends with the specification of observables describing such a system, and a characterization of the manner in which these observables are linked".

The model builder has always the choice what to observe and use as input and what to ignore. In other words, to neglect things deemed irrelevant for the purpose

G. Haag, *Modelling with the Master Equation*, DOI 10.1007/978-3-319-60300-1_1

of the model. The model should be kept as simple as possible, said Albert Einstein, but not too simple. In other words, we are interested in the ways of model building.

The quality of a model, however, depends among others on the compliance between model output and observed data. In so far, data issues are of crucial importance for any modelling decision. The purpose is not only to detect and exclude outliers, but also to determine the quality of the data, since the quality of the data constrains the quality of the model output.

Measurement errors on one hand and uncertainties and fluctuations on the other hand are intrinsic in all experiments and data. In many applications those effects are considered as small perturbations and influence the trajectory of the system only in a marginal way. If, however, the dynamics of the system is based on nonlinear interactions, phase transitions may occur and the dynamics of the system depends on or may even be dominated by fluctuations. Since fluctuations are always present, it is natural to include them in our models right from the very beginning.

Since the fundamental work of Hermann Haken (1977, 1983), a comprehensive theory, called Synergetics for the investigation of structural self-organizing space-time features of interacting multi-component systems has been provided and has demonstrated its huge modelling potential. Although the interactions and constituting units of the various systems under consideration seem to be completely incomparable on the micro-level, a close analogy between them exists on the macro-level. The interdisciplinary universality of Synergetics has its origin in the unifying concepts of model building and classification of such phenomena.

In the natural sciences, the elementary units, such as atoms or molecules, and the fundamental interactions, constituting the system are generally well known. In principle, model assumptions can directly be verified or falsified by experiments, and the reproducibility of experiments is fundamental and constitutive. Typically, one and the same experiment has to be and can be repeated under identical conditions in order to measure the value or the statistical distribution of values of an observable with a definite precision.

Much research has been done investigating self-organizing phenomena in the field of physics, chemistry and biology (Weidlich 1972; Schuster 1984; Klüver and Klüver 2011; Mainzer 2007; Arthur 1989; Rosser 2011). In these research fields, Synergetic concepts are mainly treated on the macro-level (Weidlich 2000).

The classical or quantum mechanical density matrix formalism provides a practicable framework how ensembles of interacting particles or molecules can be treated mathematically. The huge field of cooperative phenomena provides a lot of interesting examples, such as superconductivity and ferro-magnetism, to mention a few. The statistics of the laser light, namely, the phase transition from a typical lamb with stochastically emitting atoms to the high intensity laser light, characterised by coherently emitting atoms, exemplifies a self-organizing process (Sornette 2006).

Several authors were inspired by the rich field of dynamic processes of biological systems, especially by predator prey systems (see also Chap. 6). The search for analogies between economic and biological evolution was utilised in particular by Penrose (1952), Dosi (2005), Dosi et al. (1994), and Nelson and Winter (1982). The role of technological progress as an explanation of contemporary economic growth and modelling of such highly dynamic complex processes with uncertainties are important research topics up to now. Haag (1990a, b), Pyka (1999), and Erdmann (1993) have utilized the Master equation framework to formulate a dynamic theory of decision making and an evolutionary theory of innovation. The Schumpeter Clock (Mensch et al. 1991) as a micro-macro model of economic change including innovation, strategic investment, dynamic competition and short and long swings in industrial transformation belongs to the same kind of modelling framework.

In the mid-1990s the labelling "Econophysics" (Mantegna and Stanley 1999) as interdisciplinary research field was introduced by several physicists, applying theories and methods originally developed in physics, in other disciplines like economics. Especially those research problems including uncertainty or stochastic processes and nonlinear dynamics were points of interest (Soros 1994).

A framework for modelling a wide class of socio-economic phenomena has been given in the book of Weidlich and Haag (1983). In the following, we will basically proceed along the line of argumentation given in this book incorporating the results of more recent research projects related to the field named Sociodynamics. Since the definition of all these concepts is the same as in Synergetics, one can therefore consider Sociodynamics as that part of Synergetics which is devoted to social systems (Weidlich 2006).

Coming back to the difference between natural and social sciences: in social sciences the interactions between elementary units such as individuals, households, firms are rather unknown and cannot be derived from first principles. Experimental tests repeated under identical socio-economic conditions are mostly impossible. The empirical data base related to a certain subject is often rather limited and the comparability among data sets is often not guaranteed.

In view of these differences regarding modelling of socioeconomic processes some critical remarks must be made at the beginning: firstly, no direct short-cut to transfer concepts from natural to social sciences exists. Appropriate and characteristic concepts have to be developed for the quantitative description of socio-economic processes. Secondly, Synergectics, Sociodynamics and all other concepts can be applied only under certain conditions to a specific class of genuine social phenomena (Haag 1990a, b). If these conditions are fulfilled, however, a true structural relationship between natural and social sciences and not just an accidental analogy can be found. In so far, all ingredients incorporated into the models have to come from the respective sciences to avoid physicalism, namely a direct use of physical laws and a re-interpretation of those in terms of social phenomena (Müller 2012; Vega-Redondo 2007).

We have to take into account, that the evolution of any socioeconomic system is not an autonomous process but the result of human decisions occurring over time as a broad stream of concurrent, unrelated or interrelated, individual or corporate choices. The underlying mechanisms behind the millions of decisions made every day cannot be completely controlled and influenced by public authorities, at least not in a direct way (Haag 1989). Therefore, planners in charge of such systems face the difficult task of making decisions concerning a system which is largely subject to external influences in the form of national policies and entangled economies on the one hand, while the system is influenced by decisions of private firms, investors, and other individual or corporate agents on the other hand. Only limited instruments of policy are available and at their disposal, and it is of crucial importance to know in advance which of these are likely to be most effective (Fischer et al. 1988; Ball 2012).

In so far, the human society can be regarded as a multi-component system whose members, the individuals, adopt different attitudes or kinds of behaviour (Weidlich and Haag 1983). The causes of global changes in society are assumed to be correlated with the decisions of agents to change their attitudes. A complex mixture of fluctuating rational considerations, professional activities, emotional preferences and motivations finally merge into one of relatively few well demarcated resultant attitudes. Those attitudes may be related to education, politics, economic activities and consumer habit, to mention a few. The attitude space is an open one, since hitherto unknown attitudes may develop or attitudes till now considered important may disappear.

The attitudes drive the decisions of individuals. Due to individual decision processes caused by experience, emotions and thoughts based on the individual network of personal relationships, transitions from one attitude to another one are possible.

However, the detailed micro-level describing the complex interplay of rational and emotional, conscious and subconscious, genetic and environmental influences on the decisions of individuals is typically unknown. Hence, a probabilistic description instead of a deterministic account of decision behaviour is adequate.

In thermodynamics, "entropy" is a measure of the order state of matter. In a closed system, the entropy is constantly increasing and reaches a maximum of "disorder" for its equilibrium state. Therefore, the equilibrium state represents the most probable configuration of the system. In closed physical systems there is a tendency towards increasing disorder of the micro states of the system. The relation to the probability of finding a macro state through certain micro states makes it possible to transfer the entropy concept to different areas of social science (Wilson 1970). Thus the entropy concept has in principle a probabilistic background, related to the statistical distribution of events in an uncertain situation. Numerous fields of application of the entropy concept to the modelling of socioeconomic systems have been examined, such as the distribution of commodity flows and migration flows, shopping trip distribution or traffic flow assignment. The basic idea is always that the distribution of the quantities of interest can be selected as the statistically most likely distribution by means of the entropy principle, taking into account given restrictions. With regard to the statistical foundation and analysis of spatial interaction models, we follow the fundamental theoretical work of Wilson (1970) and Nijkamp and Reggiani (1998) in Sect. 8.2. However, some criticisms of the entropy concept are also appropriate. The striking elegance of the method is limited by the necessary proximity to thermodynamics. Thus, the existence of the entropy can only be shown for equilibrium systems or systems that are close to equilibrium, that is, as long as linear regression laws apply. The treatment of systems that are out of balance, and this is often the case with socioeconomic systems, is, however, rather questionable.

The Master equation is suitable for the adequate statistical treatment of non-equilibrium states. Therefore, the probability that a certain decision configuration is realized will be introduced. The Master equation is the equation of motion for this probability distribution, where transition rates between different decision configurations are the essential constituents.

On the macro-level, in turn, the decision configuration describes the distribution of attitudes of the socioeconomic system and may be considered as an appropriate set of macro variables for the system under consideration. The modelling of such transition rates in terms of variables will turn out to be the central part of the model building in natural and social sciences, respectively.

The Master equation framework can be understood as a tool for model building, where fluctuations or uncertainties are incorporated in a systematic way. The strong model building potential of the Master equation is particularly suitable for complex systems. In other words, we consider systems consisting of many interacting sub-systems, where nonlinearities are inherent, and uncertainties or fluctuations are involved and may dominate the dynamics.

The probability distribution over a given configuration contains the most detailed information about the system. In particular, not only the mean values or sometimes the most probable values can be calculated but also higher moments such as the mean square deviations. Correspondingly, the amount of mathematics required to solve the time-dependent Master equation may be considerable. However, in most practical cases, the full information contained in the configurational probability distribution cannot be exploited due to a lack of sufficiently comprehensive empiric data. Therefore it makes sense to perform a transition to a less exhaustive description in terms of quasi-closed equations of motion for mean values and variances. These dynamic equations can be derived from the Master equation in a straightforward manner. Hence, the Master equation provides the link between the micro-level of changes of single configurations (transition rates) and the macro-level of dynamic equations of motion for mean values and variances.

The non-linear form of the quasi-closed equations of motion expresses the structure of self-consistence, which is prevalent in all socio-economic systems, namely the cyclic coupling of causes and effects.

Through their cultural and economic activities, the individual members of the society contribute to what we will call a collective field related to our society with cultural, political, religious, social and economic components. This collective field acts as an order parameter of the socio-economic system and characterizes the current phase of our society. Moreover, the collective field strongly influences the decision behaviour of the individuals, by orientating their activities. The feedback between the actions of the individuals and the collective field—the cyclic coupling between causes and effects—determines the temporal development of the system.

If the outcome of the Master equation, the probability distribution function, is sharply peaked, a quasi-stable temporal development of the system characterized by a certain predictability of its trajectories may occur. However, highly divergent alternative paths of evolution of the society are possible, if the control parameters of the system attain certain critical values. Fluctuations on the micro-scale may decide into which of the divergent paths the society will bifurcate. The actions of a few influential persons or decision makers could be an example. In this case, the probability distribution for the decision configuration has lost its simple uni-modal structure (Haag 1989). Nearby the phase transition point, forecasting the system development is rather difficult if not impossible.

On the macro-level phase transitions may lead to stable attractors, periodic structures, limit cycles or even chaotic trajectories. An excellent introduction into the field of chaotic systems is given by Devaney (2003) and Gollub and Baker (1996).

In social sciences, the link between the micro-level of decisions of individuals and the macro-level of the dynamics of aggregated variables is one main target of research. However, this important link is not only of theoretical interest. It enables us to match empirical data with the outcome of the theoretical model. Of course, it is well known in this context, that it is difficult if not impossible, to give a direct and unique causal interpretation of the socio-economic situation and the behaviour of certain macro-variables of society in terms of individual motivations on the micro-level (Coleman 1992). Instead, we expect that many combinations of such motivations will merge with different intensities in the individual decision processes and will produce the observed macro-dynamics. This favours the introduction of aggregated variables (e.g. attractiveness variables), which themselves depend on a set of individual motivations.

The estimation of the parameters of the model, denoted as trend parameters, is a further important research topic in social sciences. Depending on the research issue, it is sometimes possible to introduce a cost function or penalty function, which can be used to minimize the deviations between the empirical data and the model output, taking into account various constraints. Different solution algorithms can be developed and used to optimize the parameter estimation process. In Chap. 8, we will deal with these problems.

The book is organized in three parts (see Fig. 1.1):

Part I contains the statistical fundamentals and the derivation of the Master equation and the Fokker-Planck equation. Solution methods of the Master equation, including some rather new tools for a group of special problems are presented. This part is rather technical and can be used as a toolbox.

In Chap. 2, some statistical fundamentals needed for the understanding of the modelling framework of the Master equation are presented. This includes the definition of common statistical indicators and functions. This tool box is important since stochastic processes are becoming increasingly important in many branches of physics, chemistry, biology, population dynamics, economics and social sciences. Despite the diversity of tasks and problems in these fields, there are common principles and methods which are subject of this book.

Chapter 3 is concerned with the general understanding of the fundamental aspects of the Master equation. Following the introduction of some concepts of probability theory, the Markov assumption is introduced and the Chapman-Kolmogorov

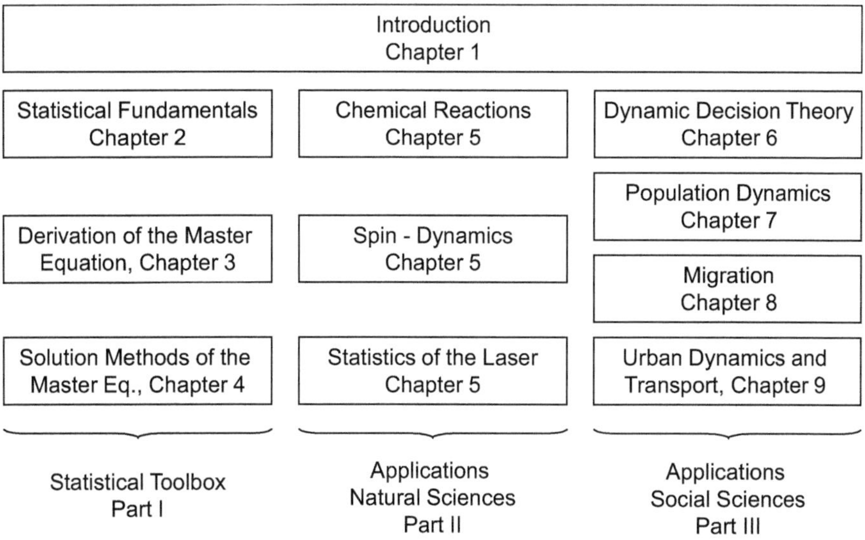

Fig. 1.1 Organization of the book

equation for conditional probabilities is derived. The Chapman-Kolmogorov equation serves as the starting point for the derivation of the Master equation. The derivation of general properties of the Master equation helps to understand the broad field of possible applications. The derivation of equations of motion for mean values and variances on both the stochastic and the quasi-deterministic level, using the method of shift operators completes this chapter.

After the derivation of the Master equation, it is logical to introduce and discuss different methods of its solution in Chap. 4. One obvious and frequently applied method consists in the approximate transformation of the discrete Master equation in a partial differential equation, namely the Fokker-Planck equation. The not so well known T-factor method is very efficient in the transformation of the Master equation into difference equations of reduced order and in continued fractions which are easier to handle. This method also provides a very elegant way to derive exact and approximate stationary solutions of the Master equation, even when detailed balance is not fulfilled. A general graph-theoretical method for the stationary solution developed by Kirchhoff for electrical networks is also presented. In case of detailed balance an exact solution method for the stationary probability distribution completes the tool box. The chapter closes with exact and approximate solution methods for one-dimensional Master equations with two particle jumps.

Part II starts with applications of the Master equation framework in the natural sciences. However, benchmarking of different solution methods is important in order to compare the Master equation framework with other modelling approaches. Therefore, we do not only apply the Master equation framework to different case studies, but also compare it with other solution methods.

In Chap. 5, the derived solution methods of Chap. 4 will be applied to some important applications out of the field of physics and chemistry in order to demonstrate the high potential of the Master equation approach in the field of natural sciences. Chemical reactions are typical examples of discrete dynamic processes obeying the mass action law. Since the various chemical reactions are considered independent of each other, a multinomial Poisson distribution is usually expected. However, in case of nonlinear chemical reactions we demonstrate that although the transition rates obey combinatorial mass-action law kinetics, a more complicated statistical distribution is obtained.

The investigation of the phase transition dynamics of spin-systems is used as an example in statistical non-equilibrium physics. The appearance of a ferromagnetic order from an initially disordered state, in other words the occurrence of a phase transition, in its conceptual simplicity is one reason for the interest in this widely applicable model type. The calculation of escape rates due to very large fluctuations will be investigated as well.

The derivation of the photon statistics of the Laserlight is selected as a typical quantum mechanical example out of the field of physics. The photon statistics of the Laser shows a typical phase transition at the so-called Laser threshold. The atoms of the Laser active material seem to be slaved: all atoms behave in a coordinated way and emit wave tracks in phase. It is interesting to apply the already introduced methods to this example and to learn more about how complicated discrete difference equations may be handled. Parts of this chapter are rather theoretical and may be skipped in a first reading.

Part III is dealing with modelling concepts in the social sciences. In Chap. 6, we derive a generalized dynamic choice model for interacting individuals using the Master equation approach. The famous multinomial-logit decision model is obtained when the system is in an equilibrium state and individuals are independent. The example shows that depending on the initial conditions of the system of agents (decision makers), and the strength of their interaction, quite different decision configurations may be obtained. Previous experience may also account for the decisions of agents, in other words decisions may depend on history. In this case the Markov assumption does no longer hold, and the Master equation fails. However, a simple trick seems to help: we fragment the whole time frame into small time sequences. The single time sequences are chosen so small that the Master equation holds within each time sequence, but not for the whole process. This provides a model of nested decision processes with memory. The emergence of conventions may be used as an example to underline the applicability of this method.

In Chap. 7, we deal with the issue of population growth. This is intended to point out ideas and thoughts behind the construction of models in population biology. I belief that this will help the reader to understand better the theoretical considerations and the outcome of the discrete Master equation approach compared with classical considerations in population dynamics based on assumed continuous population development. It is shown, that only if we take into account the discrete structure of the population, the complicated dynamics of the dying-out process can be understood. Even in the case of linear birth and death rates, the total system behaves highly nonlinear. If we ignore the discrete structure population numbers that means using differential equations such as the Fokker-Planck equation instead of the discrete Master equation, the extinction process and the life time of populations cannot be estimated appropriately.

In a further example, a Master equation City-hinterland model demonstrates how migratory phase transitions may be responsible for the tremendous growth of settlements during the last decades.

Predator-prey interaction is one of the fundamental processes of biology. A very simple but often used and discussed model for the description of predator-prey interaction is the famous Volterra-Lotka model. We will consider a special case of the Volterra-Lotka model by allowing the prey species to migrate to a habitat that the predator species cannot follow. In this case the rather unstable cycles of the Volterra-Lotka case will be replaced by a stable limit cycle.

The dynamics of a system of interacting populations closes this chapter. A sequence of phase transitions may occur and drive the system out of its stable state, via a limit cycle towards chaotic behaviour when three populations interact.

In Chap. 8, the dynamic development of spatial pattern based on human activities is investigated. This topic is one of the fundamental issues of regional science and economics. Migration processes in all their theoretical and empirical aspects are of overwhelming importance to all spheres of society. Since migration and especially the demographic composition of the stream of migrants and their social affiliation are among others factors of population growth, fundamental economic and social indicators of our society are directly influenced. After a reconsideration of the famous Gravity model the Weidlich-Haag migration model (Weidlich and Haag 1988) is derived. It is shown how the fundamental process of building up cities, in other words the instability of a homogeneous population distribution and the rank-size distribution of settlements, can be understood as a dynamic self-organization process.

As an important application of the Weidlich-Haag-model the interregional migration in Germany on the level of 402 districts is investigated. The algorithm for the estimation of all model parameters is explained in detail.

Urban dynamics is the topic of Chap. 9. We start with the development of our integrated transport and evolution model, based on the Master equation framework. The traffic subsystem as well as the urban/regional subsystem form a complex intertwined system. Its dynamics takes place on different time scales which are modelled making use of the same principles. The interactions between transport and urban/regional development are analysed for a peri-urban area, where the effects are expected to be very strong. The study area is situated in the region of Stuttgart, along the Stuttgart—Zurich corridor. The influence of the completion of the motorway A81 and the light rail system S1 that runs parallel to the A81 are considered and discussed.

In the last chapter of the book the model building framework of the Master equation is once more used for the construction of spatial-interaction models. It will be outlined how some of the usual limitations in model building can be circumvented. Our broader overall framework with full probabilistic dynamic underpinning also facilitates extensions in a number of directions. For instance, it becomes obvious how conventional spatial-interaction models can be derived and generalized, especially how fundamental dynamic aspects can be introduced. The different decision processes of developers, retailers and land owners are modelled in a rather general manner. The service sector model of Harris and Wilson (1978) will be obtained as a special case.

References

Adams D (1979) The hitchhiker's guide to the galaxy. Del Rey, Stadt
Arthur WB (1989) Competing technologies, increasing returns, and lock-in by historical events. Econ J 99:116–131
Ball P (2012) Why society is a complex matter. Meeting twenty-first century challenges with a new kind of science. Springer, Berlin
Casti JL (1992a) Reality rules: I. Wiley, New York
Casti JL (1992b) Reality rules: II. Wiley, New York
Coleman JS (1992) The economic approach to sociology. In: Radnitzky G (ed) Universal economics. Assessing the achievements. Paragon House, New York
Devaney RL (2003) An introduction to chaotic dynamical systems. Westview Press, Boulder
Dosi G (2005) Statistical regularities in the evolution of industries. A guide through some evidence and challenges for the theory. LEM working paper 17
Dosi G. Freeman C, Fabiani S (1994) The process of economic development: Introducing some stylized facts and theories on technologies, firms and institutions. Ind Corp Chang 3: 1–45
Erdmann G (1993) Elements einer evolutorischen Innovationstheorie. J.C.B. Mohr, Paul Siebeck, Tübingen
Fischer MM, Haag G, Sonis M, Weidlich W (1988) Account of different views in dynamic choice processes. In: Proceedings of the symposium of the IGU-working group on mathematical models, Canberra
Gollub JP, Baker GL (1996) Chaotic dynamics. Cambridge University Press, Cambridge
Haag G (1989) Dynamic decision theory: Applications to urban and regional topics. Kluwer, Dordrecht
Haag G (1990a) Die Beschreibung sozialwissenschaftlicher Systeme mit der Master-Gleichung. In: Ökonomie und Gesellschaft, Jahrbuch 8: Individuelles Verhalten und kollektive Phänomene. Campus, Boulder
Haag G (1990b) A master equation formulation of aggregate and disaggregate economic decision making. Sistemi Urbani 1:65–81
Haken H (1977) Synergetics: an introduction. Springer, Heidelberg
Haken H (1983) Advanced synergetics. Springer, Heidelberg
Harris B, Wilson AG (1978) Equilibrium values and dynamics of attractiveness terms in production-constrained spatial-interaction models. Environ Plan A 10:371–388
Klüver J, Klüver C (2011) Social understanding: on hermeneutics, geometrical models and artificial intelligence. Springer, Berlin
Mainzer K (2007) Thinking in complexity. The computational dynamics of matter, mind, and mankind. Springer, Berlin
Mantegna RN, Stanley HE (1999) An introduction to econophysics: correlations and complexity in finance. Cambridge University Press, Cambridge
Mensch G, Haag G, Weidlich W (1991) The Schumpeter clock. In: Technology and productivity, The Technology Economy Programme, TEP, OCDE, Paris

Müller KH (2012) The new science of cybernetics. The evolution of living research designs. Bd. 3: Research and design rules. Edition Echoraum, Wien

Nelson RR, Winter SG (1982) An evolutionary theory of economic change. Harvard University Press, Cambridge

Nijkamp P, Reggiani A (1998) Entropy, spatial interaction modelling and discrete choice analysis. Eur J Oper Res 36(1):186–196

Penrose ET (1952) Biological analogies in the theory of the firm. Am Econ Rev 58:211–221

Pyka A (1999) Der kollektive Innovationsprozess. Dunker & Humblot, Berlin

Rosser JB (2011) Complex evolutionary dynamics in urban-regional and ecologic-economic systems: from catastrophe to chaos and beyond. Springer, New York

Schuster P (ed) (1984) Stochastic phenomena and chaotic behaviour in complex systems. Springer, Berlin

Sornette D (2006) Critical phenomena in natural sciences: chaos, fractals, self-organization and disorder: concepts and tools, 2nd edn. Springer, Berlin

Soros G (1994) The alchemy of finance. Reading the mind of the market, 2nd edn. Wiley, New York

Vega-Redondo F (2007) Complex social networks. Cambridge University Press, Cambridge

Weidlich W (1972) The use of statistical methods in sociology. Collect Phenom 1:51

Weidlich W (2000) Sociodynamics. Harwood Academic Publishers, London

Weidlich W (2006) Sociodynamics: a systematic approach to mathematical modeling in the social sciences. Dover Publications, Mineola

Weidlich W, Haag G (1983) Concepts and models of a quantitative sociology: The dynamics of interaction populations, Springer series of synergetics, vol 14. Springer, New York

Weidlich W, Haag G (eds) (1988) Interregional migration: dynamic theory and comparative analysis. Springer, New York

Wilson AG (1970) Entropy in urban and regional modelling. Pion Books, London

Part I
Statistical Toolbox

Chapter 2
Statistical Fundamentals

Contents

2.1 Deterministic and Probabilistic Description of Systems

Statistics is a tool for decision making when an agent or a person is uncertain about the characteristics involved in the decision. This means that every decision involves an element of probability in it. Moreover, statistical tools are used for maintenance purposes: to forecast the life-time of machine components, to determine the length of product warranties, or to estimate the outcome of an election, to mention a few examples. The likelihood that a predefined event will occur is measured by its probability. The higher the probability of an event, the more certain is its occurrence. Statistical theory is used to describe the underlying processes and regularities of such complex systems.

All statistical problems are based on the incomplete nature of information on which statements are to be made and all statistical tasks share the problem of incomplete information. The fundamental nature of this uncertainty may be very different and depends on the definition of our system.

Models are based on rules and rules are based on experience and experiments (Casti 1992). A mathematical model should be a sufficiently accurate image or

G. Haag, *Modelling with the Master Equation*, DOI 10.1007/978-3-319-60300-1_2

picture of reality, formulated in logical symbols and rules representing the interactions of the symbols. Mathematically modelled systems may be classified as deterministic or probabilistic.

Consider a system which passes through different states in the course of time. We assume that each state is characterized by a discrete number or a set $\vec{n} = (n_1, n_2, \ldots, n_L)$ of discrete numbers, where L describes the dimension of the system. We further assume that the system is in a state $\vec{n}_0 = \vec{n}(t_0)$ at an initial time t_0.

Two possible descriptions of the evolution of the system over time are feasible (Haken 1977; Spiegel 1992):

Deterministic Evolution

In this case, information about the dynamics of the system can be considered complete. The description is fully deterministic and all states $\vec{n}(t)$ of the system at later times $t \geq t_0$ (time t is continuous, $\vec{n}(t)$ is discrete) are uniquely determined (Fig. 2.1).

Probabilistic Evolution

If the information about the system is incomplete, only probabilistic statements about the evolution of the system are possible. In other words, an exact prediction of the state $\vec{n}(t)$ reached by the system after a certain time interval $t > t_0$ is not possible. Instead the members of an ensemble of identical systems—each of them prepared in the same initial state $\vec{n}(t_0)$—will develop into different states $\vec{n}(t)$ at time t (Fig. 2.2). The information available in this particular situation is the probability to reach state $\vec{n}(t)$ at time t, given that the system has been prepared in state $\vec{n}_0 = \vec{n}(t_0)$ at time t_0. This special probability $p(\vec{n}, t | \vec{n}_0, t_0)$ is denoted as conditional probability. The Master equation is the tool to determine this central quantity in a systematic manner.

The probabilistic evolution is definitely the more general formulation, since the case of incomplete knowledge about the system comprises complete knowledge as a limiting case, whereas the converse is not true.

If the probability distribution $P(\vec{n}, t)$ to find the system in a certain state $\vec{n}$ at time t is sharply peaked around the most likely state $\vec{n}_{\max}(t)$ the limit case of almost

Fig. 2.1 Deterministic behaviour of a system

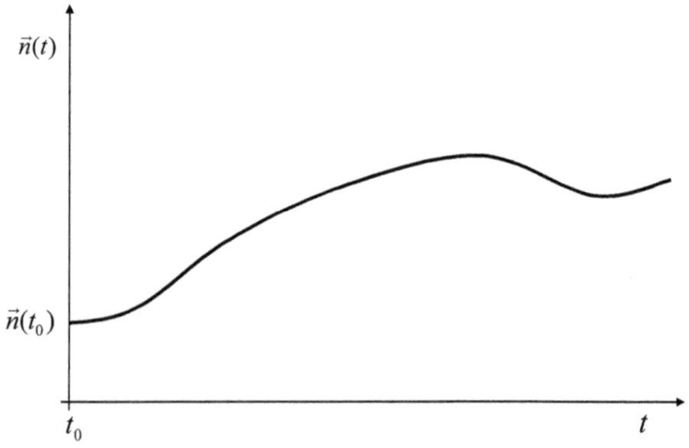

complete knowledge is revealed (Fig. 2.3). This means that the system assumes state $\vec{n} \approx \vec{n}_{\max}(t)$ with overwhelming probability at time t, whereas all other states are highly improbable at the same time.

In Fig. 2.4 the evolution of a probabilistic system is shown approaching a bi-modal distribution in the long run. If the second peak (small peak) increases over time, the system may exist in two states and phase transitions between those two states may happen, depending on the shape of the distribution $P(\vec{n}, t)$. However, the time of a switch between the two states may be very long. In Sect. 5.2.3 we will calculate the escape rate for such a bistable system.

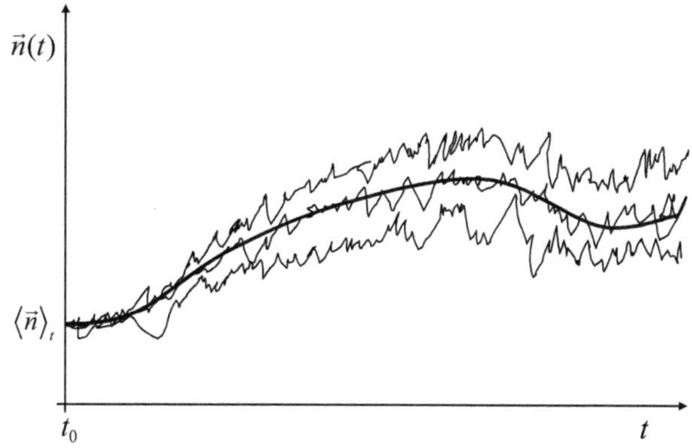

Fig. 2.2 Probabilistic evolution of a system. Three different trajectories are shown and the corresponding development of the mean value

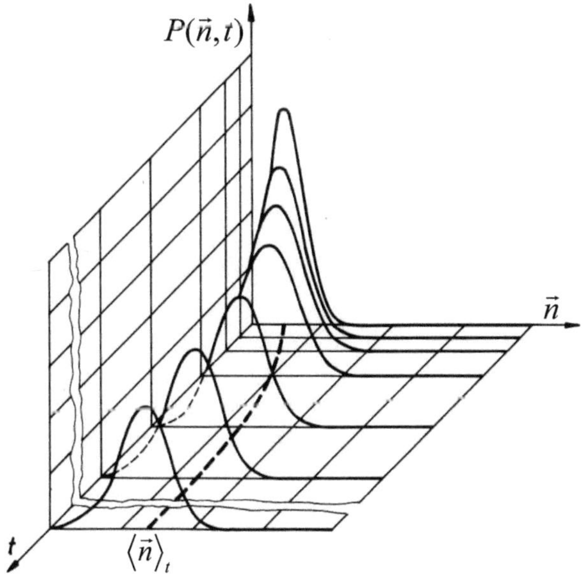

Fig. 2.3 Evolution of a typical probability distribution of a quasi-deterministic system (Source: Weidlich and Haag 1983)

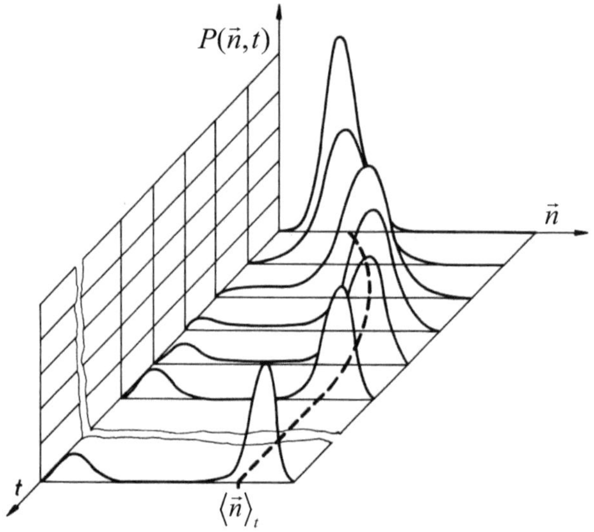

Fig. 2.4 Evolution towards a bi-modal probabilistic system. A second peak is emerging anticipating a possible phase transition in the course of time (Source: Weidlich and Haag 1983)

2.2 Mean Values, Moments, Correlation Function, Characteristic Function, Generating Function

In this part the basic tools of statistics as well as the definition of common statistical indicators and functions will be introduced. In other words, it will be explained how to use the information contained in the joint probabilities and configurational probability distribution. There are good reasons for this tool box, since stochastic processes are become increasingly important in many branches of physics, chemistry, biology, population dynamics, economics and social sciences. Despite the diversity of tasks and problems in these fields, there are common principles and methods. Let us start with a few general concepts which all have their field of applications and have been established since many years (Gut 2005; Pfeiffer and Schum 1973).

2.2.1 Mean Values and Moments

A stochastic variable is an object defined by a set of possible values and a probability distribution over this set (van Kampen 1978; Grimmet and Stirzaker 1992). The set of possible values (also called “set of states”, “sample space” or “phase space”) may be discrete and infinite, discrete and finite, continuous in a certain interval in one or many variables etc. If it is discrete and denumerable, the probability distribution will be given by a set of non-negative numbers.

The One-dimensional Case

In the one-dimensional case, if the set of possible values is discrete, we have

$$P(n,t) \geq 0 \quad \text{with} \quad \sum_n P(n,t) = 1 \tag{2.1}$$

If the range is in an interval of the x-axis ($a \leq x \leq b$), the probability distribution is determined by a non-negative function

$$P(x,t) \geq 0 \quad \text{with} \quad \int_a^b P(x,t)dx = 1 \tag{2.2}$$

Rather than developing a universal notation for all possible cases, we simply use the notation that is most appropriate or convenient.

The expectation value $E(n)$ or mean value $\bar{n}$ of a discrete random variable n will be calculated as a sum weighted with $P(n,t)$. For the *one-dimensional case*, we define for the discrete case or continuous case, respectively

$$E(n) \equiv \bar{n}(t) \equiv \sum_n nP(n,t) \quad \text{or} \quad E(x) \equiv \bar{x} \equiv \int_{-\infty}^{\infty} xP(x,t)dx \tag{2.3}$$

or in general, for any discrete or continuous function $g(n)$ or $g(x)$, respectively

$$E(g(n)) \equiv \overline{g(n)} \equiv \sum_n g(n)P(n,t) \quad \text{or} \quad E(g(x)) \equiv \overline{g(x)}$$

$$\equiv \int_{-\infty}^{\infty} g(x)P(x,t)dx \tag{2.4}$$

Especially in physics, it is convenient to indicate formally the averaging procedure by $\langle\ \rangle$ brackets

$$\langle \bullet \rangle_t = \sum_{\vec{n}} \bullet P(\vec{n},t) \quad \text{or} \quad \langle \bullet \rangle_t = \int \bullet P(\vec{x},t)d\vec{x} \tag{2.5}$$

Hence, the expectation values in (2.4), can also be described as

$$\langle g(n) \rangle_t = \sum_n g(n)P(n,t) \quad \text{or} \quad \langle g(x) \rangle_t = \int_{-\infty}^{\infty} g(x)P(x,t)dx \tag{2.6}$$

Of special interest are the first, second and third moments of a distribution. In the one-dimensional case, this means:

Moment : $\bar{n}(t)$ or $\bar{x}(t)$ or $\langle x\rangle_t$	mean value of the distribution
Moment : $\overline{n^2}(t)$ or $\overline{x^2}(t)$ or $\langle x^2\rangle_t$	information about fluctuations
Moment : $\overline{n^3}(t)$ or $\overline{x^3}(t)$ or $\langle x^3\rangle_t$	information about symmetry

Fluctuations around the mean behaviour of the stochastic system are usually described by introducing the variance $\sigma^2(t)$ of the distribution

$$\sigma^2(t) = \overline{n^2}(t) - \bar{n}(t)^2 = \sum_n (n - \bar{n}(t))^2 P(n,t) \quad \text{or} \quad \sigma^2(t) = \left\langle n^2 \right\rangle_t - \langle n\rangle_t^2$$
$$= \left\langle (n - \langle n\rangle)^2 \right\rangle_t \tag{2.7}$$

and in the continuous case

$$\sigma^2(t) = \overline{x^2}(t) - \bar{x}(t)^2 = \int_{-\infty}^{\infty} (x - \bar{x}(t))^2 P(x,t)dx \quad \text{or} \quad \sigma^2(t) = \left\langle x^2 \right\rangle_t - \langle x\rangle_t^2$$
$$= \left\langle (x - \langle x\rangle)^2 \right\rangle_t \tag{2.8}$$

where $\sigma(t) = \sqrt{\sigma^2(t)}$ is named standard deviation and describes the fluctuations around its mean.

The skewness γ is a measure of asymmetry of a probability distribution and is defined as

$$\gamma(t) = \frac{\left\langle (n - \langle n\rangle_t)^3 \right\rangle_t}{\sigma^3(t)} \tag{2.9}$$

or in the continuous case

$$\gamma(t) = \frac{\left\langle (x - \langle x\rangle_t)^3 \right\rangle_t}{\sigma^3(t)} \tag{2.10}$$

If the left tail of the distribution is more pronounced than the right tail, the skewness is negative$\gamma < 0$, the skewness is positive $\gamma > 0$ if the reverse is true.

The expectation value $E(n)$ or mean value $\vec{n}$ of a discrete random variable $\vec{n} = (n_1, n_2, \ldots, n_L)$ will be calculated as weighted mean value of $\vec{n}$.

The L-dimensional Case

For the *L-dimensional* discrete case we have

$$E(g(n_1,n_2,\ldots,n_L)) \equiv \overline{g(n_1,n_2,\ldots,n_L)} \equiv \sum_{\vec{n}} g(n_1,n_2,\ldots,n_L)P(n_1,n_2,\ldots,n_L,t) \tag{2.11}$$

or in the continuous case

$$E(g(x_1,x_2,\ldots,x_L)) \equiv \overline{g(x_1,x_2,\ldots,x_L)} \equiv \int_{-\infty}^{\infty}\cdots\int_{-\infty}^{\infty} g(x_1,x_2,\ldots,x_L)P(x_1,x_2,\ldots,x_L,t)dx_1dx_2\ldots dx_L \tag{2.12}$$

respectively. Particularly important are functions $g(n)$ or $g(x)$, respectively, defining the so-called moments (of the distribution)

$$E\left(n_1^{\nu 1}n_2^{\nu 2}\ldots n_L^{\nu L}\right) \equiv \overline{n_1^{\nu 1}n_2^{\nu 2}\ldots n_L^{\nu L}} \equiv \sum\nolimits_{n_1\ldots n_L} n_1^{\nu 1}n_2^{\nu 2}\ldots n_L^{\nu L}P(n_1,n_2,\ldots n_L,t) \tag{2.13}$$

or

$$E\left(x_1^{\nu 1}x_2^{\nu 2}\ldots x_L^{\nu L}\right) \equiv \overline{x_1^{\nu 1}x_2^{\nu 2}\ldots x_L^{\nu L}} \equiv \int_{-\infty}^{\infty}\cdots\int_{-\infty}^{\infty} x_1^{\nu 1}x_2^{\nu 2}\ldots x_L^{\nu L}P(x_1,x_2,\ldots x_L,t)dx_1dx_2\ldots dx_L \tag{2.14}$$

where the ν's are integer numbers and some of the ν's may be equal to zero.

2.2.2 *Correlation Function*

A correlation function is a statistical measure between random variables at two different points in space or time. Frequently, the correlation functions depend on spatial or temporal distance. Therefore, correlation functions are useful indicators of the assessment of interrelated functional dependencies. In model building, it is important to introduce variables, which are uncorrelated to avoid artificial dependencies. Correlation functions between the same random variable measured at two different points are often named autocorrelation functions. Correlation functions of different random variables are named cross-correlation functions to emphasise that the interrelations being considered refer to different random variables.

For the two-dimensional probability density $P(n,m,t)$ or $P(x,y,t)$ the following moments or correlation functions are of interest and will be used later in the examples:

$$\overline{n^j m^k} \equiv E\left(n^j m^k\right) = \sum_{m,n} n^j m^k P(n,m) \tag{2.15}$$

or

$$\overline{x^j y^k} \equiv E(x^j y^k) = \int_{-\infty}^{\infty} \int_{-\infty}^{\infty} x^j y^k P(x,y)dxdy \tag{2.16}$$

Correlation functions describe spatial and/or temporal dependencies. In our case, those moments (2.15), (2.16) are named usually as two-point correlation functions or two-point coherence functions.

For the *L-dimensional* case correlation functions are given by (2.13) and (2.14).

The second moment $\overline{n_1 n_2}$ is particularly important. It is often called the correlation par excellence, namely between the measurement results n_1 at point $\vec{x}_1$ at time t_1 referred as $n(\vec{x}_1, t_1)$, and n_2 at point $\vec{x}_2$ at time t_2, referred as $n(\vec{x}_2, t_2)$. Often, the results for $t_1 = t_2$, $\vec{x}_1 \neq \vec{x}_2$ or for $t_1 \neq t_2$, $\vec{x}_1 \neq \vec{x}_2$ are needed. Roughly speaking, $\overline{n_1 n_2}$ is a measure of the relationship or interaction between two events at two space-time points $\vec{x}_1, t_1$ and $\vec{x}_2, t_2$.

Sometimes $\overline{n_1 n_2} = \overline{n_1} \cdot \overline{n_2}$. Then the measurement results $\bar{n}_1$ and $\bar{n}_2$ are called uncorrelated. This means, however, only in case of a Gaussian distribution, that they must also be statistically independent. In fact, $\overline{n_1 n_2} = \overline{n_1} \cdot \overline{n_2}$ does not imply $\overline{n_1^k n_2^l} = \overline{n_1^k} \cdot \overline{n_2^l}$, which has to be fulfilled, if events are statistically independent as well. This is a consequence of the fact that the correlation function normally describes only a part of the relationships between n_1 and n_2.

It is common to write the correlation function $\overline{n_1 n_2}$ in the more detailed form

$$R(\vec{x}_1, t_1; \vec{x}_2, t_2) \equiv \overline{n(\vec{x}_1, t_1) n(\vec{x}_2, t_2)}. \tag{2.17}$$

If the positions or times in $R(\vec{x}_1, t_1; \vec{x}_2, t_2)$ occur only in the special form $\vec{x}_1 - \vec{x}_2$ or $t_1 - t_2$, we call $R(\vec{x}_1, t_1; \vec{x}_2, t_2)$ spatially or temporally homogeneous (or stationary).

Wiener-Chinchin Theorem

The Wiener–Chinchin theorem states that the autocorrelation function of a station ary random process has a spectral decomposition given by the power spectrum of that process (Engelberg 2007).

Let $\tau = t_2 - t_1$, so that $R(\vec{x}_1, \vec{x}_2, \tau)$ is temporally stationary. Then we call $R(\vec{x}_1, \vec{x}_1, \tau)$ auto-correlation function, and $R(\vec{x}_1, \vec{x}_2, \tau)$ cross-correlation function.

The (temporal) Fourier transform of the cross-correlation function

$$S(\vec{x}_1, \vec{x}_2, \nu) = \int_{-\infty}^{\infty} R(\vec{x}_1, \vec{x}_2, \tau) e^{2\pi i \nu \tau} d\tau \tag{2.18}$$

is called cross-spectral density for $\vec{x}_1 \neq \vec{x}_2$. To interpret $S(\vec{x}_1, \vec{x}_2, \nu)$, we decompose $R(\vec{x}_1, \vec{x}_2, \tau)$ directly according to Fourier

$$\begin{aligned} R(\vec{x}_1, \vec{x}_2, \tau) &= \frac{1}{T} \int_{-T/2}^{T/2} u(\vec{x}_1, t) u(\vec{x}_2, t+\tau) dt \\ &= \frac{1}{T} \int_{-T/2}^{T/2} \left[\int_{-\infty}^{\infty} \widehat{u}(\vec{x}_1, \nu) e^{-2\pi i \nu t} d\nu \right] \left[\int_{-\infty}^{\infty} \widehat{u}(\vec{x}_2, \nu') e^{-2\pi i \nu'(t+\tau)} d\nu' \right] dt \end{aligned} \tag{2.19}$$

where

$$\widehat{u}(\vec{x}_1, \nu) = \int_{-\infty}^{\infty} u(\vec{x}_1, t) e^{2\pi i \nu t} dt \tag{2.20}$$

is the Fourier transform of $u(\vec{x}_1, t)$. We consider the integrals as limit values for $T \rightarrow \infty$. Using the representation of the Dirac δ-distribution

$$\int_{-T/2}^{T/2} e^{2\pi i (\nu + \nu')t} dt = \delta(\nu + \nu') \tag{2.21}$$

with the special property of the δ-distribution

$$\int_{-\infty}^{\infty} f(x) \delta(x - x_0) dx = f(x_0)' \tag{2.22}$$

we obtain after interchanging the integrations in (2.19)

$$R(\vec{x}_1, \vec{x}_2, \tau) = \frac{1}{T} \int_{-T/2}^{T/2} \widehat{u}(\vec{x}_1, \nu) \widehat{u}(\vec{x}_2, -\nu) e^{-2\pi i \nu \tau} d\nu \tag{2.23}$$

or finally

$$S(\vec{x}_1, \vec{x}_2, \nu) = \frac{1}{T} u(\vec{x}_1, -\nu)\widehat{u}(\vec{x}_2, +\nu). \tag{2.24}$$

Hence, the Fourier transform of the autocorrelation function of a signal with frequency ν is the power spectrum of the signal. For $\vec{x}_1 = \vec{x}_2$ the relation (2.24) is called Wiener-Chinchin theorem, and the Fourier transformed of $S(\vec{x}_1, \vec{x}_1, \nu)$ equals the intensity of the signal $u(\vec{x}_1, t)$ at frequency ν.

2.2.3 *Characteristic Function*

The characteristic function is the Fourier transform of the probability density function. Therefore, the characteristic function of any real-valued random variable completely defines its probability distribution. This means that instead of working directly with the probability distribution, as an alternative way equations for the characteristic function can be treated. The transformation of equations of motion for probabilities into equations for their characteristic function is sometimes very helpful (Shephard 1991; Devore and Berk 2007).

The One-dimensional Case

The very special expectation value of $\exp(i\omega n)$

$$F_1(\omega) \equiv E\left(e^{i\omega n}\right) = \sum_n e^{i\omega n} P(n,t) \quad \text{or} \quad F_1(\omega) \equiv E\left(e^{i\omega x}\right) = \int_{-\infty}^{\infty} e^{i\omega x} P(x,t) dx \tag{2.25}$$

is named characteristic function of the probability distribution $P(n, t)$. Apparently, $F_1(\omega)$ is the Fourier transform of $P(n, t)$ or $P(x, t)$, respectively.

If $\omega = 0$, obviously

$$F_1(0) \equiv E\left(e^0\right) = \sum_n P(n,t) = 1 \quad \text{or} \quad F_1(0) \equiv E\left(e^0\right) = \int_{-\infty}^{\infty} P(x,t) dx = 1 \tag{2.26}$$

holds. It can also be proved that

$$|F_1(\omega)| \leq F_1(0) = 1 \tag{2.27}$$

is guaranteed. Why is there such an interest in the field of statistics to develop and calculate the characteristic function? To understand this question, let us consider the Taylor expansion of the exponential expression used in the definition of the characteristic function (Abramowitz and Stegun 1972):

$$e^{i\omega n} = \sum_{k=0}^{\infty} \frac{(i\omega n)^k}{k!} \tag{2.28}$$

Equation (2.28) inserted into (2.25) shows that the characteristic function serves as a generating function for the moments:

$$F_1(\omega) \equiv E\left(e^{i\omega n}\right) = \sum_{k=0}^{\infty} \frac{i^k \omega^k}{k!} \sum_n n^k P(n,t) = \sum_{k=0}^{\infty} \frac{i^k \omega^k \overline{n^k}}{k!} \tag{2.29}$$

or

$$F_1(\omega) \equiv E\left(e^{i\omega n}\right) = \sum_{k=0}^{\infty} \frac{i^k \omega^k}{k!} \int_{-\infty}^{\infty} x^k P(x,t) dx = \sum_{k=0}^{\infty} \frac{i^k \omega^k \overline{x^k}}{k!}. \tag{2.30}$$

If the power series in (2.29), (2.30) is absolutely convergent for $\omega > 0$, then $F_1(\omega)$ has a Fourier transform, namely $P(n,t)$. In other words, if all moments $\overline{n^k}$ up to the order ∞ are known, the complete probability distribution $P(n,t)$ can be constructed. Furthermore, from the definition (2.29), (2.30) it follows immediately

$$\overline{n^k} = \frac{1}{i^k} \frac{\partial^k F_1(\omega)}{\partial \omega^k}\bigg|_{\omega=0} \quad \text{or} \quad \overline{x^k} = \frac{1}{i^k} \frac{\partial^k F_1(\omega)}{\partial \omega^k}\bigg|_{\omega=0}. \tag{2.31}$$

Using the characteristic function, all moments of the distribution can be easily obtained by differentiation. Since it is often rather easy to derive directly from the Master equation an equation for the characteristic function, all moments of interest can be calculated using (2.31). This is sometimes a better strategy than to solve the Master equation for $P(n,t)$ directly and afterwards to calculate the moments via weighted integration (2.4).

The L-dimensional Case

The characteristic function in the L-dimensional case is defined accordingly

$$F_L(\omega_1, \omega_2, \ldots, \omega_L) \equiv E\left(e^{i\sum_k n_k \omega_k}\right) = \sum_{\vec{n}} e^{i\sum_k n_k \omega_k} P(n_1, n_2, \ldots, n_L, t) \tag{2.32}$$

or in the continuous case.

$$F_L(\omega_1, \omega_2, \ldots, \omega_L) \equiv E\left(e^{i\sum_k x_k \omega_k}\right)$$

$$= \int_{-\infty}^{\infty} \cdots \int_{-\infty}^{\infty} e^{i\sum_k x_k \omega_k} P(x_1, x_2, \ldots, x_L, t) dx_1 dx_2 \ldots dx_L. \tag{2.33}$$

In this case, it can also be proved that

$$F_L(0,0,\ldots,0) = 1 \quad \text{and} \quad |F_1(\vec{\omega})| \leq 1 \tag{2.34}$$

Furthermore, it is obvious that

$$F_L(\omega_1,\omega_2,\ldots,\omega_{L-1},0) = F_{L-1}(\omega_1,\omega_2,\ldots,\omega_{L-1}) \tag{2.35}$$

In analogy to the one-dimensional case, the moments can be calculated via

$$\overline{n_1^a n_2^b \ldots n_L^g} = \frac{1}{i^{a+b+\ldots g}} \frac{\partial^a}{\partial\omega_1^a} \frac{\partial^b}{\partial\omega_2^b} \cdots \frac{\partial^g}{\partial\omega_L^g} F_L(\omega_1,\omega_2,\ldots\omega_L)\bigg|_{\omega_k=0,\ k=1,\ldots,L} . \tag{2.36}$$

The integer coefficients $a \geq 0, b \geq 0, \ldots, g \geq 0$ can be zero as well.

2.2.4 *Generating Function*

The One-dimensional Case

Let the probability distribution of a discrete random variable n be given by

$$P_1(n) = a_n \quad \text{for} \quad n = 0,1,2,\ldots . \tag{2.37}$$

where $a_n \geq 0$ represents an infinite sequence of non-negative numbers. A generating function describes this infinite sequence of numbers a_n by treating them like the coefficients of a power series expansion (Papoulis 1984). The sum of this infinite series is usually called a probability generating function

$$G_1(z) = E(z^n) = \sum_{n=0}^{\infty} a_n z^n = \sum_{n=0}^{\infty} P_1(n) z^n, \tag{2.38}$$

where $P_1(n)$ is the probability mass function of n. The power series converges absolutely at least for all complex numbers z with $|z| \leq 1$; in many examples, the radius of convergence is larger.

We immediately obtain the coefficients a_n, in other words, the probability mass, with the aid of the Taylor expansion of $G_1(z)$ by taking the n-th derivative of $G_1(z)$ and dividing by $n!$

$$a_n = \frac{1}{n!} \frac{\partial^n G_1(z)}{\partial z^n}\bigg|_{z=0} . \tag{2.39}$$

The big advantage of using the generating function (2.38) and formula (2.39) trace back to the fact that in a number of important practical applications $G_1(z)$ is explicitly defined and can directly be calculated on the basis of the Master equation.

In a further step, one may easily calculate expectation values. We will apply the idea of the probability generating function to several examples in Chap. 5.

The L-dimensional Case
If $\vec{n} = (n_1, n_2, \ldots, n_L)$ is a discrete random variable taking values in the L-dimensional non-negative integer lattice, then the probability generating function of $\vec{n}$ is defined as

$$G_L(\vec{z}) = E({z_1}^{n_1}, {z_2}^{n_2}, \ldots, {z_L}^{n_L}) = \sum_{n_1, n_2, \ldots, n_L = 0}^{\infty} P_L(n_1, \ldots, n_L){z_1}^{n_1}{z_2}^{n_2} \ldots {z_L}^{n_L}, \tag{2.40}$$

where $P_L(\vec{n})$ is the probability mass function of $\vec{n} = (n_1, n_2, \ldots, n_L)$. The power series converges absolutely at least for all complex vectors

$$\vec{z} = (z_1, z_2, \ldots, z_L) \quad \text{with} \quad \max(|z_1|, \ldots, |z_1|) \leq 1. \tag{2.41}$$

Since the probability $P_L(\vec{n})$ is normalized, the probability generating function fulfils

$$G_L(\vec{1}) = 1. \tag{2.42}$$

Probability generating functions obey all properties of power series with non-negative coefficients (Johnson et al. 1993).

2.3 Examples

Five examples of important statistical distributions, namely: binomial distribution, Poisson distribution, normal distribution, Gumbel distribution and Pareto distribution will be discussed in a next step. The different distributions play an important role in the examples of the coming chapters.

2.3.1 *Binomial Distribution*

The binomial distribution is the discrete probability distribution of the number of successes n in a sequence of N independent yes/no experiments, where the probability p for success of each single experiment is the same for each outcome. In other words, the binomial distribution describes the behaviour of a random variable n if the following conditions apply:

- The number of observations N is fixed.
- Each observation is independent.
- Each observation represents one of two outcomes ("success" or "failure").

- The probability of "success" p is the same for each outcome.

The binomial distribution is frequently used to model the number of successes in a sample of size N drawn with replacement. If the experiments are carried out without replacement, the outcomes are not independent and the resulting distribution is a hypergeometric distribution (Johnson et al. 1993).

If the above mentioned conditions are met, then n has a binomial distribution with parameters N and p, and the probability that a random variable n with binomial distribution is equal to the value n, where $n = 0, 1, 2, \ldots, N$ is given by

$$P_1(n) = \binom{N}{n} p^n (1-p)^{N-n} \tag{2.43}$$

with the expression

$$\binom{N}{n} = \frac{N!}{(N-n)!n!} \tag{2.44}$$

called the binomial coefficient.

The characteristic function $F_1(\omega)$ for the binomial distribution is

$$F_1(\omega) = \langle e^{i\omega n} \rangle = \sum_{n=0}^{N} \binom{N}{n} p^n (1-p)^{N-n} e^{i\omega n} = (1 - p + pe^{i\omega})^N. \tag{2.45}$$

The Eq. (2.31) leads to the mean value

$$\langle n \rangle = Np \tag{2.46}$$

the variance of the binomial distribution

$$\sigma^2 = Np(1-p) \tag{2.47}$$

and the skewness

$$\gamma = \frac{1-2p}{\sqrt{Np(1-p)}}. \tag{2.48}$$

The probability generating function $G_1(z)$ for the binomial distribution reads

$$G_1(z) = \langle z^n \rangle = \sum_{n=0}^{N} \binom{N}{n} p^n (1-p)^{N-n} z^n = (1 - p + pz)^N. \tag{2.49}$$

In Fig. 2.5 the binomial distribution function (2.43) is shown for $N = 50$ and for two different probabilities for success: $p = 0.25$ ($\langle n \rangle = 12.5$; $\sigma^2 = 9.38$) and $p = 0.50$ ($\langle n \rangle = 25$; $\sigma^2 = 12.5$).

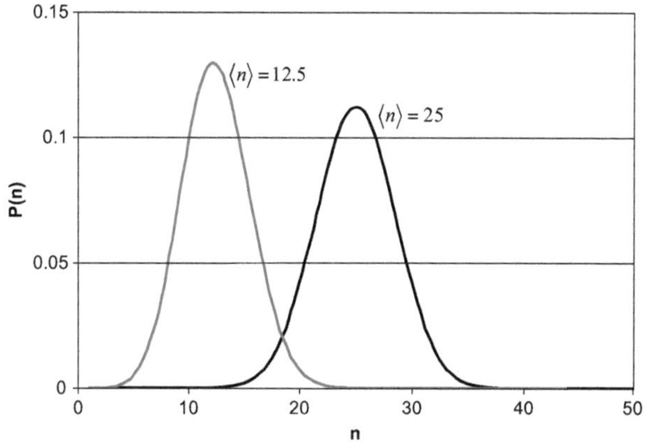

Fig. 2.5 Binomial distribution for $\langle n \rangle = 12.5$, $\sigma^2 = 9.38$ and $\langle n \rangle = 25$, $\sigma^2 = 12.5$

2.3.2 Poisson Distribution

In many practical applications one may consider N and also n as very large numbers, while the mean value $\langle n \rangle = Np$ is fixed and p tends to zero. Under the conditions: $N \to \infty$, $\langle n \rangle$ fixed and $p \to 0$, the binomial distribution may be transformed into the Poisson distribution. For this aim, suppose the sample size N becomes large, then, the distribution (2.43) becomes

$$\begin{aligned} P_1(n) &= \lim_{N\to\infty} \binom{N}{n} \left(\frac{\langle n \rangle}{N}\right)^n \left(1 - \frac{\langle n \rangle}{N}\right)^{N-n} \\ &= \lim_{N\to\infty} \frac{N(N-1)\ldots(N-n+1)}{n!} \frac{\langle n \rangle^n}{N^n} \left(1 - \frac{\langle n \rangle}{N}\right)^N \left(1 - \frac{\langle n \rangle}{N}\right)^{-n} \\ &= \lim_{N\to\infty} \frac{N(N-1)\ldots(N-n+1)}{N^n} \frac{\langle n \rangle^n}{n!} \left(1 - \frac{\langle n \rangle}{N}\right)^N \left(1 - \frac{\langle n \rangle}{N}\right)^{-n} \\ &= \lim_{N\to\infty} \frac{N}{N} \cdot \frac{N-1}{N} \cdots \frac{N-n+1}{N} \frac{\langle n \rangle^n}{n!} \left(1 - \frac{\langle n \rangle}{N}\right)^N \left(1 - \frac{\langle n \rangle}{N}\right)^{-n} \\ &= \lim_{N\to\infty} 1 \cdot \frac{\langle n \rangle^n}{n!} \left(1 - \frac{\langle n \rangle}{N}\right)^N \cdot 1 \end{aligned} \tag{2.50}$$

If we now perform the limit $N \to \infty$ keeping $\langle n \rangle$ and n fixed, the first factors in (2.50) would tend to 1, and the last factor in the expression as well. Furthermore, the second term in (2.50) tends to the exponential function (Abramowitz and Stegun 1972)

$$\lim_{N\to\infty} \left(1 - \frac{\langle n \rangle}{N}\right)^N = e^{-\langle n \rangle} \tag{2.51}$$

and finally, the Poisson distribution is obtained (Papoulis 1984; Pfeiffer and Schum 1973):

$$P_1(n) = \frac{\langle n \rangle^n}{n!} e^{-\langle n \rangle}. \tag{2.52}$$

As expected, the Poisson distribution is normalized, since

$$\sum_{n=0}^{\infty} P_1(n) = e^{-\langle n \rangle} \sum_{n=0}^{\infty} \frac{\langle n \rangle^n}{n!} = e^{-\langle n \rangle} e^{\langle n \rangle} = 1 \tag{2.53}$$

A famous example of the Poisson distribution discussed by Bortkiewicz (1898) refers to the number of people of the Prussian army killed by horse kicks. In this particular example, it is assumed that the number of trials goes to a very large number, but the probability of a horseman to be killed goes to zero.

The characteristic function $F_1(\omega)$ for the Poisson distribution reads

$$\begin{aligned} F_1(\omega) &= \langle e^{i\omega n} \rangle = \sum_{n=0}^{\infty} e^{-\langle n \rangle} \frac{\langle n \rangle^n}{n!} e^{i\omega n} = e^{-\langle n \rangle} \sum_{n=0}^{\infty} \frac{(\langle n \rangle e^{i\omega})^n}{n!} \\ F_1(\omega) &= \exp(\langle n \rangle (e^{i\omega} - 1)) \end{aligned} \tag{2.54}$$

Applying (2.41) leads to the mean value

$$\bar{n} = \langle n \rangle, \tag{2.55}$$

the variance of the Poisson distribution

$$\sigma^2 = \langle n \rangle, \tag{2.56}$$

and the skewness

$$\gamma = \frac{1}{\sqrt{\langle n \rangle}}. \tag{2.57}$$

In analogy to (2.54), the probability generating function $G_1(z)$ for the Poisson distribution can be calculated

$$\begin{aligned} G_1(z) &= \langle z^n \rangle = \sum_{n=0}^{\infty} e^{-\langle n \rangle} \frac{\langle n \rangle^n}{n!} z^n = \exp\langle n \rangle (z - 1) \\ G_1(z) &= \exp\langle n \rangle (z - 1) \end{aligned} \tag{2.58}$$

In Fig. 2.6, the Poisson distribution function (2.52) is shown for three different mean values: $\langle n \rangle = 10$, $\langle n \rangle = 20$, and $\langle n \rangle = 30$.

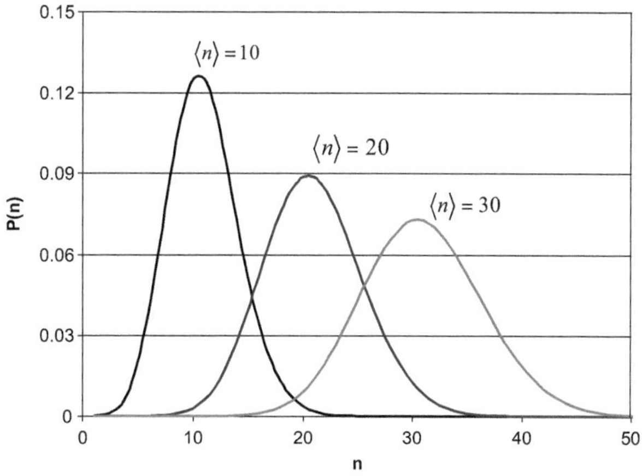

Fig. 2.6 Poisson distribution for $\langle n \rangle = 10$, $\langle n \rangle = 20$, and $\langle n \rangle = 30$

2.3.3 Normal Distribution

The normal distribution or Gaussian distribution is a very important distribution in statistics. The normal distribution can also be obtained as a limiting case of the binomial distribution for $N \to \infty$ and $p = 1/2$. Furthermore, we introduce a new variable and require that the mean value of the new variable x corresponds to $\langle x \rangle$ and the variance σ^2 remains finite. Since according to (2.47) the variance tends to infinity for $N \to \infty$, implying a transition from a discrete to a continuous variable. In the limiting case $N \to \infty$ we obtain the normal distribution (Devore and Berk 2007)

$$P_1(x) = \frac{1}{\sqrt{2\pi\sigma^2}} e^{-\frac{(x-\langle x \rangle)^2}{2\sigma^2}} \quad \text{for} \quad -\infty < x < \infty, \tag{2.59}$$

where $-\infty < \langle x \rangle < \infty$ is the mean value and $0 < \sigma^2 < \infty$ the variance of the continuous probability distribution (2.59). Normal distributions are important in statistics and are often used in the social sciences to represent real-valued random variables whose distributions are not known.

The central limit theorem states that averages of random variables independently drawn from independent distributions converge to the normal distribution. In other words, they become normally distributed when the number of random variables is sufficiently large (Engelberg 2007). Physical quantities that are expected to be the sum of many independent processes (such as measurement errors) often have distributions that are nearly normal. Moreover, many results and methods (such as propagation of uncertainty and least squares parameter fitting) can be derived analytically in explicit form when the relevant variables are normally distributed.

The great importance of the normal distribution is based on the observation that the scaled variable

$$\sqrt{n}\frac{x/n - \langle x \rangle}{\sigma} \tag{2.60}$$

is approximately normal distributed for large values of n.

As expected, the normal distribution is normalized

$$\int_{-\infty}^{\infty} P_1(x)dx = \int_{-\infty}^{\infty} \frac{1}{\sqrt{2\pi\sigma^2}} e^{-\frac{(x-\langle x \rangle)^2}{2\sigma^2}} dx = 1. \tag{2.61}$$

The characteristic function $F_1(\omega)$ for the normal distribution reads

$$\begin{aligned} F_1(\omega) &= \langle e^{i\omega x} \rangle \\ &= \int_{-\infty}^{\infty} dx \frac{1}{\sqrt{2\pi\sigma^2}} e^{-\frac{(x-\langle x \rangle)^2}{2\sigma^2}} e^{i\omega x} = \frac{1}{\sqrt{2\pi\sigma^2}} e^{-\frac{\langle x \rangle^2}{2\sigma^2}} \int_{-\infty}^{\infty} dx e^{-\frac{x^2}{2\sigma^2} + \left(\frac{\langle x \rangle}{\sigma^2} + i\omega\right)x} \\ &= \frac{1}{\sqrt{2\pi\sigma^2}} e^{-\frac{\langle x \rangle^2}{2\sigma^2}} \sqrt{2\pi\sigma^2} e^{+\left(\frac{\langle x \rangle}{\sigma^2} + i\omega\right)^2 \frac{\sigma^2}{2}} \\ F_1(\omega) &= \exp(i\langle x \rangle \omega - \sigma^2\omega^2/2) \end{aligned} \tag{2.62}$$

For the derivation of (2.62) we used the following result for complex parameters α, β with $\mathrm{Re}(\alpha) > 0$

$$\int_{-\infty}^{\infty} dx e^{-\alpha x^2 + \beta x} = \sqrt{\frac{\pi}{\alpha}} e^{\beta^2/(4\alpha)} \tag{2.63}$$

Applying (2.31) with (2.62) leads to the mean value.

$$\bar{x} = \langle x \rangle, \tag{2.64}$$

the variance of the normal distribution

$$\sigma^2 = \left\langle (x - \langle x \rangle)^2 \right\rangle \tag{2.65}$$

and the skewness.

$$\gamma = 0. \tag{2.66}$$

In Fig. 2.7, the normal distribution function (2.59) is shown for $\langle n \rangle = 75$ and for three different variance values: $\sigma^2 = 10$, $\sigma^2 = 20$ and $\sigma^2 = 40$.

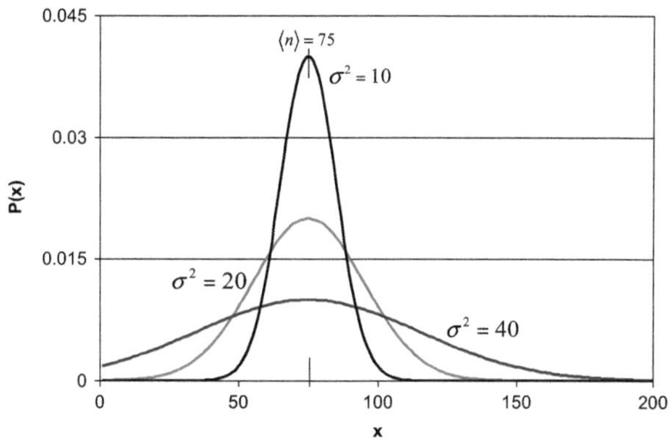

Fig. 2.7 Normal distribution for $\langle n \rangle = 75$ and $\sigma^2 = 10$, $\sigma^2 = 20$, and $\sigma^2 = 40$

2.3.4 Gumbel Distribution

The Gumbel distribution (Gumbel 1941, 1954) is a standard distribution for the modelling of extremal events (Embrechts et al. 1991), namely the distribution of the maximum (or the minimum) of a number of samples of various distributions. For example, it is used for the estimation of the distribution of an extreme river level in a particular year, given a set of maxima values for past years, or for predicting an extreme earthquake or other natural disasters. It can be shown that the Gumbel distribution is likely to be useful if the distribution of the underlying sample data is of normal or exponential type. The Gumbel distribution is a particular case of the generalized extreme value distribution (also known as the Fisher-Tippett distribution).

The probability density function of the Gumbel distribution reads

$$P_1(z) = \frac{1}{\beta} e^{-(z+e^{-z})}, \quad \text{where} \quad z = \frac{x-\mu}{\beta}, \tag{2.67}$$

where μ (mode) is the maximum of the distribution function and β is related to the variance of the distribution.

The mean value is given by

$$\langle x \rangle = \mu + \beta\gamma, \tag{2.68}$$

where $\gamma \approx 0.5772$ is the Euler–Mascheroni constant (Lehmer 1975).

The variance of this Gumbel distribution is

$$\sigma^2 = \frac{\pi^2}{6}\beta^2. \tag{2.69}$$

The characteristic function $F_1(\omega)$ for the Gumbel distribution reads

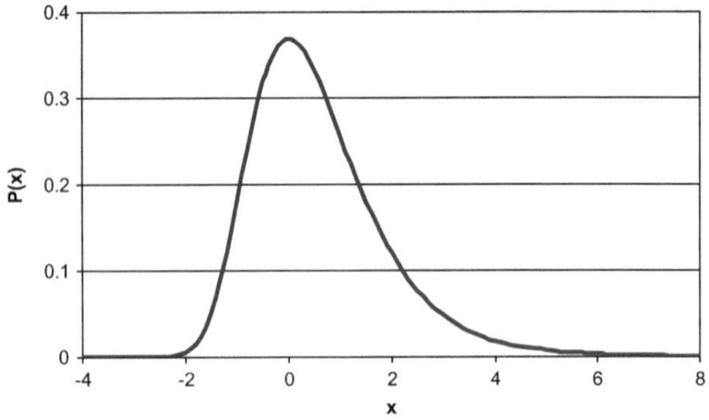

Fig. 2.8 Gumbel distribution for $\beta=1, \mu=0$ and $\langle x\rangle=\gamma$ and $\sigma^2=\pi^2/6$

$$F_1(\omega) = \left\langle e^{i\omega x}\right\rangle = \Gamma(1 - i\beta\omega)\exp(i\beta\omega), \tag{2.70}$$

where $\Gamma(x)$ is the Gamma function.

In many practical applications the standard form of the Gumbel distribution is used, characterized by the parameters $\mu=0$ and $\beta=1$ (see Fig. 2.8), since β has only an influence on the variance of the distribution, but not on the position of its maximum.

In this case the Gumbel probability density function reads

$$P_1(x) = e^{-(x+e^{-x})} \tag{2.71}$$

with the cumulative double-exponential distribution function

$$\int_{-\infty}^{x} P_1(x')dx' = e^{-e^{-x}} \tag{2.72}$$

In this particular case, the maximum of the distribution (mode) is located at zero, and the mean value is given by

$$\langle x\rangle = \gamma \approx 0.5772 \tag{2.73}$$

In Fig. 2.8, the Gumbel distribution (2.67) is shown for $\beta=1, \mu=0$, leading to $\langle x\rangle=0.5772$ and to $\sigma^2=\pi^2/6 = 1.65$.

2.3.5 *Pareto Distribution*

The Pareto distribution is often presented as a tail function, which gives the probability of finding larger values than a given value x. The Italian economist Alfredo Pareto (1848–1923) proposed this log-linear relationship for modelling the number of people N with income larger than the specific value x

$$\log N = \log A - \alpha \log x \quad \text{or} \quad N = Ax^{-\alpha} \tag{2.74}$$

This power law distribution (2.74) is often used for the description of scale-invariant social and physical phenomena.

Let x be a random variable, then the Pareto probability density function is given by

$$P_1(x) = \begin{cases} \dfrac{\alpha x_m^\alpha}{x^{\alpha+1}} & \text{for} \quad x \geq x_m \\ 0 & \text{for} \quad x < x_m \end{cases} \tag{2.75}$$

where $x_m > 0$ represents the possible minimum value of x, and $\alpha > 0$ is the Pareto coefficient, sometimes also called Pareto index or tail index.

All segments of the distribution are self-similar (scale invariant) and represent a straight line when plotted in a log-log diagram.

The probability density function is normalized

$$\int_{x_m}^{\infty} P_1(x)dx = \alpha x_m^\alpha \int_{x_m}^{\infty} x^{-(\alpha+1)}dx = 1 \tag{2.76}$$

in the interval $x \geq x_m \geq 0$.

The characteristic function $F_1(\omega)$ of the Pareto distribution (2.75) reads

$$\begin{aligned} F_1(\omega) &= \langle e^{i\omega x} \rangle = \int_{x_m}^{\infty} e^{i\omega x} P_1(x) = \int_{x_m}^{\infty} e^{i\omega x} \alpha \frac{x_m^\alpha}{x^{\alpha+1}} dx \\ &= \alpha(-i\omega x_m)^\alpha \Gamma(-\alpha, -i\omega x_m) \end{aligned} \tag{2.77}$$

where $\Gamma(a,x) = \int_x^\infty e^{-t} t^{a-1} dt$ is the incomplete Gamma function (Abramowitz and Stegun 1972).

The mean value of the Pareto distribution is given by

$$\langle x \rangle = \begin{cases} \infty & \text{for} \quad \alpha \leq 1 \\ \dfrac{\alpha x_m}{\alpha - 1} & \text{for} \quad \alpha > 1 \end{cases} \tag{2.78}$$

and the variance

$$\sigma^2(x) = \begin{cases} \infty & \text{for} \quad \alpha \in (1, 2] \\ \left(\frac{x_m}{\alpha-1}\right)^2 \dfrac{\alpha}{\alpha - 2} & \text{for} \quad \alpha > 2 \end{cases} \tag{2.79}$$

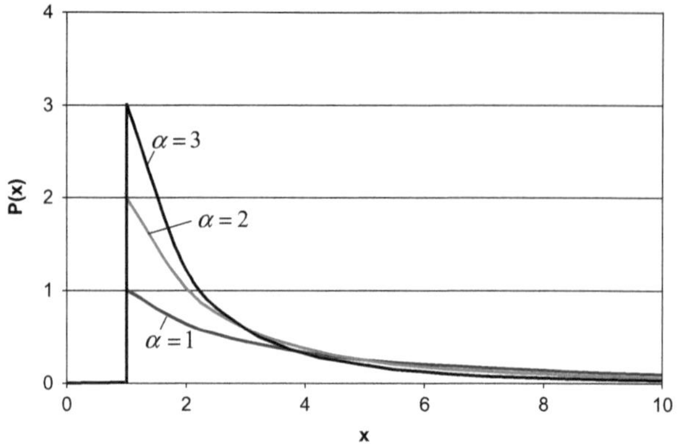

Fig. 2.9 Pareto distribution for $x_m = 1$ and $\alpha = 1$, $\alpha = 2$, and $\alpha = 3$

For $\alpha \leq 1$ the variance does not exist.

In Fig. 2.9, the Pareto distribution (2.75) is shown for $x_m = 1$, and $\alpha = 1$, $\alpha = 2$, and $\alpha = 3$.

References

Abramowitz M, Stegun IA (1972) Handbook of mathematical functions. Dover Publications, New York

Casti JL (1992) Reality rules: I. Wiley, New York

Devore J, Berk K (2007) Modern mathematical statistics with applications. Duxberry Press, Pacific Grove

Embrechts P, Klüppelberg C, Mikosch T (1991) Modelling extremal events. Springer, New York

Engelberg S (2007) Random signals and noise: a mathematical introduction. CRC Press, Boca Raton

Grimmett G, Stirzaker D (1992) Probability and random processes. Oxford University Press, Oxford

Gumbel EJ (1941) The return period of flood flows. Ann Math Stat 12:163–190

Gumbel EJ (1954) Statistical theory of extreme values and some practical applications, Applied mathematics series, vol 33. U.S. Department of Commerce, National Bureau of Standards, Washington, DC

Gut A (2005) Probability: a graduate course. Springer, New York

Haken H (1977) Synergetics: an introduction. Springer, Heidelberg

Johnson NL, Kotz S, AW K (1993) Univariate discrete distributions. Wiley, New York

Lehmer DH (1975) Euler constant for arithmetical progressions. Acta Arithm 27(1):125–142

Papoulis A (1984) Probability, random variables and stochastic processes. McGraw-Hill, New York

Pfeiffer PE, Schum DA (1973) Introduction to applied probability. Academic, New York

Shephard NG (1991) From characteristic function to distribution function: a simple framework fort he theory. Economet Theory 7:519–529

Spiegel MR (1992) Theory and problems of probability and statistics. McGraw-Hill, New York

Van Kampen NG (1978) An introduction to stochastic processes for physicists. In: Garido L, Seglar P, Shepherd PJ (eds) Stochastic processes in nonequilibrium systems. Proceedings, Sitges 1978. Springer, New York

von Bortkiewicz L (1898) Das Gesetz der kleinen Zahlen. B.G. Teubner, Leipzig

Weidlich W, Haag G (1983) Concepts and models of a quantitative sociology: the dynamics of interaction populations, Springer series of synergetics, vol 14. Springer, New York

Chapter 3
Derivation of the Chapman–Kolmogorov Equation and the Master Equation

Contents

3.1 The Chapman–Kolmogorov Equation

Before deriving the Master equation, it is useful to introduce some fundamental concepts of probability theory (van Kampen 1978, 1981; Gardiner 1983; Honerkamp 1998).

As before, it is assumed that the system can be in one of mutually exclusive discrete states, which are characterized by the vector $\vec{n} = (n_1, n_2, \ldots, n_L)$ consisting of one or a multiple of discrete numbers. In the course of time, transitions between the different states occur. The probability distribution function of finding the system in state or configuration $\vec{n}$ at time t is defined as

$$P(\vec{n}, t) \geq 0 \tag{3.1}$$

This configurational probability has the following statistical interpretation: in an ensemble consisting of a large number of identically prepared systems, so that each of them belongs to the same probability distribution, one may find systems in state $\vec{n}$ at time t with approximately the relative frequency $P(\vec{n}, t)$. Approaching the limit

G. Haag, *Modelling with the Master Equation*, DOI 10.1007/978-3-319-60300-1_3

of an infinite ensemble its systems are found with exactly the relative frequency $P(\vec{n},t)$.

As the system can be found with certainty in any one of its states (configurations) $\vec{n}$ at any time t, the probability distribution over all possible configurations has to satisfy the normalization condition:

$$\sum_{\vec{n}} P(\vec{n},t) = 1 \tag{3.2}$$

where the sum extends over all configurations $\vec{n}$.

Coming back to the conditional probability

$$p(\vec{n}_2,t_2|\vec{n}_1,t_1) \geq 0 \tag{3.3}$$

to find the system in state $\vec{n}_2$ at time t_2, given that it was with certainty in state $\vec{n}_1$ at time t_1. The conditional probability is fundamental for the dynamics of the system, since it describes how the probability spreads out in the time interval $(t_2 - t_1) \geq 0$, on condition that it was concentrated on state $\vec{n}_1$ at time t_1. The conditional probability may also depend on the precious history of the system, in other words on the states the system traversed before arriving in state $\vec{n}_1$ at time t_1. In this general case the probability evolution process may become very complicated (see Sect. 6.5.4).

Fortunately, in many cases the so-called Markov assumption holds, at least as a good approximation. This Markov assumption postulates that the evolution within time of the conditional probability $p(\vec{n}_2,t_2|\vec{n}_1,t_1)$ only depends on the initial state $\vec{n}_1$ at time t_1, but not on states of the system prior to t_1. In other words, after arriving at state $\vec{n}_1$, the system has lost its historic memory and previous states do not matter in the process of further evolution.

The following relations are a consequence of the definition of the conditional probability:

$$p(\vec{n}_2,t_1|\vec{n}_1,t_1) = \delta_{\vec{n}_2\vec{n}_1} \tag{3.4}$$

where

$$\delta_{\vec{n}_2\vec{n}_1} = \begin{cases} 1 & \text{for } \vec{n}_2 = \vec{n}_1 \\ 0 & \text{for } \vec{n}_2 \neq \vec{n}_1 \end{cases} \tag{3.5}$$

and

$$\sum_{\vec{n}_2} p(\vec{n}_2,t_2|\vec{n}_1,t_1) = 1 \tag{3.6}$$

Equation (3.4) follows because at time $t_2 = t_1$ the state $\vec{n}_1$ is taken with certainty. Equation (3.6) holds, since the system at time t_2 must be in one of the states $\vec{n}_2$ of the system.

Furthermore, let us recall the so-called joint probability:

$$p(\vec{n}_k, t_k; \vec{n}_{k-1}, t_{k-1}; \ldots; \vec{n}_2, t_2; \vec{n}_1, t_1) \tag{3.7}$$

This n-fold function is the joint probability to find the system in state $\vec{n}_1$ at time t_1, in state $\vec{n}_2$ at time t_2 and ... in state $\vec{n}_k$ at time t_k. From this definition, it follows that the lower order joint probabilities can be obtained from the higher order ones by the following reduction formula:

$$p(\vec{n}_3, t_3; \vec{n}_1, t_1) = \sum_{\vec{n}_2} p(\vec{n}_3, t_3; \vec{n}_2, t_2; \vec{n}_1, t_1) \tag{3.8}$$

or in the general case

$$\begin{aligned} &p(\vec{n}_k, t_k; \ldots; \vec{n}_{l+1}, t_{l+1}; \vec{n}_{l-1}, t_{l-1}; \ldots; \vec{n}_1, t_1) \\ &\quad = \sum_{\vec{n}_l} p(\vec{n}_k, t_k; \ldots; \vec{n}_{l+1}, t_{l+1}; \vec{n}_l, t_l; \vec{n}_{l-1}, t_{l-1}; \ldots; \vec{n}_1, t_1) \end{aligned} \tag{3.9}$$

Clearly, the summation in (3.8) over all possible states at time t_2 leads to the probability of being in state $\vec{n}_1$ at time t_1 and in state $\vec{n}_3$ at time t_3 irrespective of the intermediate state at time t_2.

By introducing the Markov assumption, all joint probabilities can be expressed in terms of the conditional probability (3.3) and the configurational probability distribution (3.1). In particular, the two-fold joint probability clearly reads:

$$p(\vec{n}_2, t_2; \vec{n}_1, t_1) = p(\vec{n}_2, t_2 | \vec{n}_1, t_1) P(\vec{n}_1, t_1). \tag{3.10}$$

Since the probability to find the system in state $\vec{n}_1$ at time t_1 and in state $\vec{n}_2$ at time t_2 is synonymous with the probability to find it in state $\vec{n}_1$ at time t_1 multiplied with the conditional probability to find it in state $\vec{n}_2$ at time t_2, given that it was in $\vec{n}_1$ at t_1. Generalizing this consideration and assuming that the conditional probability does not depend on the previous history—the Markov assumption—we obtain:

$$\begin{aligned} p(\vec{n}_k, t_k; \vec{n}_{k-1}, t_{k-1}; \ldots; \vec{n}_2, t_2; \vec{n}_1, t_1) = p(\vec{n}_k, t_k | \vec{n}_{k-1}, t_{k-1}) \ldots p(\vec{n}_2, t_2 | \vec{n}_1, t_1) \\ \times P(\vec{n}_1, t_1) \end{aligned} \tag{3.11}$$

The composition formulas for the joint probabilities (3.10) and (3.11) may now be combined with reduction formula (3.8). Taking the sum over $\vec{n}_1$ in (3.10) and using (3.7) yield:

$$P(\vec{n}_2,t_2) = \sum_{\vec{n}_1} p(\vec{n}_2,t_2;\, \vec{n}_1,t_1) = \sum_{\vec{n}_1} p(\vec{n}_2,t_2|\vec{n}_1,t_1)P(\vec{n}_1,t_1) \tag{3.12}$$

Equation (3.12) shows that the probability distribution $P(\vec{n}_1,t_1)$ is propagated in the time interval $(t_2 - t_1) \geq 0$ by means of the conditional probability $p(\vec{n}_2,t_2|\vec{n}_1,t_1)$ also called the propagator to the distribution $P(\vec{n}_2,t_2)$. In other words, if one knows the conditional probability, the whole dynamics of the system, the whole process is well known. The Master equation which will be considered next is nothing but a tool to calculate the conditional probability based on the internal structure of the system in a systematic way.

Inserting (3.10 and 3.11) in (3.8) yield

$$p(\vec{n}_3,t_3|\vec{n}_1,t_1)P(\vec{n}_1,t_1) = \sum_{\vec{n}_2} p(\vec{n}_3,t_3|\vec{n}_2,t_2)p(\vec{n}_2,t_2|\vec{n}_1,t_1)P(\vec{n}_1,t_1) \tag{3.13}$$

Since (3.13) must hold for an arbitrary initial distribution $P(\vec{n}_1,t_1)$ one can conclude that

$$p(\vec{n}_3,t_3|\vec{n}_1,t_1) = \sum_{\vec{n}_2} p(\vec{n}_3,t_3|\vec{n}_2,t_2)p(\vec{n}_2,t_2|\vec{n}_1,t_1) \tag{3.14}$$

also holds. Equation (3.14) is the well-known *Chapman–Kolmogorov* equation. The propagator of the probability distribution from t_1 to t_3 namely $p(\vec{n}_3,t_3|\vec{n}_1,t_1)$ can be decomposed into a product of propagators from the initial state at t_1 over all possible intermediate states $\vec{n}_2$ at t_2 to the final state $\vec{n}_3$ at t_3. The *Chapman–Kolmogorov* equation can also be seen as definition of a Markov process.

3.2 Derivation of the Master Equation

According to (3.12) and (3.14), the conditional probability is the crucial quantity determining the evolution with time of any probability distribution $P(\vec{n},t)$. The Master equation is nothing but a differential equation in time for the propagator or the probability distribution itself. For its derivation let us consider (3.12) for times $t_1 = t$ and $t_2 = t + \tau$, where τ is an infinitesimally short time interval. Proceeding in this way, we obtain the short-time evolution equation

$$P(\vec{n}_2,t+\tau) = \sum_{\vec{n}_1} p(\vec{n}_2,t+\tau|\vec{n}_1,t)P(\vec{n}_1,t) \tag{3.15}$$

The short-time propagator is now being expanded in a Taylor series around t with respect to $t_2 = t + \tau$ yielding

$$p(\vec{n}_2, t+\tau|\vec{n}_1, t_1) = p(\vec{n}_2, t|\vec{n}_1, t) + \tau \left. \frac{\partial p(\vec{n}_2, t_2|\vec{n}_1, t)}{\partial t_2} \right|_{t_2=t} + O(\tau^2) \tag{3.16}$$

Making use of (3.4) and (3.5) in (3.16), we obtain

$$p(\vec{n}_2, t|\vec{n}_1, t) = \delta_{\vec{n}_2 \vec{n}_1}$$

and

$$\sum_{\vec{n}_2} \left. \frac{\partial p(\vec{n}_2, t_2|\vec{n}_1, t)}{\partial t_2} \right|_{t_2=t} = 0 \tag{3.17}$$

where the sum extends over all states $\vec{n}_2$ of the system. Re-inserting (3.17) in (3.16) results in

$$p(\vec{n}_2, t+\tau|\vec{n}_1, t_1) = \tau \ w_t(\vec{n}_2, \vec{n}_1) \quad \text{for } \vec{n}_1 \neq \vec{n}_2 \tag{3.18}$$

and

$$p(\vec{n}_2, t+\tau|\vec{n}_1, t_1) = 1 - \tau \sum_{\vec{n}_1} w_t(\vec{n}_2, \vec{n}_1) \quad \text{for } \vec{n}_1 = \vec{n}_2 \tag{3.19}$$

where the probability transition rate

$$w_t(\vec{n}_2, \vec{n}_1) = \left. \frac{\partial p(\vec{n}_2, t_2|\vec{n}_1, t)}{\partial t_2} \right|_{t_2=t} \geq 0 \tag{3.20}$$

has been introduced and higher order terms in τ can be neglected in the limiting case $\tau \to 0$. Equation (3.18) states on the one hand, that given the system was in state $\vec{n}_1$, the probability to reach state $\vec{n}_2$ in the infinitesimally short time interval τ will be proportional to that time interval and to the transition rate $w_t(\vec{n}_2, \vec{n}_1) \geq 0$ from $\vec{n}_1$ to $\vec{n}_2$. On the other hand, the probability to remain in the same state during the interval τ is 1 minus the probability transferred to all other states within the time interval (3.19).

The transition rates are sometimes confused with probabilities. However, the $w(\vec{n}_2, \vec{n}_1)$ are rates, in other words, transitions per time unit. The physical dimension of the conditional probability is [1] and the transition rates $w(\vec{n}_2, \vec{n}_1)$ have the dimension [1/*time*].

By inserting (3.18) and (3.19) in (3.15), dividing the left and right-hand sides by τ, performing some trivial arrangements and taking the limit $\tau \to 0$ with

$$\frac{dP(\vec{n},t)}{dt} = \lim_{\tau \to 0} \frac{P(\vec{n},t+\tau) - P(\vec{n},t)}{\tau} \tag{3.21}$$

the general form of the Master equation is obtained

$$\frac{dP(\vec{n},t)}{dt} = \sum_{\vec{m}} w_t(\vec{n},\vec{m})P(\vec{m},t) - \sum_{\vec{m}} w_t(\vec{m},\vec{n})P(\vec{n},t) \tag{3.22}$$

where the sum extends over all states (configurations) $\vec{m}$. To avoid subscripted indices $\vec{n}$ and $\vec{m}$ are now introduced instead of $\vec{n}_1$ and $\vec{n}_2$, respectively.

The Master equation (3.22) can be interpreted in an illustrative way. The quantities $w_t(\vec{n},\vec{m})$ are transition probabilities per time unit or transition rates in the following sense: $w_t(\vec{n},\vec{m})P(\vec{m},t)$ is the probability transferred from state $\vec{m}$ to the state $\vec{n}$ per unit of time, which is also denoted as the probability flux from $\vec{m}$ to $\vec{n}$. Then (3.22) can be read as a rate equation for probabilities and the Master equation as a balance equation for probability flows. The change per unit of time of the probability of state $\vec{n}$ [left-hand side of (3.22)] is the sum of two terms with opposite effects. Firstly, there is a probability flux from all other states $\vec{m}$ into the state $\vec{n}$ [first term of the right-hand side of (3.22)]. Secondly, there is a probability flux out of state $\vec{n}$ into all other states $\vec{m}$ [second term of the right-hand side of (3.22)]. The change per time unit of the probability $P(\vec{n},t)$ is caused by the difference of those probability fluxes.

The Master equation provides the most detailed description about the evolution of a system under conditions of uncertainty or restricted information. The quantities representing the knowledge about the system are the transition rates, namely the transition probabilities per unit of time $w_t(\vec{n},\vec{m})$ to reach $\vec{n}$ from $\vec{m}$.

The transition rates contain all information we have about the process. In social sciences, the transition rates can often be inferred from phenomenological considerations (see Chaps. 6, 7, 8, and 9) based on plausibility arguments. In natural sciences, fundamental and substantive considerations may be the starting point in the construction of the transition rates (see Chap. 5) or the transition rates may be derived from first principles.

On the one hand, using the transition rates $w_t(\vec{n},\vec{m})$ the evolution with time of the configurational probability $P(\vec{n},t)$ is obtained. On the other hand, the propagation of $P(\vec{n},t)$ is given by the conditional probability (3.15). Therefore, via the Master equation and the transition rates the conditional probability, the Greens-function of the Master equation (Haag and Hänggi 1979, 1980), can be computed.

The Master equation represents a set of homogeneous first-order differential equations for the evolution with time of the configurational probability distribution $P(\vec{n},t)$. The configurational probability $P(\vec{n},t)$ represents the occupation probability of a cell $\vec{n} \in C$ in a highly dimensional configuration space $C = \{0,1,2,\ldots\}^L$ in analogy to the phase space in statistical physics.

The Master equation (3.22) is valid for all probability distributions especially for the conditional probability $p(\vec{n}, t, \vec{m}, t')$. The configuration space comprises all possible states of the considered system.

For many applications of the Master equation, it is convenient to introduce $\vec{m} = \vec{n} + \vec{k}$ in (3.22), where $\vec{k}$ extends over all neighbouring states with non-vanishing transition rates $w_t\left(\vec{n}, \vec{n} + \vec{k}\right) > 0$ and $w_t\left(\vec{n} + \vec{k}, \vec{n}\right) > 0$.

$$\begin{aligned}\frac{dP(\vec{n},t)}{dt} &= \sum_{\vec{k}} w_t\left(\vec{n}, \vec{n} + \vec{k}\right)P\left(\vec{n} + \vec{k}, t\right) - \sum_{\vec{k}} w_t\left(\vec{n} + \vec{k}, \vec{n}\right)P(\vec{n}, t) \\ &= \sum_{\vec{k}} \tilde{w}_t\left(-\vec{k}, \vec{n} + \vec{k}\right)P\left(\vec{n} + \vec{k}, t\right) - \sum_{\vec{k}} \tilde{w}_t\left(+\vec{k}, \vec{n}\right)P(\vec{n}, t)\end{aligned} \tag{3.23}$$

where $\vec{k} = (k_1, k_2, \ldots k_L)$ and $k_j = 0, \pm 1, \pm 2, \ldots$ In (3.23) another notation for the transition rates has been introduced as well, which is used for some mathematical treatments:

$$w_t\left(\vec{n} + \vec{k}, \vec{n}\right) = \tilde{w}_t\left(+\vec{k}, \vec{n}\right) \tag{3.24}$$

The first and the second line in (3.23) are identical, since the summation over $\vec{k}$ extends over all positive and negative values of $\vec{k}$ related to non-vanishing transition rates.

In case of a continuous configuration space, the Master equation can be appropriately modified, leading to

$$\frac{dP(\vec{x},t)}{dt} = \int w_t(\vec{x}, \vec{x}')P(\vec{x}', t)d\vec{x}' - \int w_t(\vec{x}', \vec{x})P(\vec{x}, t)d\vec{x} \tag{3.25}$$

Since, the Master equation approach generates a complex system of equations for the probability distribution, both exact and approximate methods of solution are highly valuable. A variety of such methods, involving more or less restrictive assumptions will be treated in the next chapter.

3.3 General Properties of the Master Equation

In this chapter, important theorems of the solutions of the Master equation are presented (Haken 1977). The proofs are purely mathematical and can be dropped in a first reading.

The following assumptions are made:

(a) time-independent transition rates

Suppose, the transition rates depend not explicitly on time

$$w_t(\vec{n}, \vec{m}) = w(\vec{n}, \vec{m}) \tag{3.26}$$

but only on its initial configuration state $\vec{m}$ and its final configuration state $\vec{n}$

(b) connectiveness of all configuration states

Presume each state (cell) in configuration space is connected with each other state (cell) by at least one chain of non-vanishing transition rates. Thus by writing $\vec{m} = \vec{n}_{j+1}, \vec{n} = \vec{n}_j$ we may pass from any initial state (cell) $\vec{n}_0$ by a chain of intermediate states (cells) $\vec{n}_k$ via transitions with transition rates $w(\vec{n}_{j+1}, \vec{n}_j) > 0$ to any state (cell) $\vec{n}_L$ in configuration space, so that

$$w(\vec{n}_L, \vec{n}_{L-1})\, w(\vec{n}_{L-1}, \vec{n}_{L-2}) \ldots w(\vec{n}_2, \vec{n}_1) w(\vec{n}_1, \vec{n}_0) > 0. \tag{3.27}$$

The following proofs are based on theorems of linear algebra, collected in mathematical text books.

3.3.1 *Normalization and Positiveness of the Probability Distribution*

The normalization property and the positiveness of the configurational probability $P(\vec{n}, t)$ are guaranteed for all times.

Proof

Suppose, at an initial time $t = 0$, the configurational probability $P(\vec{n}, t)$ is positive. Thus, it fulfils

$$0 \leq P(\vec{n}, 0) \leq 1 \tag{3.28}$$

for all $\vec{n}$ and $P(\vec{n}, t)$ is normalized

$$\sum_{\vec{m}} P(\vec{n}, 0) = 1 \tag{3.29}$$

then for all later times $t \geq 0$ the positiveness

$$0 \leq P(\vec{n}, t) \leq 1 \tag{3.30}$$

and normalization

$$\sum_{\vec{m}} P(\vec{n}, t) = 1 \tag{3.31}$$

of $P(\vec{n}, t)$ are guaranteed.

To prove this statement, we firstly sum up over all configurations (states) $\vec{n}$ in (3.22)

$$\sum_{\vec{n}} \frac{dP(\vec{n},t)}{dt} = \frac{d}{dt}\sum_{\vec{n}} P(\vec{n},t) = \sum_{\vec{n},\vec{m}} w(\vec{n},\vec{m})P(\vec{m},t) - \sum_{\vec{n},\vec{m}} w(\vec{m},\vec{n})P(\vec{n},t) = 0 \tag{3.32}$$

The normalization condition is guaranteed and remains normalized, if it has been normalized at the beginning.

Secondly, suppose that at time t_0 the probability of a certain configuration state $\vec{n}_0$ reaches zero

$$P(\vec{n}_0, t_0) = 0 \text{ and } P(\vec{n}, t_0) \geq 0 \quad \text{for } \vec{n} \neq \vec{n}_0 \tag{3.33}$$

Then the Master equation (3.22) exhibits

$$\frac{dP(\vec{n}_0,t)}{dt} = \sum_{\vec{m}} \underbrace{w(\vec{n}_0,\vec{m})}_{\geq 0} \underbrace{P(\vec{m},t_0)}_{\geq 0} - \sum_{\vec{m}} \underbrace{w(\vec{m},\vec{n}_0)}_{\geq 0} \underbrace{P(\vec{n}_0,t_0)}_{=0} \geq 0 \tag{3.34}$$

and $P(\vec{n}_0, t \geq t_0)$ cannot become negative.

3.3.2 *The Liouville Representation of the Master Equation and Eigen Values, Eigen States and Symmetrization*

The real part of the eigen values is always positive, $\mathrm{Re}(\lambda) \geq 0$, *and* $\mathrm{Im}(\lambda) = 0$ *holds in addition if detailed balance is fulfilled.*

Proof

The Master equation represents a linear system of difference-differential equations, which is discrete in configuration space and continuous in time. Formally the Master equation can be written as

$$\frac{dP(\vec{n},t)}{dt} = \sum_{\vec{m}} L(\vec{n},\vec{m})P(\vec{m},t) \text{ or } \frac{dP}{dt} = LP \tag{3.35}$$

with the Liouville operator

$$L(\vec{n},\vec{m}) = w(\vec{n},\vec{m}) - \delta_{\vec{n}\vec{m}} \sum_{\vec{m}'} w(\vec{m}',n) = L_{\vec{n}\vec{m}} \tag{3.36}$$

Inserting

$$P(\vec{m}, t) = \varphi_{\vec{m}} e^{-\lambda t} \tag{3.37}$$

into the Master equation (3.35) yields a set of linear algebraic equations for the eigen functions $\varphi_{\vec{m}}$ with eigen values λ

$$\sum_{\vec{m}} L_{\vec{n}\vec{m}} \varphi_{\vec{m}}^{(\alpha)} = -\lambda_\alpha \varphi_{\vec{m}}^{(\alpha)} \tag{3.38}$$

where the additional index α indicates a possible set of eigen values and eigen states.

Since the Liouville operator is not symmetric

$$L_{\vec{n}\vec{m}} \neq L_{\vec{m}\vec{n}} \tag{3.39}$$

the eigenvectors of the adjoint problem

$$\sum_{\vec{m}} \chi_{\vec{n}}^{(\alpha)} L_{\vec{n}\vec{m}} = -\lambda_\alpha \chi_{\vec{n}}^{(\alpha)} \tag{3.40}$$

are different from the eigenvectors of (3.38). The eigenvectors φ and χ form a bi-orthogonal system so that we can decompose $L_{\vec{n}\vec{m}}$ into

$$L_{\vec{n}\vec{m}} = -\sum_{\alpha} \lambda_\alpha \varphi_{\vec{n}}^{(\alpha)} \chi_{\vec{m}}^{(\alpha)} \tag{3.41}$$

In the next step, we symmetries the Liouville matrix $L_{\vec{n}\vec{m}}$ using the stationary solution $P_{st}(\vec{n})$ of the Master equation:

$$L_{\vec{n}\vec{m}}^{s} = w(\vec{n}, \vec{m}) \frac{P_{st}^{1/2}(\vec{m})}{P_{st}^{1/2}(\vec{n})} \tag{3.42}$$

for $\vec{n} \neq \vec{m}$ and the form (3.36) for $\vec{n} = \vec{m}$. Furthermore, it is assumed that detailed balance holds, namely

$$w(\vec{n}, \vec{m}) P_{st}(\vec{m}) = w(\vec{m}, \vec{n}) P_{st}(\vec{n}) \tag{3.43}$$

To prove that (3.42) is a symmetric matrix is very easy.

The meaning of this symmetrization for the eigen values λ becomes obvious, if we redefine the eigen states accordingly

$$\varphi_{\vec{n}}^{(\alpha)} = P_{st}^{1/2}(\vec{n}) \widetilde{\varphi}_{\vec{n}}^{(\alpha)} \tag{3.44}$$

and

$$\chi_{\vec{m}}^{(\alpha)} = P_{st}^{1/2}(\vec{m})\widetilde{\chi}_{\vec{m}}^{(\alpha)} \tag{3.45}$$

Inserting (3.44) in (3.38) finally yields

$$\sum_{\vec{m}} L_{\vec{n}\vec{m}}^{s} \widetilde{\varphi}_{\vec{m}}^{(\alpha)} = -\lambda_{\alpha}\widetilde{\varphi}_{\vec{m}}^{(\alpha)} \tag{3.46}$$

and in analogy inserting (3.45) in (3.40)

$$\sum_{\vec{m}} \widetilde{\chi}_{\vec{n}}^{(\alpha)} L_{\vec{n}\vec{m}}^{s} = -\lambda_{\alpha}\widetilde{\chi}_{\vec{n}}^{(\alpha)} \tag{3.47}$$

Since the Liouville matrix (3.42) $L_{\vec{n}\vec{m}}^{s}$ is symmetric, the eigenvectors $\widetilde{\varphi}$ may be identified with the $\widetilde{\chi}$. Thus, using (3.44) and (3.45) results into

$$\varphi_{\vec{n}}^{(\alpha)} = P_{st}(\vec{n})\chi_{\vec{n}}^{(\alpha)} \tag{3.48}$$

Properties of the eigen values λ can now be derived by using a well-known theorem of linear algebra for symmetric matrices.

$$-\lambda = Extr\left\{\frac{(\widetilde{\chi}L^{s}\widetilde{\chi})}{(\widetilde{\chi}\widetilde{\chi})}\right\} = Extr\left\{\frac{(\chi L\varphi)}{(\chi\varphi)}\right\} \tag{3.49}$$

with the abbreviation

$$(\chi L\varphi) = \sum_{\vec{n},\,\vec{m}} \chi_{\vec{n}} L_{\vec{n}\vec{m}}^{s} \varphi_{\vec{m}} \tag{3.50}$$

Finally, we obtain after some cumbersome calculations

$$\lambda = Extr\left\{\frac{1}{2}\frac{\sum\limits_{\vec{n},\,\vec{m}} (\chi_{\vec{m}} - \chi_{\vec{n}})^{2} w(\vec{n},\vec{m}) P_{st}(\vec{m})}{\sum\limits_{\vec{m}} \chi_{\vec{m}}^{2} P_{st}(\vec{m})}\right\} \geq 0 \tag{3.51}$$

From (3.51) it is evident that the eigen values are non-negative $\lambda \geq 0$ and $\mathrm{Im}(\lambda) = 0$. The eigen value $\lambda = 0$ belongs to the stationary state.

3.3.3 Existence of a Stationary State

There always exists one stationary solution $P_{st}(\vec{n})$ of the Master equation.

Proof
Suppose, $P_{st}(\vec{n})$ exists: then, the time derivative of the Master equation vanishes for all states enumerated by $\vec{n} = 1, 2, \ldots, c$, where c is the total number of states in configuration space. Then the stationary solution of the Master equation is determined by the following set of linear algebraic equations for $P_{st}(\vec{n})$:

$$0 = \sum_{\vec{m}} w_t(\vec{n}, \vec{m}) P_{st}(\vec{m}) - \sum_{\vec{m}} w_t(\vec{m}, \vec{n}) P_{st}(\vec{n}) \equiv \sum_{\vec{m}} L(\vec{n}, \vec{m}) P_{st}(\vec{m}). \quad (3.52)$$

The algebraic system of Eq. (3.52) has at least one nontrivial solution $P_{st}(\vec{n}) \neq 0$ if and only if the determinant of the coefficients (3.36) vanishes. This fact is obvious, since, the Eq. (3.52) are not independent of each other

$$\sum_{\vec{n}, \vec{m}} L(\vec{n}, \vec{m}) P_{st}(\vec{m}) = 0 \quad (3.53)$$

and also

$$\sum_{\vec{n}} L(\vec{n}, \vec{m}) = 0 \quad (3.54)$$

holds. The solution is only determined up to an arbitrary constant which is finally fixed by the normalization condition

$$\sum_{\vec{n}} P_{st}(\vec{n}) = 1. \quad (3.55)$$

The stationary solution is unique provided (3.27) is valid.

3.3.4 *Convergence of the Distribution to its Stationary State*

Any time-dependent solution $P(\vec{n}, t)$ of the Master equation approaches for $t \to \infty$ the same stationary solution $P_{st}(\vec{n})$, provided (3.26) *and* (3.27) *are fulfilled.*

Proof
For the proof, we introduce the information "Entropy"-function, which acts as a Lyapunov-function in configuration space

$$S(t) = -\sum_{\vec{n}} P(\vec{n}, t) \log\left(\frac{P(\vec{n}, t)}{P_{st}(\vec{n})}\right). \quad (3.56)$$

From (3.56) it is obvious that

$$S(t) = 0 \quad \text{for} \quad P(\vec{n}, t) = P_{st}(\vec{n}) \tag{3.57}$$

is true, and we will prove, that

$$S(t) \leq 0 \quad \text{for} \quad P(\vec{n}, t) \neq P_{st}(\vec{n}) \tag{3.58}$$

also holds. The proof of the convergence

$$\lim_{t \to \infty} P(\vec{n}, t) = P_{st}(\vec{n}) \tag{3.59}$$

is a little bit tricky and cumbersome. First of all, we introduce

$$\begin{aligned} S(t) &= -\sum_{\vec{n}} \left\{ P(\vec{n}, t) \log\left(\frac{P(\vec{n}, t)}{P_{st}(\vec{n})}\right) - P(\vec{n}, t) + P_{st}(\vec{n}) \right\} \\ &= -\sum_{\vec{n}} P_{st}(\vec{n}) \Big\{ \underbrace{X(\vec{n}, t) \log X(\vec{n}, t) - X(\vec{n}, t) + 1}_{\geq 0} \Big\} \end{aligned} \tag{3.60}$$

with the abbreviation

$$X(\vec{n}, t) = \frac{P(\vec{n}, t)}{p_{st}(\vec{n})}. \tag{3.61}$$

In the next step, the temporal evolution of $S(t)$ is considered. The time derivative of (3.60) leads to

$$\begin{aligned} \frac{dS(t)}{dt} &= -\sum_{\vec{n}} \left\{ \frac{dP(\vec{n}, t)}{dt} \log\left(\frac{P(\vec{n}, t)}{p_{st}(\vec{n})}\right) + P(\vec{n}, t) \frac{P_{st}(\vec{n})}{P(\vec{n}, t)} \frac{dP(\vec{n}, t)/dt}{p_{st}(\vec{n})} - \frac{dP(\vec{n}, t)}{dt} \right\} \\ &= -\sum_{\vec{n}} \frac{dP(\vec{n}, t)}{dt} \log\left(\frac{P(\vec{n}, t)}{p_{st}(\vec{n})}\right) \end{aligned} \tag{3.62}$$

The Master equation (3.22) is now inserted into (3.62) leading to

$$\frac{dS(t)}{dt} = -\sum_{\vec{n}, \vec{m}} \left(w(\vec{n}, \vec{m}) P(\vec{m}, t) - w(\vec{m}, \vec{n}) P(\vec{n}, t) \right) \log\left(\frac{P(\vec{n}, t)}{P_{st}(\vec{n})}\right) \tag{3.63}$$

After some tricky rearrangements, we obtain the final form of the derivative of $S(t)$:

$$\frac{dS(t)}{dt} = -\sum_{\vec{n},\,\vec{m}} w(\vec{n},\vec{m})P(\vec{m},t)\big(\log X(\vec{n},\vec{m}) - X\big(\vec{n},\vec{m}\big) + 1\big) \leq 0 \tag{3.64}$$

Since $w(\vec{n},\vec{m})P(\vec{m},t) > 0$, the derivative of $S(t)$ is always negative. The stationary state is reached if and only if $S(t) = 0$, that means for $X(\vec{n},\vec{m}) = 1$, or in other words, if (3.59) is approached. For $t \to \infty$ the entropy reaches its maximum value related to the stationary solution $P_{st}(\vec{n})$. Moreover, this statement holds even for time-dependent transition rates $w_t(\vec{n},\vec{m})$.

3.4 Equations of Motion for Mean Values and Variances

In this section, we derive equations of motion for mean values and variances on both the stochastic and the quasi-deterministic level. However, only the stochastic or probabilistic level is the fully exact and conceptually consistent one. Although the method of the Master equation is conceptually satisfying and consistent, it has the disadvantage that the Master equation must be solved numerically in most cases and that the huge amount of information contained in the configurational probability distribution can often only be compared with relatively poor empiric material (Grimmett and Stirzaker 1992).

In most cases, however, the mean square deviations from the most probable path of the system will be very small because of mutual cancellation of fluctuations. This holds in particular if the probability distribution can be expected to be unimodal and sharply peaked. In case of small numbers, the discrete structure of the Master equation has to be considered in detail and approximate mean value equations may lead to completely wrong results (see Chap. 7.1).

However, if the system behaves quasi-continuous, for example for huge population numbers in population dynamics, a quasi-deterministic or mean value description is adequate. Those approximate dynamic equations of the mean values and corresponding variances can be directly derived from the Master equation.

The mean value and mean deviations of a set of stochastic variables characterized by the vector $\vec{n} = (n_1, n_2, \ldots, n_L)$, consisting of one or a multiple of discrete numbers can be calculated via the normalized configurational probability distribution function $P(\vec{n},t) \geq 0$

$$\bar{n}_k(t) = \sum_{\vec{n}} n_k P(\vec{n},t) = \langle n_k \rangle_t \tag{3.65}$$

where the sum extends over all possible configurations in configuration space. Equations of motion for moments of the distribution can directly be derived from the Master equation (3.23) using

$$\frac{d}{dt}\langle n_l\rangle_t = \sum_{\vec{n}} n_l \frac{dP(\vec{n},t)}{dt} \tag{3.66}$$

Insertion of the Master equation (3.23) into (3.66) yields

$$\begin{aligned}
\frac{d}{dt}\langle n_l\rangle_t &= \sum_{\vec{n}} n_l \frac{dP(\vec{n},t)}{dt} \\
&= \sum_{\vec{n}} n_l \left(\sum_{\vec{k}} w_t\left(\vec{n}, \vec{n}+\vec{k}\right) P\left(\vec{n}+\vec{k},t\right) - \sum_{\vec{k}} w_t\left(\vec{n}+\vec{k}, \vec{n}\right) P(\vec{n},t) \right) \\
&= \sum_{\vec{k}} \left(\sum_{\vec{n}} n_l w_t\left(\vec{n}, \vec{n}+\vec{k}\right) P\left(\vec{n}+\vec{k},t\right) - \sum_{\vec{n}} n_l w_t\left(\vec{n}+\vec{k}, \vec{n}\right) P(\vec{n},t) \right) \\
&= \sum_{\vec{k}} \left\langle n_l w_t\left(\vec{n}, \vec{n}+\vec{k}\right)\right\rangle - \sum_{\vec{k}} \left\langle n_l w_t\left(\vec{n}+\vec{k}, \vec{n}\right)\right\rangle
\end{aligned} \tag{3.67}$$

The exact mean value Eqs. (3.67) still have the disadvantage to be not self-contained, because one needs the probability distribution $P(\vec{n},t)$ to calculate the right-hand side of (3.67). In analogy to (3.67), the exact equations of motion for the higher order moments and correlation functions can be formally obtained

$$\begin{aligned}
\frac{d}{dt}\left\langle n_l^p n_s^q \right\rangle_t &= \sum_{\vec{n}} n_l^p n_s^q \frac{dP(\vec{n},t)}{dt} \\
&= \sum_{\vec{k}} \left\langle n_l^p n_s^q w_t\left(\vec{n}, \vec{n}+\vec{k}\right)\right\rangle - \sum_{\vec{k}} \left\langle n_l^p n_s^q w_t\left(\vec{n}+\vec{k}, \vec{n}\right)\right\rangle
\end{aligned} \tag{3.68}$$

Depending on the specific mathematical form of the transition rates and/or the shape of the distribution function, different approximations can be introduced leading finally to quasi-closed mean value equations.

A frequently used, and often well justifiable assumption assumes that the probability distribution is a well behaved and sharply peaked unimodal distribution. In this special case, it can be assumed that the mean value of a function is comparable with the function of its mean value:

$$\langle f(\vec{n})\rangle_t \approx f\left(\langle\vec{n}\rangle_t\right) \tag{3.69}$$

We will deal with such assumptions in some of the subsequent applications (Chaps. 6, 7, and 8) of the book. Another frequently used assumption presumes for example that higher order moments and/or correlation functions can be factorized into products of lower order moments, such as:

$$\left\langle n_i^2 m_j \right\rangle_t \approx \left\langle n_i^2 \right\rangle_t \left\langle m_j \right\rangle_t \tag{3.70}$$

Those assumptions, if justifiable, help to obtain quasi-closed mean value and variance equations, as well as approximate equations for higher order moments. If, however, the probability distribution does not have a well behaved structure, for instance in the case of bistable systems or if small numbers—that means the discrete structure of the system becomes important, as in case of dying-out processes of population—the full complexity of the Master equation has to be taken into account (see Chap. 7).

The Eqs. (3.67), (3.68) with the assumptions described in (3.69), (3.70) are a system of autonomous differential equations for the quasi mean values. Therefore, they may have several attractors describing stationary states, limit cycles or even strange attractors related to deterministic chaotic systems. Since each attractor has its own basin of attraction, the initial state $\vec{n}(t = t_0)$, in other words, the state in which the system is prepared at $t = t_0$ determines which attractor will be approached in the course of time $t \to \infty$.

Contrary to the temporal behaviour of the quasi mean values, the probability distribution $P(\vec{n}, t)$ approaches unambiguously its stationary state $P_{st}(\vec{n})$. In consequence, the exact mean value equations will follow and approach the uniquely determined stationary value $\langle \vec{n} \rangle_{st}$.

3.4.1 A Specific Structure of the Transition Rates

The mathematical structure of the configurational transition rates $w_t\left(\vec{n} + \vec{k}, \vec{n}\right)$ is completely determined by the system under consideration, and the transition rates determine the complexity of the Master equation. In many applications, the (total) transition rate may be decomposed in a sum of transition rates related to specific effects or interactions within the system. In the following chapters, we shall discuss systems from different fields of research. Each application stands for a very specific structure of the transition rates and has an influence on the appropriate mathematical method of solution of the Master equation (Weidlich and Haag 1983).

In a rather general case, we consider a decomposition of the (total) transition rate $w_t\left(\vec{n} + \vec{k}, \vec{n}\right)$ into different specific contributions

$$\begin{aligned} w_t\left(\vec{n} + \vec{k}, \vec{n}\right) &= \sum_{i,j} w_{ji}\left(\vec{n} + \vec{k}, \vec{n}\right) + \sum_i w_{i+}\left(\vec{n} + \vec{k}, \vec{n}\right) \\ &\quad + \sum_i w_{i-}\left(\vec{n} + \vec{k}, \vec{n}\right) \end{aligned} \tag{3.71}$$

where the index t indicates a possible explicit time dependency of the transition rates. Let us remind that the configurational transition rate is the probability per time unit to find a certain configuration $\vec{n}+\vec{k}=\{n_1+k_1,\ldots,n_i+k_i,\ldots,n_L+k_L\}$ at time $t+\tau$, given that the configuration $\vec{n}=\{n_1,\ldots,n_L\}$ was realized at time t.

The first term in (3.71) describes all possible transitions between different states $i \leftrightarrow j$, such as migration acts between two different regions, leading to a change of configuration $\vec{n} \rightarrow \vec{n}+\vec{k}$. The second and third terms are related to birth and death events, respectively, leading also to a respective change of configuration $\vec{n} \rightarrow \vec{n}+\vec{k}$.

This specific structure of the (total) configurational transition rate is typical for socio-economic processes, but may occur in physical, biological or chemical systems as well.

(a) Birth- and Death Processes

The second and the third terms in (3.71) are related to birth-processes, where $w_{i+}\left(\vec{n}+\vec{k},\vec{n}\right)$ describes a single birth-event at position i and $w_{i-}\left(\vec{n}+\vec{k},\vec{n}\right)$ describes a single death-event at the position i. The single transition rates have to be summed up to contribute to the configurational transition rate $w_t\left(\vec{n}+\vec{k},\vec{n}\right)$ via

$$\sum_i w_{i+}\left(\vec{n}+\vec{k},\vec{n}\right) \geq 0 \quad \text{for} \quad \vec{k}=\{\ldots,0,\ldots,1_i,\ldots,0\} \tag{3.72}$$

and

$$\sum_i w_{i-}\left(\vec{n}+\vec{k},\vec{n}\right) \geq 0 \quad \text{for} \quad \vec{k}=\{\ldots,0,\ldots,-1_i,\ldots,0\} \tag{3.73}$$

and where $w_{i+}\left(\vec{n}+\vec{k},\vec{n}\right)=0$ and $w_{i-}\left(\vec{n}+\vec{k},\vec{n}\right)=0$ for all other $\vec{k}$. The vector $\vec{k}$ for non-vanishing transition rates w_{i+}, w_{i-} contain zeros except for +1, (-1) in the respective position of a birth or death event.

(b) Migration/Transfer Processes

Migration processes such as population or transfer-processes from one state to another are described by $w_{ji}\left(\vec{n}+\vec{k},\vec{n}\right)$

$$\sum_{i,j} w_{ji}\left(\vec{n}+\vec{k},\vec{n}\right) \geq 0 \quad \text{for} \quad \vec{k}=\left\{0,\ldots,1_j,\ldots,0,\ldots,-1_i,\ldots,0\right\} \tag{3.74}$$

and where $w_{ji}\left(\vec{n}+\vec{k},\vec{n}\right)=0$ for all other $\vec{k}$. This means, that the vector $\vec{k}$ for non-vanishing transition rates $w_{ji}\left(\vec{n}+\vec{k},\vec{n}\right)$ contains zeros except for the integers (+1) and (-1) in the positions j, i , respectively. The transitions (3.74) conserve the number of population if e.g. population dynamics would be considered.

As already mentioned, the mathematical structure of the transition rates is determined by the system to be modelled and thus sets the explicit mathematical structure of the Master equation. Therefore, in the following we present tools for a systematic development of dynamic equations of motion, based on the Master equation. This is the task of the shift operators, introduced in the next section.

3.4.2 Shift Operators

The shift operators sometimes also called translation operators help to simplify the forthcoming derivations of dynamic mean value and variance equations, taking into account the specific structure of the above introduced transition rates (van Kampen 1978).

Let us introduce shift operators $E_{i1}^{\pm 1}$ acting on a function$f(\vec{n})$ of the configuration space as follows

$$E_i^{\pm 1} f(n_1, \ldots, n_i, \ldots, n_L) = f(n_1, \ldots, (n_i \pm 1), \ldots, n_L). \tag{3.75}$$

This definition implies the further relations

$$n_k E_i^{+1} E_j^{-1} f(n_1, \ldots, n_i, \ldots, n_L) = E_i^{+1} E_j^{-1} \left(n_k - \delta_{ki} + \delta_{kj}\right) f(n_1, \ldots, n_i, \ldots, n_L) \tag{3.76}$$

$$n_k E_i^{\pm 1} f(n_1, \ldots, n_i, \ldots, n_L) = E_i^{\pm 1} (n_k \mp \delta_{ki}) f(n_1, \ldots, n_i, \ldots, n_L) \tag{3.77}$$

$$E_i^{+1} E_j^{-1} f(n_1, \ldots, n_i, \ldots, n_L) = f\left(n_1, \ldots, (n_i + 1), \ldots, \left(n_j - 1\right), \ldots, n_L\right) \tag{3.78}$$

where δ_{ik} is the Kronecker symbol

$$\delta_{ik} = \begin{cases} 1 & \text{for} \quad i = k \\ 0 & \text{for} \quad i \neq k \end{cases} \tag{3.79}$$

Furthermore, we introduce the so-called general configurations (GC) with $\vec{n} = \{n_1, \ldots, n_i, \ldots, n_L\}$, where the n_i could be positive or zero $n_i \geq 0$ or even negative $n_i < 0$. Of course, a realisable (true) configuration (DC) consists only of such components, which are composed of positive integers or zeros.

The functions $P(\vec{n}, t)$, $w_t\left(\vec{n} + \vec{k}, \vec{n}\right)$ introduced so far are only defined for true configurations (DC). However, we can easily extend their definition to general configurations (GC) by the requirement

$$f(\vec{n}) = \begin{cases} f(\vec{n}) & \text{for } \vec{n} \in DC \\ 0 & \text{otherwise} \end{cases} \tag{3.80}$$

In all forthcoming calculations, we make use of this formal extension. It implies that sums over $f(\vec{n})$ with $\vec{n}$ running through all true decision configurations, DC, can be formally extended to sums over all general configurations, GC. It is evident, that then the following relations hold

$$\sum_{\vec{n}}^{GC} E_i^{\pm 1} f(\vec{n}) = \sum_{\vec{n}}^{GC} E_i^1 E_j^{-1} f(\vec{n}) = \sum_{\vec{n}}^{GC} f(\vec{n}) \tag{3.81}$$

because the summation includes all contributions $-\infty < n_i < \infty$, $i = 1, 2, \ldots, L$. Equation (3.81) means, that if we shift the operators $E_i^{\pm 1}$ to the left-hand side (lhs) the sum over all contributions will remain constant and we can formally neglect the operators.

3.4.3 Exact and Approximate Mean Value Equations

Inserting (3.71) in (3.22) and making use of the above introduced formalism (3.75)–(3.81), the Master equation can be written in an equivalent form:

$$\begin{aligned} \frac{dP(\vec{n},t)}{dt} &= \sum_{j,i} \left(E_i E_j^{-1} - 1\right) w_{ji}\left(\vec{n} + \vec{k}, \vec{n}\right) P(\vec{n},t) \\ &+ \sum_i \left(E_i^{-1} - 1\right) w_{i+}\left(\vec{n} + \vec{k}, \vec{n}\right) P(\vec{n},t) \\ &+ \sum_i \left(E_i^{+1} - 1\right) w_{i-}\left(\vec{n} + \vec{k}, \vec{n}\right) P(\vec{n},t) \end{aligned} \tag{3.82}$$

The compact formulation of the Master equation (3.82) will be the starting point for the derivation of appropriate mean value equations in the next step.

In principle, the Master equation contains such a tremendous amount of information compared with the standard empirical situation (available empirical data), that a less exhaustive description in terms of mean values seems to be adequate, at least in most socio-economic applications. Moreover, we expect, that the mean value, $\langle \vec{n}(t) \rangle$ of the system under consideration practically coincides with the realized configuration. Therefore, it is highly desirable to derive self-contained equations of motion as an acceptable approximation of the dynamics of the system.

Let us first start with the definition of the mean value $\langle f(\vec{n}) \rangle$ of an arbitrary function $f(\vec{n})$ in configuration space, according to Eq. (2.6)

$$\langle f(\vec{n}) \rangle = \sum_{\vec{n}}^{DC} f(\vec{n}) P(\vec{n},t) = \sum_{\vec{n}}^{GC} f(\vec{n}) P(\vec{n},t) \tag{3.83}$$

In particular, the mean value of $\langle n_k \rangle$ is of general importance. In analogy to (3.67) we multiply the Master equation with n_k and sum up over all configurations $\vec{n}$

$$\frac{d}{dt}\langle n_k \rangle = \sum_{\vec{n}}^{GC} n_k \frac{dP(\vec{n},t)}{dt}$$

$$= \sum_{\vec{n}}^{GC} \sum_{j,i} n_k \left(E_i E_j^{-1} - 1 \right) w_{ji} \left(\vec{n} + \vec{k}, \vec{n} \right) P(\vec{n},t)$$

$$+ \sum_{\vec{n}}^{GC} \sum_{i} n_k \left(E_i^{-1} - 1 \right) w_{i+} \left(\vec{n} + \vec{k}, \vec{n} \right) P(\vec{n},t)$$

$$+ \sum_{\vec{n}}^{GC} \sum_{i} n_k \left(E_i^{+1} - 1 \right) w_{i-} \left(\vec{n} + \vec{k}, \vec{n} \right) P(\vec{n},t) \quad (3.84)$$

By making use of (3.75)–(3.81) and the definition (3.83), and taking the shift the operators to the left-hand side of (3.84) we obtain:

$$\frac{d}{dt}\langle n_k \rangle = \sum_{j,i}^{L} \sum_{\vec{n}}^{GC} E_i E_j^{-1} \left(\delta_{jk} - \delta_{ik} \right) w_{ji} \left(\vec{n} + \vec{k}, \vec{n} \right) P(\vec{n},t)$$

$$+ \sum_{i=1}^{L} \sum_{\vec{n}}^{GC} E_i^{-1} \delta_{ki} w_{i+} \left(\vec{n} + \vec{k}, \vec{n} \right) P(\vec{n},t)$$

$$- \sum_{i=1}^{L} \sum_{\vec{n}}^{GC} E_i^{+1} \delta_{ki} w_{i-} \left(\vec{n} + \vec{k}, \vec{n} \right) P(\vec{n},t) \quad (3.85)$$

and after performing the sums over i ,j in (3.85)

$$\frac{d}{dt}\langle n_k \rangle = \sum_{j=1}^{L} \sum_{\vec{n}}^{GC} w_{kj} \left(\vec{n} + \vec{k}, \vec{n} \right) P(\vec{n},t) - \sum_{i=1}^{L} \sum_{\vec{n}}^{GC} w_{ik} \left(\vec{n} + \vec{k}, \vec{n} \right) P(\vec{n},t)$$

$$+ \sum_{\vec{n}}^{GC} w_{k+} \left(\vec{n} + \vec{k}, \vec{n} \right) P(\vec{n},t) - \sum_{\vec{n}}^{GC} w_{k-} \left(\vec{n} + \vec{k}, \vec{n} \right) P(\vec{n},t) \quad (3.86)$$

This finally leads along with the definition of the mean values to the exact equations of motion

$$\frac{d}{dt}\langle n_k \rangle = \sum_{j=1}^{L} \left\langle w_{kj} \left(\vec{n} + \vec{k}, \vec{n} \right) \right\rangle - \sum_{j=1}^{L} \left\langle w_{jk} \left(\vec{n} + \vec{k}, \vec{n} \right) \right\rangle$$

$$+ \left\langle w_{k+} \left(\vec{n} + \vec{k}, \vec{n} \right) \right\rangle - \left\langle w_{k-} \left(\vec{n} + \vec{k}, \vec{n} \right) \right\rangle \quad (3.87)$$

The intuitive interpretation of (3.87) is obvious: the evaluation with time of the mean value $\langle n_k \rangle$ is due to migratory transitions from all other states $j \neq k$ into the state k as well as transitions from state k to all other states and in addition birth and death event related to state k. As already described at the beginning of this chapter [see (3.67)], the exact Eqs. (3.87) are still not self-contained. It depends on the explicit mathematical structure of the transition rates $w_{ji}\left(\vec{n}+\vec{k},\vec{n}\right)$, $w_{i+}\left(\vec{n}+\vec{k},\vec{n}\right)$, $w_{i-}\left(\vec{n}+\vec{k},\vec{n}\right)$ whether or not a closed system or hierarchical system of differential equations can be derived starting with (3.87).

If, however, the probability distribution is well behaved, sharply peaked and unimodal, according to (3.69), (3.70), namely, if it can be assumed that the mean value of a function is approximately equal to that function of the mean value, and higher order moments factorize, a self-contained equation of motion for the mean value of the system can be obtained. We will apply this procedure in some of the following examples.

The self-contained equations of motion establish a set of ordinary (nonlinear) coupled differential equations for the $\langle n_k \rangle$.

Since the mean values by definition are averages over paths with fluctuating deviations, their evolution is described by deterministic equations. In principle, deterministic chaos may be also possible if the nonlinearities depict a specific structure. The trajectories (single paths belonging to a specific initial condition) may have one or several stationary states. It will be shown, that they coincide with the maximum (the maxima) of the stationary probability distribution[1] $P_{st}(\vec{n})$. All time-dependent solutions approach for $t \to \infty$ one of these stationary states, but the initial starting point determines which of the stationary states will be approached.

3.4.4 Exact and Approximate Equations of Motion for the Variance

The variance matrix $\sigma_{kl}^2(t)$ is the mean value of bilinear deviations of the system variables from their mean value, as

$$\sigma_{kl}^2(t) = \langle (n_k - \langle n_k \rangle)(n_l - \langle n_l \rangle) \rangle = \langle n_k n_l \rangle - \langle n_k \rangle \langle n_l \rangle \tag{3.88}$$

We further introduce the notation

$$\Delta n_k(t) = (n_k - \langle n_k \rangle), \quad \text{where} \quad \langle \Delta n_k(t) \rangle = 0 \quad \text{for all } k \tag{3.89}$$

for the derivation of the random variables n_k from their mean values. According to Chap. 2 the definition of the variance matrix is given by

[1]In case of a limit cycle, the limit cycle corresponds to a hill structure of the probability distribution (see…).

$$\sigma_{kl}^2(t) = \sum_{\vec{n}} \Delta n_k \Delta n_l P(\vec{n}, t). \tag{3.90}$$

We now precede in analogy to the derivation of the mean value equations in the previous chapter and derive equations of motion for the variances $\sigma_{kl}^2(t)$

$$\frac{d\sigma_{kl}^2(t)}{dt} = \sum_{\vec{n}}^{GC} \Delta n_k(t) \Delta n_l(t) \frac{dP(\vec{n}, t)}{dt} \tag{3.91}$$

where we have taken into account (3.89), namely $\langle \Delta n_k(t) \rangle = \langle \Delta n_l(t) \rangle = 0$

Inserting the Master equation (3.82) in (3.91) yields (Weidlich 2006)

$$\begin{aligned}
\frac{d\sigma_{kl}^2(t)}{dt} &= \sum_{i,j}^{L} \sum_{\vec{n}}^{GC} \Delta n_k(t) \Delta n_l(t) \left(E_j^{+1} E_i^{-1} - 1\right) w_{ij}\left(\vec{n} + \vec{k}, \vec{n}\right) P(\vec{n}, t) \\
&+ \sum_{i}^{L} \sum_{\vec{n}}^{GC} \Delta n_k(t) \Delta n_l(t) \left(E_i^{-1} - 1\right) w_{i+}\left(\vec{n} + \vec{k}, \vec{n}\right) P(\vec{n}, t) \\
&+ \sum_{i}^{L} \sum_{\vec{n}}^{GC} \Delta n_k(t) \Delta n_l(t) \left(E_i^{+1} - 1\right) w_{i-}\left(\vec{n} + \vec{k}, \vec{n}\right) P(\vec{n}, t)
\end{aligned} \tag{3.92}$$

Now we make use of (3.75)–(3.80) and shift the operators $E_i^{\pm 1}$ to the lhs in the equations and apply (3.81)

$$\begin{aligned}
\frac{d\sigma_{kl}^2(t)}{dt} &= \sum_{i,j}^{L} \sum_{\vec{n}}^{GC} \left(\left(\delta_{il} - \delta_{jl}\right) \Delta n_k(t) + \left(\delta_{ik} - \delta_{jk}\right) \Delta n_l(t) + \left(\delta_{il} - \delta_{jl}\right)\right. \\
&\quad \left.\left(\delta_{ik} - \delta_{jk}\right)\right) w_{ij}\left(\vec{n} + \vec{k}, \vec{n}\right) P(\vec{n}, t) \\
&\quad + \sum_{i}^{L} \sum_{\vec{n}}^{GC} \left(\delta_{ik} \Delta n_l(t) + \delta_{il} \Delta n_k(t) + \delta_{ik} \delta_{il}\right) w_{i+}\left(\vec{n} + \vec{k}, \vec{n}\right) P(\vec{n}, t) \\
&\quad - \sum_{i}^{L} \sum_{\vec{n}}^{GC} \left(\delta_{ik} \Delta n_l(t) + \delta_{il} \Delta n_k(t) - \delta_{ik} \delta_{il}\right) w_{i-}\left(\vec{n} + \vec{k}, \vec{n}\right) P(\vec{n}, t)
\end{aligned} \tag{3.93}$$

After performing the sums over the Kronecker symbols, we finally obtain the exact equations of motion for the variances

$$\begin{aligned}\frac{d\sigma_{kl}^2(t)}{dt} &= \sum_{j=1}^{L}\left\langle \Delta n_l(t) w_{kj}\left(\vec{n}+\vec{k},\vec{n}\right)\right\rangle - \sum_{i=1}^{L}\left\langle \Delta n_l(t) w_{ik}\left(\vec{n}+\vec{k},\vec{n}\right)\right\rangle \\ &+ \sum_{j=1}^{L}\left\langle \Delta n_k(t) w_{lj}\left(\vec{n}+\vec{k},\vec{n}\right)\right\rangle - \sum_{i=1}^{L}\left\langle \Delta n_k(t) w_{il}\left(\vec{n}+\vec{k},\vec{n}\right)\right\rangle \\ &+ \sum_{j=1}^{L}\delta_{kl}\left\langle w_{kj}\left(\vec{n}+\vec{k},\vec{n}\right)\right\rangle + \sum_{i=1}^{L}\delta_{lk}\left\langle w_{ik}\left(\vec{n}+\vec{k},\vec{n}\right)\right\rangle \\ &- \left\langle w_{lk}\left(\vec{n}+\vec{k},\vec{n}\right)\right\rangle - \left\langle w_{kl}\left(\vec{n}+\vec{k},\vec{n}\right)\right\rangle \\ &+ \left\langle \Delta n_l(t) w_{k+}\left(\vec{n}+\vec{k},\vec{n}\right)\right\rangle + \left\langle \Delta n_k(t) w_{l+}\left(\vec{n}+\vec{k},\vec{n}\right)\right\rangle \\ &+ \delta_{kl}\left\langle w_{k+}\left(\vec{n}+\vec{k},\vec{n}\right)\right\rangle - \left\langle \Delta n_l(t) w_{k-}\left(\vec{n}+\vec{k},\vec{n}\right)\right\rangle \\ &- \left\langle \Delta n_k(t) w_{l-}\left(\vec{n}+\vec{k},\vec{n}\right)\right\rangle + \delta_{lk}\left\langle w_{k-}\left(\vec{n}+\vec{k},\vec{n}\right)\right\rangle \end{aligned} \tag{3.94}$$

The variance Eq. (3.94) is very heavy to solve, at least analytically. We still have the problem that for the solution of the different required mean values $\langle f\left(\vec{n},t\right\rangle$ the probability distribution has to be known. However, Eq. (3.94) can be an excellent starting point for appropriate approximations [see (3.69) and (3.70)].

References

Gardiner CW (1983) Handbook of stochastic methods for physics, chemistry and the natural sciences, Springer Series Synergetics, vol 13. Springer, Heidelberg

Grimmett G, Stirzaker D (1992) Probability and random processes. Oxford University Press, Oxford

Haag G, Hänggi P (1979) Exact solutions of discrete master equations in terms of continued fractions. Z Phys B 34:411–417

Haag G, Hänggi P (1980) Continued fraction solutions of discrete master equations not obeying detailed balance II. Z Phys B 39:269–279

Haken H (1977) Synergetics: an introduction. Springer, Heidelberg

Honerkamp J (1998) Statistical physics: an advanced approach with applications. Springer, Berlin

Van Kampen NG (1978) An introduction to stochastic processes for physicists. In: Garido L, Seglar P, Shepherd PJ (eds) Stochastic processes in nonequilibrium systems. Proceedings, Sitges 1978. Springer, New York

Van Kampen NG (1981) Stochastic processes in physics and chemistry. North-Holland, Amsterdam

Weidlich W (2006) Sociodynamics: a systematic approach to mathematical modeling in the social sciences. Dover, Mineola

Weidlich W, Haag G (1983) Concepts and models of a quantitative sociology: the dynamics of interacting populations, Springer series of synergetics, vol 14. Springer, New York

Chapter 4
Solution Methods of Master Equations

Contents

4.1 The Fokker–Planck Approximation

The Master equation (3.22) is a difference-differential equation, since it describes the evolution with time of the occupation probability $P(\vec{n}, t)$ of discrete states $\vec{n}$. The Kramers-Moyal expansion of the Master equation leads to a partial differential equation of infinite order for the probability $P(\vec{x}, t)$ for continuous states $\vec{x}$. Since $P(\vec{x}, t) \geq 0$ has to be positive, one can prove that all terms of the expansion $s \geq 3$ have to be neglected to avoid negative values of $P(\vec{x}, t)$ for specific configurations, or all derivatives have to be taken into account. The remaining partial differential equation is the so-called Fokker–Planck equation. Therefore, the Fokker–Planck equation is an approximate equation of the Master equation.

For the derivation of the Fokker–Planck equation it is supposed that $P(\vec{n}, t)$ and the transition rates $\tilde{w}_t\left(\vec{k}, \vec{n}\right)$ are differentiable functions of the L variables $\vec{n} = (\ldots, n_i, \ldots)$. Expanding the first term on the rhs (right-hand side) of the Master equation (3.23) in a Taylor series with respect to $\vec{n} + \vec{k}$ around $\vec{n}$:

G. Haag, *Modelling with the Master Equation*, DOI 10.1007/978-3-319-60300-1_4

$$\frac{\partial P(\vec{n},t)}{\partial t} = \sum_{\vec{k}} \tilde{w}_t\left(-\vec{k}, \vec{n}+\vec{k}\right)P\left(\vec{n}+\vec{k},t\right) - \sum_{\vec{k}} \tilde{w}_t\left(-\vec{k}, \vec{n}\right)P(\vec{n},t)$$

$$= \sum_{\vec{k}} \sum_{s_1}^{\infty} \cdots \sum_{s_L}^{\infty} \frac{k_1^{s_1} \ldots k_L^{s_L}}{s_1! \ldots s_L!} \frac{\partial^{s_1}}{\partial n_1^{s_1}} \cdots \frac{\partial^{s_L}}{\partial n_L^{s_L}} \tilde{w}_t\left(-\vec{k}, \vec{n}\right)P(\vec{n},t). \tag{4.1}$$

Introducing the moments of the distribution of order $s = s_1 + s_2 + \cdots s_L$ defined by

$$m_{s_1 \ldots s_L}(\vec{n},t) = \sum_{\vec{k}} k_1^{s_1} \ldots k_L^{s_L} \tilde{w}_t\left(\vec{k}, \vec{n}\right)$$

$$= \lim_{\tau \to 0} \frac{1}{\tau} \sum_{\vec{k}} k_1^{s_1} \ldots k_L^{s_L} p\left(\vec{n}+\vec{k}, t+\tau, |\vec{n}, t\right)$$

$$= \lim_{\tau \to 0} \frac{1}{\tau} \langle [n_1(t+\tau) - n_1(t)]^{s_1} \cdots [n_L(t+\tau) - n_L(t)]^{s_L} \rangle \tag{4.2}$$

and inserting (4.2) in (4.1) derives the stochastic partial differential equation of infinite order:

$$\frac{\partial P(\vec{n},t)}{\partial t} = \sum_{s=1}^{\infty} \frac{(-1)^s}{s!} \sum_{s_1, s_2, \ldots, s_L}^{s_1 + \ldots s_L = s} \frac{s!}{s_1! \ldots s_L!} \frac{\partial^{s_1}}{\partial n_1^{s_1}} \cdots \frac{\partial^{s_L}}{\partial n_L^{s_L}} \left(m_{s_1 \ldots s_L}(\vec{n},t) P(\vec{n},t)\right) \tag{4.3}$$

or after appropriate scaling of the variables

$$\frac{\partial P(\vec{x},t)}{\partial t} = \sum_{s=1}^{\infty} \frac{(-1)^s \varepsilon^{s-1}}{s!} \sum_{s_1, s_2, \ldots, s_L}^{s_1 + \ldots s_L = s} \frac{s!}{s_1! \ldots s_L!} \frac{\partial^{s_1}}{\partial x_1^{s_1}} \cdots \frac{\partial^{s_L}}{\partial x_L^{s_L}} \left(M_{s_1 \ldots s_L}(\vec{x},t) P(\vec{x},t)\right) \tag{4.4}$$

where

$$x_i = \frac{n_i}{N} \text{ and } \Delta x_i = \frac{1}{N} \equiv \varepsilon \text{ and } m_{s_1 \ldots s_L}(\vec{n},t) = N \cdot M_{s_1 \ldots s_L}(\vec{x},t) \tag{4.5}$$

and the normalization is introduced:

$$\sum_{\vec{n}} P(\vec{n},t) = \sum_{\vec{x}} P(\vec{x},t) \Delta x_1 \ldots \Delta x_L \approx \int_0^{\infty} P(\vec{x},t) d\vec{x} = 1 \tag{4.6}$$

The Fokker–Planck equation is the truncated version of the partial differential Eq. (4.4) up to the second order, that means all terms with $s \geq 3$ are neglected. This approach is justified for $N \gg 1$ or in other words, if $\varepsilon \ll 1$.

Then the Fokker–Planck equation reads (Weidlich and Haag 1983):

$$\begin{aligned} \frac{\partial P(\vec{x},t)}{\partial t} &= -\sum_{i=1}^{L} \partial_i \left[K_i(\vec{x},t)P(\vec{x},t)\right] + \frac{\varepsilon}{2}\sum_{i,j=1}^{L} \partial_i \partial_j \left[Q_{ij}(\vec{x},t)P(\vec{x},t)\right] \\ &= -\sum_{i=1}^{L} \partial_i I_i(\vec{x},t) \end{aligned} \tag{4.7}$$

with the so-called drift-coefficient

$$K_i(\vec{x},t) = \varepsilon \sum_{\vec{k}} k_i \tilde{w}_t\left(\vec{k},\vec{n}\right) = \lim_{\tau \to 0} \frac{1}{\tau} \langle x_i(t+\tau) - x_i(t) \rangle, \tag{4.8}$$

fluctuation coefficient

$$\begin{aligned} Q_{ij}(\vec{x},t) &= \varepsilon \sum_{\vec{k}} k_i k_j \tilde{w}_t\left(\vec{k},\vec{n}\right) \\ &= \lim_{\tau \to 0} \frac{1}{\tau} \left\langle (x_i(t+\tau) - x_i(t))\left(x_j(t+\tau) - x_j(t)\right) \right\rangle, \end{aligned} \tag{4.9}$$

and the abbreviation for the probability flux

$$I_i(\vec{x},t) = K_i(\vec{x},t)P(\vec{x},t) - \frac{\varepsilon}{2}\sum_{j=1}^{L} \partial_j\left(Q_{ij}(\vec{x},t)P(\vec{x},t)\right) \tag{4.10}$$

The drift coefficient $K_i(\vec{x},t)$ encompasses all those effects influencing the mean behaviour of the trajectory of the system. The fluctuation coefficient comprises the impacts of fluctuations, namely the stochastic elements leading to fluctuations around the mean pass of the system.

The Fokker–Planck equation is an exact equation for Gauss–Markoff processes, since it only contains information about the first and the second moment of the distribution. Methods of solution of Fokker–Planck equations are well known and described in detail by Risken (1972, 1996).

When we deal with: atoms or molecules in physics or chemistry, or agents, decision maker, consumer, households or individuals in social sciences, the discrete structure of the Master equation is most appropriate. Therefore, my intension is to deal especially with discrete stochastic processes and assigned solution methods.

4.2 T-factor Method

Exact solutions of Master equations in discrete or continuous representations however, are often too complicated for explicit evaluation (Schnakenberg 1976) or are available only in relatively simple cases (Haken 1975; Haag et al. 1977). Therefore, methods for the determination of approximate stationary and non-stationary solutions have been developed (van Kampen 1965; Kubo et al. 1973; Görtz and Walls 1976). The presented T-factor method, developed by Haag (1977), is suitable for the dynamic and stationary treatment of discrete Master equations. This method is particularly adapted to physical problems. With the help of this method it is easily possible to derive a continued fraction representation of the master equation. Based on this continued fraction solution a reduced and often more appropriate equation for mathematical solution can be deduced (Haag and Hänggi 1979, 1980). In Chaps. 5 and 7 different examples out of the field of physics, chemistry and biology will underline the importance and usefulness of this treatment via benchmark tests of different solution techniques. This method, therefore, represents a useful complement to the commonly used methods of generating functions and eigenvector analysis.

4.2.1 *Fundamentals of the T-factor Method*

In the following we consider positive, normalized and uniquely determined probability distributions

$$P(\vec{n},t) > 0 \ \forall_{\vec{n}\in C} \text{ and } \sum_{\vec{n}} P(\vec{n},t) = 1 \tag{4.11}$$

satisfying the Pauli master equation (3.22)

$$\frac{dP(\vec{n},t)}{dt} = \sum_{\vec{m}} w_t(\vec{n},\vec{m})P(\vec{m},t) - \sum_{\vec{m}} w_t(\vec{m},\vec{n})P(\vec{n},t) \tag{4.12}$$

The time index t is omitted for simplification in this section, so that $P(\vec{n},t) \equiv P(\vec{n})$. We further assume that the initial distribution has been established via a physical process leading to $P(\vec{n}) > 0$ for all $\vec{n} \in C$ at time $t = t_0$. In this sense, we exclude systems with a δ-like initial distribution, characterized by only one occupied cell in configuration space.

We define a set of transition factors (T-factors) at time t by

$$g(\vec{m}|\vec{n}) = P(\vec{m})/P(\vec{n}) \tag{4.13}$$

As we can see from the definition (4.13), the T-factors contain no normalization factors. They are therefore completely determined by the properties of the system under consideration, namely the transition rates and their dynamics.

From the definition (4.13) and assumptions (4.11) the important properties can be deduced

$$g(\vec{m}|\vec{n}) > 0 \tag{4.14}$$

$$g(\vec{m}|\vec{n}) = g\left(\vec{m}|\vec{k}\right)g\left(\vec{k}|\vec{n}\right) \tag{4.15}$$

$$\lim_{\vec{n}, m_i \to \infty} g(\vec{m}|\vec{n}) < 1 \tag{4.16}$$

$$g(\vec{m}|\vec{m}) = 1 \tag{4.17}$$

$$g(\vec{m}|\vec{n})^{-1} = g(\vec{n}|\vec{m}) \tag{4.18}$$

Equations (4.14)–(4.18) yield for all $\vec{m} \in C, \vec{n} \in C, \vec{k} \in C$.

Relation (4.15) states the uniqueness of the probability distribution, (4.14) and (4.16) reflect that the distribution has to be positive and normalized. It is remarkable that the $g(\vec{m}|\vec{n})$ have the structure of a propagator in configuration space for a fixed time t.

From (4.11) and the definition (4.13) we immediately obtain

$$P(\vec{n}) = \left(\sum_{\vec{m}} g(\vec{m}|\vec{n})\right)^{-1}. \tag{4.19}$$

In particular, we conclude that the set of all transition factors $g(\vec{m}|\vec{n})$ determines the set of all $P(\vec{n})$ and vice versa.

If we consider (4.15) there exists a great number of possible paths in configuration space to proceed from any cell $\vec{n}_o$ with occupation probability $P(\vec{n}_o)$ to cell $\vec{n}_M$ with occupation probability $P(\vec{n}_M)$. Therefore, we may select a mathematically appropriate path $\vec{n}_o \to \vec{n}_M$ characterized by a one-dimensional parameter j and non-vanishing transition rates. For many systems, it is practical to introduce $g\left(\vec{n}_{j+1}|\vec{n}_j\right) > 0$ which act between adjacent cells in configuration space and pass from $\vec{n}_o$ to $\vec{n}_M$ by a chain of intermediate cells $\left\{\vec{n}_0, \vec{n}_1, \ldots, \vec{n}_j, \ldots, \vec{n}_M\right\}$. Introducing the T-factors $g\left(\vec{n}_{j+1}|\vec{n}_j\right) > 0$ acting between neighbouring cells, we obtain the representation of the distribution function

$$P(\vec{n}_M) = P(\vec{n}_0) \prod_{j=0}^{M-1} g\left(\vec{n}_{j+1}|\vec{n}_j\right). \tag{4.20}$$

The T-factors $g\left(\vec{n}_{j+1}|\vec{n}_j\right) > 0$ are often smooth functions of j especially in the interesting region $j >> 1$, we can define extreme values of the distribution for

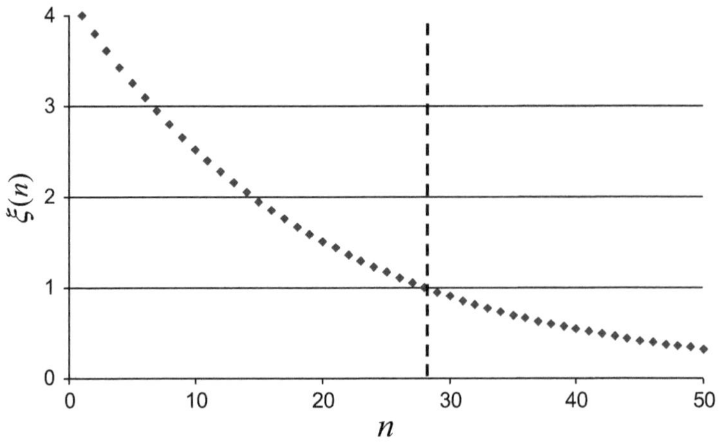

Fig. 4.1 T-factor for a typical uni-modal distribution function

$$g(\vec{n}_{l+1}|\vec{n}_l)+=1, \text{ and } \underset{l}{\Delta}\, g(\vec{n}_{l+1}|\vec{n}_l) < 0 \quad \text{(maximum)} \tag{4.21}$$

$$g(\vec{n}_{l+1}|\vec{n}_l)+=1, \text{ and } \underset{l}{\Delta}\, g(\vec{n}_{l+1}|\vec{n}_l) > 0 \quad \text{(minimum)} \tag{4.22}$$

where l indicates the extreme point (maxima or minima) of the distribution along the selected path, and the symbol $\stackrel{+}{=}$ indicates the very next value l to $g\left(\vec{n}_{j+1}|\vec{n}_j\right) \approx 1$.

Figure 4.1 shows a typical distribution of the T-factors $\xi(n) = P(n)/P(n-1)$, related to a unimodal one-dimensional probability distribution (Fig. 4.2). It is obvious that the shape of the T-factor distribution $\xi(n)$ is varying slowly compared to that of the probability distribution $P_{st}(n)$, respectively.

4.2.2 Exact Transformation of the Equation of Motion

L-dimensional case

We now want to transform the Master equation using the definition of the T-factors to an exact equation of motion for the T-factors. Therefore, we differentiate the definition Eq. (4.13) with respect to time and perform some simple manipulations

$$\frac{\dot{g}\,(\vec{m}|\vec{n})}{g(\vec{m}|\vec{n})} = J(\vec{m}) - J(\vec{n}) \text{ or } \frac{d}{dt}\ln g(\vec{m}|\vec{n}) = \underset{\vec{m}-\vec{n}}{\Delta} J(\vec{m}), \tag{4.23}$$

where

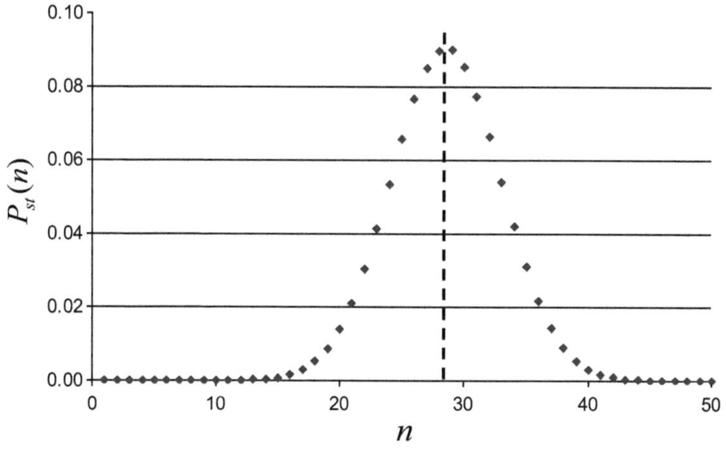

Fig. 4.2 Probability distribution related to the T-factors above

$$J(\vec{m}) = \frac{\dot{P}(\vec{m})}{P(\vec{m})} = \sum_{\vec{n}} (w(\vec{m}, \vec{n})\, g(\vec{n}|\vec{m}) - w(\vec{n}, \vec{m})), \tag{4.24}$$

and

$$\underset{\vec{m}-\vec{n}}{\Delta} J(\vec{m}) = J(\vec{m}) - J(\vec{n}). \tag{4.25}$$

Equation (4.24) inserted into (4.23) leads to the exact equation of motion for the transition factors. However, (4.23) is still a differential—difference equation as well as the Master equation.

If, in particular, the transition rates decompose into a sum of terms, the flow term $J(\vec{m})$ also decomposes into a corresponding sum of terms

$$w(\vec{m}|\vec{n}) = \sum_i w_i(\vec{m}|\vec{n}) \Leftrightarrow J(\vec{m}) = \sum_i J_i(\vec{m}), \tag{4.26}$$

where

$$J_i(\vec{m}) = \sum_{\vec{n}} (w_i(\vec{m}, \vec{n})\, g(\vec{n}|\vec{m}) - w_i(\vec{n}, \vec{m})). \tag{4.27}$$

One-dimensional case

Let us now consider the one-dimensional case of the Master equation. Therefore, we replace the multi indices $\vec{m}$ by a one-dimensional index $m = 0, 1, 2, \ldots$ In this case, the equation of motion for the transition factors reads

$$\frac{d}{dt}\ln g(m|n) = \underset{m-n}{\Delta} J(m) \tag{4.28}$$

with

$$J(m) = \sum_{r}^{\pm R} (w(m, m-r)g(m-r|m) - w(m-r, m)), \tag{4.29}$$

where we take into account transition rates $w(m, m-r)$ not only between neighbouring states, characterized by $r = \pm 1$, but also between all cells in configuration space distinguished by

$$w(m, m-r) \geq 0 \quad \text{for} \quad r = \pm 1, \pm 2, \ldots \pm R. \tag{4.30}$$

In the next step, we substitute $n = m - 1$ in (4.28) and introduce as a transition factor acting between neighbouring states

$$\xi(m) = g(m|m-1), \tag{4.31}$$

which is often a slowly varying function even if the probability distribution $P(m)$ varies rapidly (compare Figs. 4.1 and 4.2). Then we find by inserting (4.31) in (4.28)

$$\frac{d\xi(m)}{dt} = \xi(m)\Delta\sum\nolimits_{r>0}\left\{\left(w(m, m+r)\prod\nolimits_{j=1}^{r}\xi(m+j) - w(m+r, m)\right)\right.$$
$$\left. + \left(w(m, m-r)\prod\nolimits_{j=0}^{r-1}\xi^{-1}(m-j) - w(m-r, m)\right)\right\} \tag{4.32}$$

for all m. Equation (4.32) describes a set of coupled nonlinear differential difference equations for the transition factors $\xi(m)$, which determine the probability distribution

$$P(m) = P(0)\prod_{j=1}^{m}\xi(j) = P(0)\exp\sum_{j=1}^{m}\ln\xi(j), \tag{4.33}$$

where again $P(0)$ is fixed by the normalization condition of the probability distribution.

According to (4.21), (4.22) extreme values of the probability distribution can be found if

$$\xi(m_0) \overset{+}{=} 1 \text{ and } \underset{1}{\Delta}\xi(m_0) < 0 \quad \text{(maximum)}$$

$$\xi(m_0) \overset{+}{=} 1 \text{ and } \underset{1}{\Delta}\xi(m_0) > 0 \quad \text{(minimum)} \tag{4.34}$$

The treatment of the nonlinear equation of motion (4.32) is very cumbersome and (4.32) represents only an interstate in the derivation of a more appropriate equation of motion. Therefore, let us consider the transition to a continuous variable by replacing

$$\xi(m) \to q(x) \text{ and } P(m) \to P(x) \text{ as well as } J(m) \to J(x). \tag{4.35}$$

Using the Taylor expansion

$$f(x+j) = e^{j\frac{\partial}{\partial x}} f(x), \tag{4.36}$$

we obtain the still exact partial differential equation

$$\dot{q}(x) = q(x)\left(1 - e^{-\frac{\partial}{\partial x}}\right)J(x) \tag{4.37}$$

with

$$\begin{aligned} J(x) = &\sum_{r>0}^{R}\left(w(x,x+r)\prod_{j=1}^{r}\left(e^{j\frac{\partial}{\partial x}}q(x)\right) - w(x+r,x)\right) \\ &+\sum_{r>0}^{R}\left(w(x,x-r)\prod_{j=0}^{r-1}\left(e^{-j\frac{\partial}{\partial x}}q(x)\right)^{-1} - w(x-r,x)\right) \end{aligned} \tag{4.38}$$

Equations (4.37), (4.38) correspond to the Kramers-Moyal expansion of the Master equation and represent a generalization of the already known Görtz and Walls (1976) equation. The probability distribution $P(x)$ is again obtained via

$$P(x) = P(0)\exp\sum_{j=1}^{m} \ln q(j) \approx P(0)\exp\int_{1}^{m} \ln q(x)dx, \tag{4.39}$$

where we can perform the sum in (4.39) via Eulers sum rule (Abramowitz and Stegun 1972).

$$\sum_{j=1}^{m} \ln q(j) = \int_{1}^{m} \ln q(x)dx + \frac{1}{2}\ln\left(\frac{q(m)}{q(1)}\right) + \cdots. \tag{4.40}$$

Suppose, $q(x)$ and the transition rates $w(x, x+r)$ for $r = \pm 1, \pm 2, \ldots \pm R$ are smooth functions of x then a "local" approximation concerning $q(x)$ can be assumed. This assumption strongly simplifies the equations of motion and we finally obtain the approximate equation for the continuous transition factors

$$\dot{q}(x) = q(x)\frac{\partial}{\partial x}\sum_{r}^{\pm R} w(x, x+r)(q^{r}(x) - 1). \tag{4.41}$$

This Eq. (4.41) describes a first order quasi-linear partial differential equation (Görtz and Walls 1976). Hence, the method of characteristics as a technique for solving partial differential equations can be applied. The mathematical procedure is to reduce a partial differential equation to a family of ordinary differential equations. Once the solutions of the ordinary differential equations are found, it can be solved along the characteristic curves and transformed into a solution for the original partial differential equation.

Suppose the probability distribution function $P(x)$ is sharply peaked on the site $y(t) = \langle x(t) \rangle$, then the difference between the most probable value of $P(x)$ and its mean value is negligible. In this case, we may expand the continuous transition factor around its maximum

$$q(y(t)) = 1 \tag{4.42}$$

in a Taylor expansion

$$q(x) = 1 + \left(\frac{\partial q}{\partial x}\right)_{\langle x \rangle}(x - y(t)) + \cdots, \tag{4.43}$$

and insert (4.43) in (4.39), (4.40) and perform the integration. This leads to a normal distribution around its mean value $y(t) = \langle x \rangle$

$$\begin{aligned} P(x) &= P(y(t))\exp\int_{y(t)}^{x} \ln\left(1 + \left(\frac{\partial q}{\partial x}\right)_{y}(x' - y(t))dx'\right) \\ &\approx P(y(t))\exp\left\{-\frac{(x - y(t))}{2\sigma^2(t)}\right\} \end{aligned}$$

and finally after normalization to the time-dependent Gauss-distribution function

$$P(x) = \frac{1}{\sqrt{2\pi\sigma^2(t)}}\exp\left\{-\frac{(x - y(t))^2}{2\sigma^2(t)}\right\}. \tag{4.44}$$

where the variance $\sigma^2(t)$ of the distribution is given by

$$\sigma^2(t) = -\left[\left(\frac{\partial q}{\partial x}\right)_y\right]^{-1}. \tag{4.45}$$

Hence, the slope of the transition factor $q(x)$ at (4.42) directly determines the variance. The higher the slope, the smaller the value of the variance. The introduced assumptions lead to normal distributed fluctuations according to the central limit theorem. In so far (4.41) corresponds in its physical content to the Fokker–Planck equation, but represents a first order partial differential equation instead of a second order partial differential equation. As a consequence, different mathematical solution algorithms can be applied. It depends on the mathematical structure of the transition rates which solution algorithm—Fokker–Planck equation or Görtz equation—seems to be most appropriate for the treatment of the analytical solution.

Equations of motion for the mean value $\langle x\rangle$ of the probability distribution and the variance $\sigma^2(t)$ follow in a straight forward manner via implicit differentiation of (4.42) and using (4.41)

$$\dot{y}(t) = -\frac{\left(\frac{\partial q}{\partial t}\right)_y}{\left(\frac{\partial q}{\partial x}\right)_y} = -\frac{q(y)}{\left(\frac{\partial q}{\partial x}\right)_y}\frac{\partial}{\partial x}J(x)\Big|_y \tag{4.46}$$

and

$$\dot{\sigma}^2(t) = -\frac{\partial}{\partial t}\left(\left[\frac{\partial q}{\partial x}\right]_y^{-1}\right) = \frac{1}{\left(\frac{\partial q}{\partial x}\right)_y^2}\left(\frac{\partial \dot{q}}{\partial x}\right)_y = \frac{1}{\left(\frac{\partial q}{\partial x}\right)_y^2}\frac{\partial}{\partial x}\left(q(x)\frac{\partial J(x)}{\partial x}\right)_y. \tag{4.47}$$

For further evaluation, we need the first and second derivatives of $J(x)$ at the point of the maximum $y(t)$:

$$\begin{aligned}
J(x)|_y &= \sum_r^{\pm R} w(x, x+r)(q^r(x) - 1)\Big|_y = 0 \\
\frac{\partial J(x)}{\partial x}\Big|_y &= \left(\frac{\partial q}{\partial x}\right)_y \sum_r^{\pm R} r w(x, x+r)\Big|_y \\
\frac{\partial^2 J(x)}{\partial x^2}\Big|_y &= 2\left(\frac{\partial q}{\partial x}\right)_y \frac{\partial}{\partial x}\sum_r^{\pm R} r w(x, x+r)\Big|_y \\
&\quad + \left(\frac{\partial q}{\partial x}\right)_y^2 \sum_r^{\pm R} r(r-1) w(x, x+r)\Big|_y
\end{aligned} \tag{4.48}$$

Inserting (4.48) in (4.46) and (4.47) we obtain the well-known, coupled equations of motion for the first and second moment of the distribution

$$\frac{dy(t)}{dt} = K(y(t)) \tag{4.49}$$

and

$$\frac{d\sigma^2(t)}{dt} = 2\sigma^2(t)\frac{\partial K(y(t))}{\partial y(t)} + Q(y(t)) \tag{4.50}$$

with the abbreviations

$$K(y(t)) = \sum_r^{\pm R} rw(y+r,y) \text{ and } Q(y(t)) = \sum_r^{\pm R} r^2 w(y+r,y). \tag{4.51}$$

Equation (4.49) describes the movement of the system along the most likely path $y(t) \approx \langle x(t) \rangle$, caused by the driving force $K(y(t))$. The fluctuations $\sigma^2(t)$ around this most likely path are given by the dynamic Eq. (4.50), where $Q(y(t))$ is the fluctuation coefficient. Since the transition probabilities $w(x, x+r) > 0$ do not explicitly depend on time, we can integrate the equation of motion for the variance (4.50) formally using (4.49)

$$\sigma^2(t) = \sigma_0^2 \left[\frac{K(y)}{K(y_0)}\right]^2 + K^2(y) \int_{y_0}^{y} \frac{Q(y')}{K^3(y')} dy', \tag{4.52}$$

where we used the initial conditions

$$\sigma^2 = \sigma_0^2, y = y_0 \tag{4.53}$$

at time $t = 0$. The solution (4.52) is non-divergent only for $K(y_0) \neq 0$. In the special case, close to an unstable point y_m, where $K(y_m) \approx 0$ an exponential fluctuation enhancement may occur.

4.3 Stationary Solution of the Master Equation

If the transition rates $w(\vec{n}, \vec{m})$ do not explicitly depend on time, the Master equation describes an equilibration process, starting with an arbitrary initial distribution $P(\vec{n}, t)$ and ending up with an unique final probability distribution $P_{st}(\vec{n})$ for $t \to \infty$. In the stationary state the left-hand side of the Master equation (3.22) equals zero:

$$\sum_{\vec{m}} w(\vec{n},\vec{m})P_{st}(\vec{m}) - \sum_{\vec{m}} w(\vec{m},\vec{n})P_{st}(\vec{n}) = 0 \quad \text{for all } \vec{n}. \tag{4.54}$$

Applying the T-factor method yields:

$$J(\vec{n}) = \sum_{\vec{m}} (w(\vec{n},\vec{m})g_{st}(\vec{m}|\vec{n}) - w(\vec{m},\vec{n})) = 0 \quad \text{for all } \vec{n}. \tag{4.55}$$

Equations (4.54, 4.55) express the fact that under stationary conditions, the sum of all transitions per unit of time from any state $\vec{m}$ into any state $\vec{n}$ of the configuration space must be balanced by the sum of all transitions from $\vec{n}$ into all other states $\vec{m}$. Therefore, (4.54) is nothing but a balance equation for probability fluxes. Moreover, a nontrivial solution of (4.54) can always be found, since (4.54) describes a set of linear algebraic equations for $P_{st}(\vec{n})$. However, in general, the solution of the stationary Master equation (4.54) has a very cumbersome form and is not easy to calculate, as we can see in the next chapter.

4.3.1 Kirchhoff's Exact Solution for Stationary Systems

In this section, a graph-theoretical solution based on a theorem developed by Kirchhoff (1844) for electrical networks will be presented. We follow Haken (1977) in the representation of this solution path.

Firstly, a graph *G* associated with the Eq. (4.54) is defined. This graph *G* contains all vertices and edges characterized by $w(\vec{n},\vec{m}) > 0$.

Secondly, certain sub graphs are defined. They are obtained from *G* by omitting certain edges. This sub graph, called maximal tree *T*(*G*), is defined by the following rules:

$$\begin{aligned} &\text{(a) } T(G) \text{ covers sub graph so that} \\ &\qquad \text{all edges of } T(G) \text{ are edges of } G \\ &\qquad T(G) \text{ contains all vertices of } G \\ &\text{(b) } T(G) \text{ is connected} \\ &\text{(c) } T(G) \text{ contains no circuits(cyclic sequence of edges)} \end{aligned} \tag{4.56}$$

From the definition (4.56) it follows immediately that one has to drop a certain minimum number of edges of *G*.

Thirdly, we define a directed maximal tree with the index *n*, $T_n(G)$. The directed tree is obtained from *T*(*G*) by directing all edges of *T*(*G*) towards the vertex with index *n*. After these preliminaries, the stationary solution $P_{st}(n)$ can be constructed.

Fourthly, we assign a numerical value $A(T_n(G))$, to each directed maximal graph $T_n(G)$. This value is obtained as a product of all transition rates $w(\vec{n},\vec{m}) > 0$ whose edges occur in $T_n(G)$ in the corresponding direction.

Fifthly, we define S_n as the sum over all maximal trees with the same index *n*

$$S_n = \sum_{allT_n(G)} A(T_n(G)). \tag{4.57}$$

Sixthly, Kirchhoff's formula for the probability distribution is then given by

$$P_{st}(n) = \frac{S_n}{\sum_{k=1}^{N} S_k}. \tag{4.58}$$

Unfortunately, for higher numbers of vertices this procedure becomes rather tedious. However, at least in many practical cases it allows for a deep insight into the construction of the stationary solution. Furthermore, it permits the decomposition of the problem into several smaller parts in many examples.

As Kirchhoff's solution seems rather abstract, let us consider the method by means of a simple example.

Example: Relaxation Processes in a Four-level Atom

We consider a four-level atom as depicted in Fig. 4.3. We assume that the atom can be excited by an external source, e.g. optical pumping by a Laser source. After the excitation of the fourth energy level, a step-by-step relaxation process starts e.g. via emission of photons or phonons. If the Laser operates under stationary conditions, the system will approach its stationary state after a short settling time. The question is, what will be the occupation probability $P_{st}(k)$ of the different eigen states E_k, $k = 1, 2, \ldots, 4$? Of course, the probability of occupation of a specific state depends on the transition probabilities $w(i,j) \geq 0$ for $i, j = 1, \ldots, 4$ between the different energy levels. Since we have circular transitions between the different energy levels, the principle of detailed balance is not fulfilled.

We now apply the graph-theoretical solution method of Kirchhoff, as described above and define a graph G, containing all vertices and edges characterized by $w(\vec{n}, \vec{m}) > 0$ according to Fig. 4.3. Secondly, Fig. 4.4 the maximal trees $T(G)$ belonging to the four-level system are extracted and based on that the directed maximal trees, Fig. 4.5.

As described, the values for the directed maximal trees of our example are obtained as product of all nonvanishing transition rates

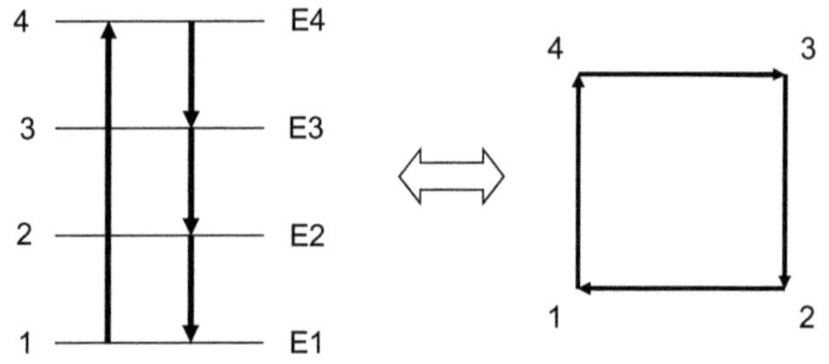

Fig. 4.3 Four-level atom (*left*) and its corresponding graph (*right*). Circular transitions violating the principle of detailed balance. 1→4: excitation by external sources, 4→3; 3→2; and 2→1 recombination processes

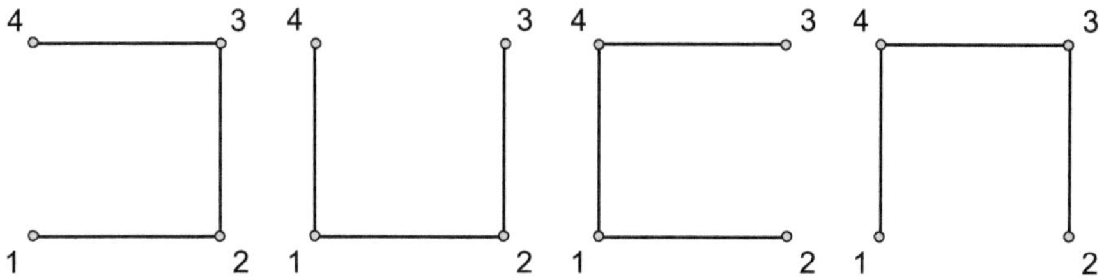

Fig. 4.4 The maximal trees $T(G)$ belonging to the graph of the four-level atom

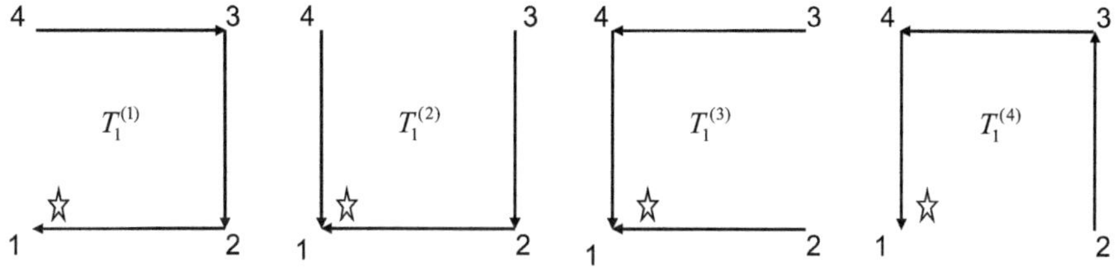

Fig. 4.5 The directed maximal trees $T_1(G)$ belonging to (Fig. 4.4)

$$
\begin{array}{ll}
A\left(T_1^{(1)}\right) = w(1,2)w(2,3)w(3,4) \neq 0 & A\left(T_2^{(1)}\right) = w(2,1)w(2,3)w(3,4) = 0 \\
A\left(T_1^{(2)}\right) = w(1,4)w(1,2)w(2,3) = 0 & A\left(T_2^{(2)}\right) = w(1,4)w(2,1)w(2,3) = 0 \\
A\left(T_1^{(3)}\right) = w(1,2)w(1,4)w(4,3) = 0 & A\left(T_2^{(3)}\right) = w(2,1)w(1,4)w(4,3) = 0 \\
A\left(T_1^{(4)}\right) = w(3,2)w(4,3)w(1,4) = 0 & A\left(T_2^{(4)}\right) = w(2,3)w(3,4)w(4,1) \neq 0 \\
S_1 = w(1,2)w(2,3)w(3,4) & S_2 = w(2,3)w(3,4)w(4,1)
\end{array}
$$

and

$$
\begin{array}{ll}
A\left(T_3^{(1)}\right) = w(2,1)w(3,2)w(3,4) = 0 & A\left(T_4^{(1)}\right) = w(2,1)w(3,2)w(4,3) = 0 \\
A\left(T_3^{(2)}\right) = w(1,4)w(2,1)w(3,2) = 0 & A\left(T_4^{(2)}\right) = w(4,1)w(1,2)w(2,3) \neq 0 \\
A\left(T_3^{(3)}\right) = w(1,2)w(4,1)w(3,4) \neq 0 & A\left(T_4^{(3)}\right) = w(1,2)w(4,1)w(4,3) = 0 \\
A\left(T_3^{(4)}\right) = w(3,2)w(3,4)w(4,1) = 0 & A\left(T_4^{(4)}\right) = w(3,2)w(4,3)w(4,1) = 0 \\
S_3 = w(1,2)w(4,1)w(3,4) & S_4 = w(4,1)w(1,2)w(2,3)
\end{array}
$$

In our example, this finally leads to the stationary probabilities, using (4.57), (4.58)

$$P_{st}(1) = \frac{w(1,2)w(2,3)w(3,4)}{w(1,2)w(2,3)w(3,4) + w(2,3)w(3,4)w(4,1) + w(1,2)w(4,1)w(3,4) + w(4,1)w(1,2)w(2,3)}$$

and accordingly, for the other probabilities:

$$P_{st}(2) = \frac{w(2,3)w(3,4)w(4,1)}{w(1,2)w(2,3)w(3,4) + w(2,3)w(3,4)w(4,1) + w(1,2)w(4,1)w(3,4) + w(4,1)w(1,2)w(2,3)}$$
$$P_{st}(3) = \frac{w(1,2)w(4,1)w(3,4)}{w(1,2)w(2,3)w(3,4) + w(2,3)w(3,4)w(4,1) + w(1,2)w(4,1)w(3,4) + w(4,1)w(1,2)w(2,3)}$$
$$P_{st}(4) = \frac{w(4,1)w(1,2)w(2,3)}{w(1,2)w(2,3)w(3,4) + w(2,3)w(3,4)w(4,1) + w(1,2)w(4,1)w(3,4) + w(4,1)w(1,2)w(2,3)}$$

It can be easily verified that the normalization condition of the probability is satisfied.

This example also demonstrates the limitations of Kirchhoff's method.

4.3.2 *Exact Stationary Solution for Systems with Detailed Balance*

Detailed balance is a much stronger requirement than (4.54, 4.55). Instead of a balance of the total probability in- and out-flows of each cell in configuration space, in case of detailed balance the in- and out-flows between each pair of cells $\vec{m}, \vec{n}$ in configuration space vanish:

$$w(\vec{n}, \vec{m})P_{st}(\vec{m}) - w(\vec{m}, \vec{n})P_{st}(\vec{n}) = 0 \quad \text{for all } \vec{n}, \vec{m}. \tag{4.59}$$

In other words, each flow must have a counter flow of the same size. Consequently, a necessary condition for detailed balance (4.59) requires a reverse transition rate $w(\vec{m}, \vec{n}) > 0$ to each non-vanishing transition rate $w(\vec{n}, \vec{m}) > 0$

If detailed balance (4.59) holds, the stationary transition factors are easily determined by the transition rates

$$g_{st}(\vec{m}|\vec{n}) = \frac{w(\vec{m}, \vec{n})}{w(\vec{n}, \vec{m})} \quad \text{or} \quad g\left(\vec{n}_{j+1}|\vec{n}_j\right) = \frac{w\left(\vec{n}_{j+1}|\vec{n}_j\right)}{w\left(\vec{n}_j|\vec{n}_{j+1}\right)} > 0. \tag{4.60}$$

We introduce a possible path $\left\{\vec{n}_0, \vec{n}_1, \ldots, \vec{n}_j, \ldots, \vec{n}_M\right\}$ in configuration space between adjacent cells, so that $g\left(\vec{n}_{j+1}|\vec{n}_j\right) > 0$. According to this path, we proceed from any cell $\vec{n}_o$ with occupation probability $P(\vec{n}_o)$ to cell $\vec{n}_M$ with occupation probability $P(\vec{n}_M)$ via

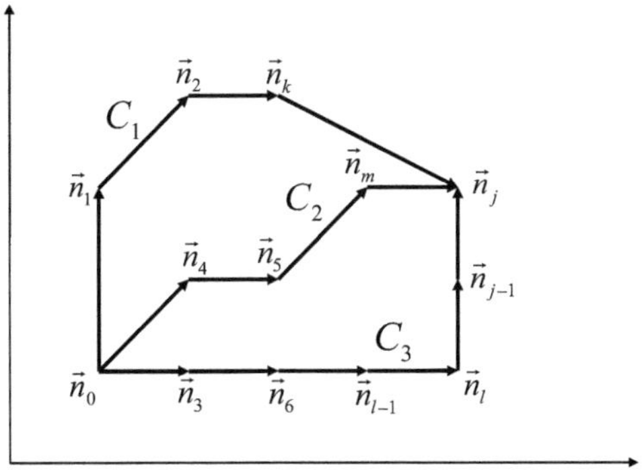

Fig. 4.6 Different equivalent possible chains for the calculation of the stationary probability distribution in case of detailed balance $C_1 = (\vec{n}_0, \vec{n}_1, \vec{n}_2, \vec{n}_k, \vec{n}_j)$, $C_2 = (\vec{n}_0, \vec{n}_4, \vec{n}_5, \vec{n}_m, \vec{n}_j)$, $C_3 = (\vec{n}_0, \vec{n}_3, \vec{n}_6, \vec{n}_{l-1}, \vec{n}_l, \vec{n}_{j-1}, \vec{n}_j)$

$$P(\vec{n}_M) = P(\vec{n}_0) \prod_{j=0}^{M-1} g(\vec{n}_{j+1} | \vec{n}_j) = P(\vec{n}_0) \prod_{j=0}^{M-1} \frac{w(\vec{n}_{j+1}, \vec{n}_j)}{w(\vec{n}_j, \vec{n}_{j+1})}. \tag{4.61}$$

Finally, the value of $P_{st}(\vec{n}_0)$ is determined by the normalization condition

$$\sum_{\vec{n}_M} P_{st}(\vec{n}_M) = 1. \tag{4.62}$$

The exact stationary solution, if detailed balance holds, was first developed by Haken (1974). However, since condition (4.59) is not directly appropriate to check the existence of detailed balance, it implies that the stationary solution has to be known beforehand.

Therefore, it was important to derive conditions for the transition rates which are equivalent to (4.59) but not containing the unknown stationary solution $P_{st}(\vec{n})$ (Haag 1978). Figure 4.6 shows three different possible chains for the calculation of the stationary probability distribution in case of detailed balance for a two-dimensional system.

Since (4.59) implies (4.61) for any chain C_r and for any final state $\vec{n}$, we may also apply (4.61) to any closed loop chain Γ (Fig. 4.7). Therefore, as a necessary and sufficient condition for detailed balance (Haag 1978, 1989)

$$(\Gamma) \prod_{j} \frac{w(\vec{n}_{j+1}, \vec{n}_j)}{w(\vec{n}_j, \vec{n}_{j+1})} = 1 \tag{4.63}$$

has to be fulfilled, for every closed loop $\Gamma = \{\vec{n}_0, \vec{n}_1, \ldots, \vec{n}_j, \ldots, \vec{n}_0\}$. The Eq. (4.63) imply conditions between the forms the transition rates and the involved parameters. In Sect. 6.5 this test is used to prove detailed balance for specific transition rates related to migratory processes.

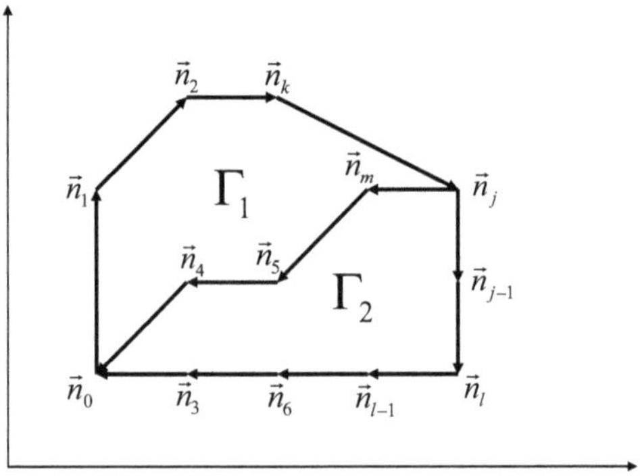

Fig. 4.7 Two closed loops fulfilling (4.63): $\Gamma_1 = \left(\vec{n}_0, \vec{n}_1, \vec{n}_2, \vec{n}_k, \vec{n}_j, \vec{n}_m, \vec{n}_5, \vec{n}_4, \vec{n}_0\right)$, and $\Gamma_2 = \left(\vec{n}_0, \vec{n}_1, \vec{n}_2, \vec{n}_k, \vec{n}_j, \vec{n}_{j-1}, \vec{n}_l, \vec{n}_{l-1}, \vec{n}_6, \vec{n}_3, \vec{n}_0\right)$

4.4 Continued Fraction Solutions for Two Particle Jumps

In this chapter, we confine ourselves to a discussion of the stationary solution of discrete Master equations of one-variable processes. We follow the papers of Haag and Hänggi (1979, 1980) in the presentation of results. One-dimensional Master equations occur in many fields, such as in quantum optics, spin-relaxation, chemical reactions or population dynamics when the statistical system under consideration can be assumed to be spatially uniform. Firstly, spatial uniformity can arise, for example, since the system is small, e.g. in case of some biological systems, so it cannot exhibit any phase boundaries or, secondly, it can be imposed by external boundary conditions such as thorough stirring in chemical reactions.

In contrast to a one-dimensional Fokker–Planck equation where the stationary solution is easily obtained by a simple integration procedure, the solutions of discrete Master equations are generally of more complicated structure. This is due to the fact, that except for simple birth and death processes the Master equation with multiple transitions generally does not obey a detailed balance relation. In addition, the discrete structure of the phase space leads to complicated solutions, especially for small numbers. Those very specific and important details cannot be represented by the continuous structure of Fokker–Planck equation.

In order to elucidate the basic ideas, we start the investigation of the stationary stochastic of systems where two-particle jumps may occur, not obeying detailed balance. For these cases the stationary birth- and death Master equation reads

$$0 = \sum_m w(n,m)P_{st}(m) - \sum_m w(m,n)P_{st}(n) \tag{4.64}$$

$$w(m,.n) = 0 \quad \text{for}\, |m-n| > 2 \tag{4.65}$$

with the abbreviations

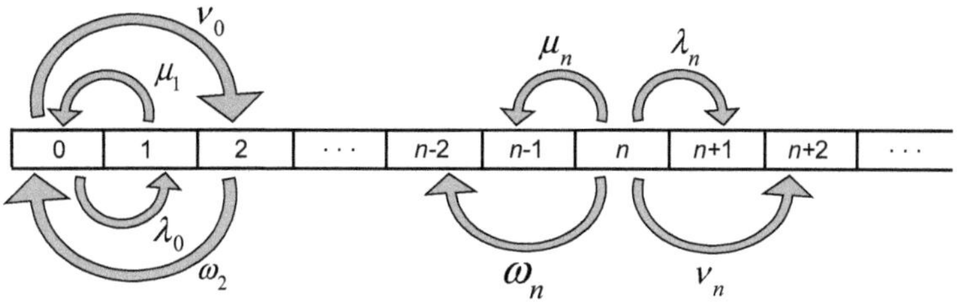

Fig. 4.8 Possible one- and two-particle jumps

$$w(n+1,n) \equiv \lambda_n \geq 0 \quad \text{for } n = 0, 1, \ldots \tag{4.66}$$
$$w(n-1,n) \equiv \mu_n \geq 0 \quad \text{for } n = 1, 2, \ldots \tag{4.67}$$
$$w(n+2,n) \equiv \nu_n \geq 0 \quad \text{for } n = 0, 1, \ldots \tag{4.68}$$
$$w(n-2,n) \equiv \omega_n \geq 0 \quad \text{for } n = 2, 3, \ldots \tag{4.69}$$

In Fig. 4.8 the one- and two-particle jumps between cells of the one-dimensional Master equation are illustrated.

The following abbreviations are introduced for simplicity

$$[0] = \lambda_0 + \nu_0 \tag{4.70}$$
$$[1] = \lambda_1 + \mu_1 + \nu_1 \tag{4.71}$$
$$[n] = \lambda_n + \nu_n + \mu_n + \omega_n. \tag{4.72}$$

Insertion of the chosen abbreviations into the Master equation (4.64) leads to

$$0 = -[0]P_{st}(0) + \mu_1 P_{st}(1) + \omega_2 P_{st}(2) \tag{4.73}$$
$$0 = \lambda_0 P_{st}(0) - [1]P_{st}(1) + \mu_2 P_{st}(2) + \omega_3 P_{st}(3)$$
$$\ldots \tag{4.74}$$
$$0 = \nu_{n-2}P_{st}(n-2) + \lambda_{n-1}P_{st}(n-1) - [n]P_{st}(n) + \mu_{n+1}P_{st}(n+1) + \omega_{n+2}P_{st}(n+2) \quad \text{for } n = 2, 3, \ldots \tag{4.75}$$

In the next step, the nearest-neighbour transition factors

$$\xi(n) = P_{st}(n)/P_{st}(n-1) \geq 0 \tag{4.76}$$

are introduced and inserted into (4.73). This yields

$$\xi(1) = \frac{[0]}{\mu_1 + \omega_2 \xi(2)} \tag{4.77}$$

and from the boundary condition (4.74), we obtain

$$\xi(2) = \frac{[1][0] - \mu_1\lambda_0}{\omega_2\lambda_0 + [0](\mu_2 + \omega_3\xi(3))}. \tag{4.78}$$

This result suggests the following ansatz for a recursion relation of the transition factors

$$\xi(n) = \frac{a_{n-1}}{b_n + a_{n-2}(\mu_n + \omega_{n+1}\xi(n+1))} \geq 0, \tag{4.79}$$

where the coefficients $a_n \geq 0$ and $b_n \geq 0$ are determined by the recursion relations

$$a_n = [n]a_{n-1} - \omega_n\nu_{n-2}\frac{a_{n-1}a_{n-3}}{a_{n-2}} - (b_n + \mu_n a_{n-2})\left(\nu_{n-2}\frac{b_{n-1} + \mu_{n-1}a_{n-3}}{a_{n-2}} + \lambda_{n-1}\right) \tag{4.80}$$

$$b_n = \omega_n(\nu_{n-3}(b_{n-2} + \mu_{n-2}a_{n-4}) + \lambda_{n-2}a_{n-3}) \tag{4.81}$$

with the following values of the recursion expressions (4.80) and (4.81)

$$\begin{aligned} a_{-1} &= 1, \quad a_0 = [0], \quad a_1 = [1][0] - \mu_1\lambda_0 \\ b_1 &= 0, \quad b_2 = \omega_2\lambda_0. \end{aligned} \tag{4.82}$$

This recursion relations can be proved using (4.75) and by complete induction on n. In terms of the transition factors (4.76), the stationary solution of (4.64) can be written as

$$P_{st}(n) = P_{st}(0)\prod_{k=1}^{n}\xi(k) \tag{4.83}$$

with $P_{st}(0)$ determined by normalization of the distribution. By inserting (4.79) into (4.83) the distribution function (4.83) can be also recast in the form of a continued fraction

$$P_{st}(n) = P_{st}(0)\frac{a_0 a_1 \ldots a_{n-1}}{d_n + d_{n-1}a_{n-2}\omega_{n+1}\xi(n+1)} \tag{4.84}$$

with coefficients d_n satisfying the recursion relation

$$d_n = (b_n + \mu_n a_{n-2})d_{n-1} + \omega_n a_{n-1}a_{n-3}d_{n-2} \tag{4.85}$$

together with the initial values

$$d_0 = 1, \; d_1 = \mu_1. \tag{4.86}$$

The above formulated mathematical relations are the starting point for the development of different solution methods

(a) continued fraction solutions
(b) exact reduced difference equation
(c) approximation method.

and it depends on the mathematical structure and complexity of the system, which of those solution methods seems to be the most appropriate one.

(a) The continued fraction solution

With an equivalence transformation, we can cast (4.79) in the form of an infinite continued fraction (Abramowitz and Stegun 1972, Perron 1913)

$$\xi(n) = \alpha_n \cfrac{1}{1+\cfrac{\beta_n}{1+\cfrac{\beta_{n+1}}{1+}\cdots}} \\ = \alpha_n \frac{1}{1+}\frac{\beta_n}{1+}\frac{\beta_{n+1}}{1+}\cdots \tag{4.87}$$

where the parameters α_n and β_n are given by

$$\alpha_n = \frac{a_{n-1}}{b_n + \mu_n a_{n-2}} \geq 0 \text{ and } \beta_n = \omega_{n+1} \frac{a_n a_{n-2}}{(b_n + \mu_n a_{n-2})(b_{n+1} + \mu_{n+1} a_{n-1})} \geq 0. \tag{4.88}$$

Because the number of terms of (4.87) is infinite, the continued fraction is called an infinite continued fraction. Obviously the structure of the continued fraction (4.87) disappears if $\beta_n = 0$ or, in other words, if no two-step down jumps exist.

Since the application of the Master equation to different systems may lead to different continued fractions, we will introduce a short review of some theorems, related to continued fractions.

Mathematics related to continued fractions

Let us consider a continued fraction of the following structure:

$$f = b_0 + \cfrac{a_1}{b_1 + \cfrac{a_2}{b_2 + \cfrac{a_3}{b_3+}}\cdots} \\ = b_0 + \frac{a_1}{b_1+}\frac{a_2}{b_2+}\frac{a_3}{b_3+}\cdots \tag{4.89}$$

Continued fractions are very appropriate for computational purposes, because they converge much faster than Taylor-series of the same order. For this reason, we consider so called r-th terminating fraction of f (Abramowitz and Stegun 1972):

$$f_r = \frac{A_r}{B_r} = b_0 + \frac{a_1}{b_1+} \frac{a_2}{b_2+} \cdots \frac{a_r}{b_r} \tag{4.90}$$

If $\lim_{r\to\infty} \frac{A_r}{B_r}$ exists, the infinite continued fraction is called convergent (Abramowitz and Stegun 1972).

If $a_i = 1$ and b_i are integers there is always convergence.

If a_i and b_i are positive, then $f_{2r} < f_{2r+2}$ and $f_{2r-1} > f_{2r+1}$.

The factors A_r and B_r can easily be computed via the following two recursion equations:

$$\begin{aligned} A_r &= b_r A_{r-1} + a_r A_{r-2} \\ B_r &= b_r B_{r-1} + a_r B_{r-2} \end{aligned} \tag{4.91}$$

where $A_{-1} = 1, A_0 = b_0, B_{-1} = 0, B_0 = 1$.

The truncated form of the continued fraction (4.89), prepared optimal for numerical calculations, reads

$$\begin{aligned} \xi_r(n) &= \alpha_n \frac{1}{1+} \frac{\beta_n}{1+} \frac{\beta_{n+1}}{1+} \cdots \frac{\beta_{n+r-2}}{1} \\ &= \alpha_n \frac{A_r}{B_r} \end{aligned} \tag{4.92}$$

Comparing (4.92) with (4.90), we can assign the coefficients

$$\begin{aligned} b_0 &= 0, a_1 = 1, a_2 = \beta_n, \ldots, a_r = \beta_{n+r-2} \\ b_1 &= 1, b_2 = 1, \ldots, b_r = 1. \end{aligned} \tag{4.93}$$

Together with the initial conditions

$$A_{-1} = 1, A_0 = b_0, B_{-1} = 0, B_0 = 1, \tag{4.94}$$

we obtain the final recursion relations

$$\begin{aligned} A_r &= A_{r-1} + \beta_{n+r-2} A_{r-2} \\ B_r &= B_{r-1} + \beta_{n+r-2} B_{r-2} \end{aligned} \tag{4.95}$$

for the continued fraction of the one-dimensional stationary Master equation (4.64) not obeying detailed balance.

(b) Exact reduced difference equation

Using the definition of the transition function (4.76) in (4.79) ultimately yields a reduced difference equation for the stationary solution of the Master equation (4.64)

$$a_{n-2}\omega_{n+1} P_{st}(n+1) + (b_n + a_{n-2}\mu_n) P_{st}(n) - a_{n-1} P_{st}(n-1) = 0 \quad \text{for } n = 1, 2, \ldots. \tag{4.96}$$

Compared with the difference equation, given by the Master equation (4.64), the difference Eq. (4.96) is reduced from order 4 in (4.64) to order 2 in (4.96). Moreover, the coefficients of the difference equation (4.96) are given by recursion relations and not simply by modified transition rates between next neighbours. Therefore, Eq. (4.96) is not a Master equation. The difference equation (4.96) has two fundamental solutions, but only the physical relevant solution characterised by $P_{st}(n) \geq 0$ is of interest in our case.

Textbooks about solution methods of difference equations have a long tradition (Meschkowski 1959, Rommelfanger 1986) but are currently difficult to find. In the next Chap. 5, we will deal with a number of important examples applying the method of Laplace (see Sect. 5.1).

(c) Approximation methods

In the previous sections, we described methods of solution for the stationary transition rates (4.76) casted in a very appropriate form for computer evaluation. Analytical solutions of the continued fraction representations cannot be obtained easily, except for special cases, which will be our focus in the next chapter. Therefore, the investigation of approximate solutions is of vast interest. Moreover, the approximate solution could be a favoured starting point for deep going analytical and numerical research.

Based on the observation that the transition factor $\xi(n)$ varies generally on a slower scale than the probabilities $P_{st}(n)$ themselves, we may set in (4.79) approximately $\xi(n) \approx \xi(n+1)$ and solve the quadratic equation, yielding

$$\xi(n) \approx \frac{b_n + a_{n-2}\mu_n}{2a_{n-2}\omega_{n+1}} \left\{ \sqrt{1 + \frac{4\omega_{n+1}a_{n-1}a_{n-2}}{(b_n + a_{n-2}\mu_n)^2}} - 1 \right\} > 0. \tag{4.97}$$

We have chosen the positive sign in the quadratic equation, since $\xi(n) \geq 0$ has to be satisfied.

Suppose, we can state that

$$\left| \frac{4\omega_{n+1}a_{n-1}a_{n-2}}{(b_n + a_{n-2}\mu_n)^2} \right| \ll 1. \tag{4.98}$$

Then we obtain for the approximate transition factor

$$\xi(n) \approx \alpha_n \tag{4.99}$$

and the approximate probability distribution

$$P_{st}(n) = P_{st}(0) \prod_{k=1}^{n} \xi(k) = P_{st}(0) \prod_{k=1}^{n} \alpha_k. \tag{4.100}$$

Moreover, the extreme values $\{\hat{n}\}$ of the probability distribution are approximately given by setting

$$\xi(\bar{n}) \approx 1 \tag{4.101}$$

and solving (4.97) with (4.101)

$$a_{\hat{n}-2}\omega_{\hat{n}+1} + (b_{\hat{n}} + a_{\hat{n}-2}\mu_{\hat{n}}) - a_{\hat{n}-1} \approx 0. \tag{4.102}$$

Equation (4.102) determines the extreme values of the distribution.

Suppose $\hat{n}$ to be an extreme value of the distribution, we may expand $\xi(n)$ around $\hat{n}$ in a Taylor series

$$\xi(n) \approx 1 + \left(\frac{\partial \xi(n)}{\partial n}\right)_{n=\hat{n}} (n - \hat{n}) + \cdots \tag{4.103}$$

and by use of (4.83) and the Euler MacLaurin summation formula, we obtain the Gaussian approximation in lowest order

$$P_{st}(n) \approx C \exp\left(-\frac{(n-\bar{n})^2}{2\sigma^2}\right), \tag{4.104}$$

where the variance of the distribution is given by

$$\sigma^2 = -\left[\left(\frac{\partial \xi(n)}{\partial n}\right)_{n=\bar{n}}\right]^{-1}. \tag{4.105}$$

or in more detail:

$$\sigma^2 = \frac{2a_{\bar{n}}\omega_{\bar{n}} + a_{\bar{n}}\mu_{\bar{n}} + b_{\bar{n}}}{\frac{\partial}{\partial n}\left[a_n\omega_n + a_n\mu_n + b_n - a_n\right]_{n=\bar{n}}}. \tag{4.106}$$

Using (4.106) the evolution of the variance related to the evolution of the extreme values (4.102) can be calculated.

Three special cases

1. Transition rates $\nu_n = 0\,,\omega_n = 0$

In Fig. 4.9, the Master equation for next-neighbour transitions is illustrated.

In case of next-neighbour transitions only, the regression formulas (4.80, 4.81) yield after some minor calculations

$$a_n = \prod_{k=0}^{n} \lambda_k \ \text{ and } \ b_n = 0. \tag{4.107}$$

For this special case, the continued fraction solution (4.79) reduces to a single quotient

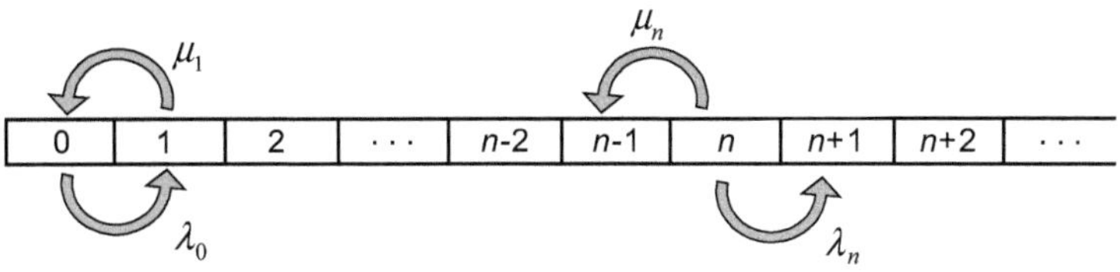

Fig. 4.9 Transitions between next-neighbours only

$$\xi(n) = \frac{\lambda_{n-1}}{\mu_n}, \tag{4.108}$$

and we obtain for the exact stationary probability distribution

$$P_{st}(n) = P_{st}(0)\prod_{k=1}^{n}\xi(k) = P_{st}(0)\prod_{k=1}^{n}\frac{\lambda_{k-1}}{\mu_k}. \tag{4.109}$$

Extreme values of the probability distribution are approximately defined by

$$\xi(\hat{n}) \approx 1, \tag{4.110}$$

or in other words, if the average birth rate equals the average death rate,

$$\lambda_{\hat{n}} \approx \mu_{\hat{n}}\,. \tag{4.111}$$

holds.

2. *Transition rates* $\nu_n = 0$

In Fig. 4.10, the special case of a Master equation with next-neighbour transitions and two-particle down jumps is illustrated.

For this special case, in which all the two-particle jump birth rates ν_n vanish, the recursion relations for a_n (4.80) and b_n (4.81) simplify considerably

$$a_n = (\lambda_n + \mu_n + \omega_n)a_{n-1} - (b_n + \mu_n a_{n-2})\lambda_{n-1} \tag{4.112}$$
$$b_n = \omega_n \lambda_{n-2} a_{n-3} \tag{4.113}$$

with the solution:

$$a_n = \prod_{k=0}^{n}\lambda_k, \quad \text{for } n = 0, 1, \ldots \quad \text{and } b_n = \omega_n \prod_{k=0}^{n-2}\lambda_k, \quad \text{for } n = 2, 3, \ldots \tag{4.114}$$

This leads to a simplified structure of the continued fraction (4.79)

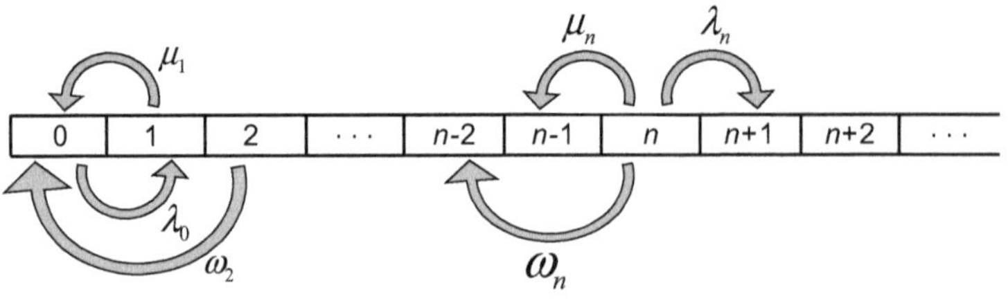

Fig. 4.10 Next-neighbour transitions and two-particle down jumps

$$\xi(n) = \frac{\lambda_{n-1}}{(\omega_n + \mu_n) + \omega_{n+1}\xi(n+1))}. \tag{4.115}$$

Extreme values of the probability distribution $\hat{n}$ are approximately defined by

$$\xi(\hat{n}) \approx 1. \tag{4.116}$$

If we insert (4.116) in (4.115), this finally leads to the condition

$$2\omega_{\hat{n}} + \mu_{\hat{n}} - \lambda_{\hat{n}-1} \approx 0, \text{ and } \sigma^2 = \frac{3\omega_{\hat{n}} + \mu_{\hat{n}}}{\frac{\partial}{\partial n}(2\omega_n + \mu_n - \lambda_n)|_{n=\hat{n}}} \tag{4.117}$$

In other words (4.117) stays, that the average birth rate equals the average death rate. The Gaussian approximation is known to give very satisfactory results when the transition rates correspond to a large system size (van Kampen 1981).

Furthermore, we obtain for the reduced difference equation (from order 3 to order 2)

$$\omega_{n+1}P_{st}(n+1) + (\omega_n + \mu_n)P_{st}(n) - \lambda_{n-1}P_{st}(n-1) = 0 \quad \text{for} \quad n = 1, 2, \ldots \tag{4.118}$$

This reduced difference equation is appropriate as starting point for analytical considerations (see Chap. 5). Despite the fact that (4.118) is not a Master equation as already mentioned, the transition rates enter (4.118) in a linear way.

For comparison reasons consider the exact Master equation (4.119) for this special case $\nu_n = 0$:

$$\begin{aligned}&\omega_{n+2}P_{st}(n+2) + \mu_{n+1}P_{st}(n+1) - (\omega_n + \mu_n + \lambda_n)P_{st}(n)\\&\quad + \lambda_{n-1}P_{st}(n-1) = 0 \quad \text{for } n = 1, 2, \ldots\end{aligned} \tag{4.119}$$

The straightforward calculation of transition factors $\xi(n)$ in form of infinite continued fractions and their subsequent backward transformation maintained a simplified but equivalent difference equation for the probability distribution $P_{st}(n)$.

3. Transition rate $\omega_n = 0$

In Fig. 4.11, the special case of a Master equation with next-neighbour transitions and two-particle up jumps is illustrated.

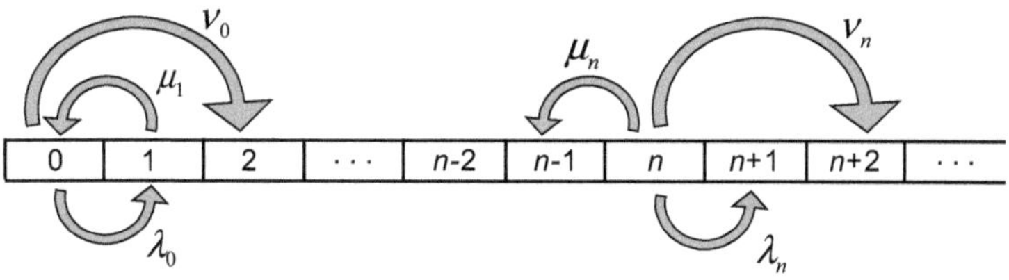

Fig. 4.11 Next-neighbour transitions and two-particle up jumps

In the case of identically vanishing two-particle jump death rates $\omega_n = 0$ the continued fraction coefficients in (4.81) yield

$$b_n = 0. \tag{4.120}$$

From (4.79) and (4.80), we obtain for the T-factors

$$\xi(n) = \frac{a_{n-1}}{\mu_n a_{n-2}} \tag{4.121}$$

and therefore for the probability distribution

$$P_{st}(n) = P_{st}(0) \prod_{k=1}^{n} \xi(k) = P_{st}(0) \frac{a_{n-1}}{\mu_1 \mu_2 \ldots \mu_n}. \tag{4.122}$$

The evaluation of the stationary probability distribution is reduced to the calculation of the $(n-1)$-th order continued fraction coefficient a_{n-1} which obeys the recursion relation

$$a_n = (\lambda_n + \mu_n + \nu_n) a_{n-1} - \mu_n \mu_{n-1} \nu_{n-2} a_{n-3} - \mu_n \lambda_{n-1} a_{n-2} \tag{4.123}$$

Using the transformation

$$a_n = f_n \prod_{i=1}^{n} \mu_i, \tag{4.124}$$

we recast (4.123) in the form

$$f_n - f_{n-1} = \left(\frac{\lambda_n}{\mu_n} + \frac{\nu_n}{\mu_n} \right) f_{n-1} - \frac{\lambda_{n-1}}{\mu_{n-1}} f_{n-2} - \frac{\nu_{n-2}}{\mu_{n-2}} f_{n-3} \tag{4.125}$$

with

$$f_{-3} = f_{-2} = 0, f_{-1} = 1, f_0 = \nu_0 + \lambda_0. \tag{4.126}$$

Summation of the relation in (4.125) from $n = 0$ to $n = i$ equals

$$f_i = \left(\frac{\lambda_i}{\mu_i} + \frac{\nu_i}{\mu_i}\right) f_{i-1} + \frac{\nu_{i-2}}{\mu_{i-2}} f_{i-2}. \tag{4.127}$$

The recursion relation (4.127), written in terms of the coefficients a_n, yields for (4.123) the considerably simplified recursion relation

$$a_n = (\lambda_n + \nu_n) a_{n-1} + \mu_n \nu_{n-1} a_{n-2} \tag{4.128}$$

for $n = 1, 2, \ldots$ or finally for the reduced difference equation using (4.122) for the probability distribution

$$\mu_{n+1} P_{st}(n+1) - (\lambda_n + \nu_n) P_{st}(n) - \nu_{n-1} P_{st}(n-1) = 0 \quad \text{for } n = 1, 2, \ldots \tag{4.129}$$

For comparison reasons, consider the exact Master equation (4.119) for this special case $\omega_n = 0$:

$$\begin{aligned} &\mu_{n+1} P_{st}(n+1) - (\mu_n + \lambda_n + \nu_n) P_{st}(n) + \lambda_{n-1} P_{st}(n-1) \\ &\quad + \nu_{n-2} P_{st}(n-2) = 0 \quad \text{for } n = 2, 3, \ldots \end{aligned} \tag{4.130}$$

Again, we found that the straightforward calculation of transition factors $\xi(n)$ in form of infinite continued fractions and their subsequent backward transformation maintained a simplified but equivalent difference equation for the probability distribution $P_{st}(n)$.

References

Abramowitz M, Stegun IA (1972) Handbook of mathematical functions. Dover, New York

Görtz R, Walls DF (1976) Steady state solutions of master equations without detailed balance. Z Phys B 25:423–427

Haag G (1977) Transition factor method for discrete master equations and applications to chemical reactions. Z Phys B 29:153–159

Haag G (1978) Neue Lösungsmethoden diskreter Mastergleichungen. Theses University of Stuttgart, Stuttgart

Haag G (1989) Dynamic decision theory: applications to urban and regional topics. Kluwer, Dordrecht

Haag G, Hänggi P (1979) Exact solutions of discrete master equations in terms of continued fractions. Z Phys B 34:411–417

Haag G, Hänggi P (1980) Continued fraction solutions of discrete master equations not obeying detailed balaance II. Z Phys B 39:269–279

Haag G, Weidlich W, Alber P (1977) Approximation methods for stationary solutions of discrete master equations. Z Phys B 26:207–215

Haken H (1974) Exact stationary solution of the master equation for systems far from thermal equilibrium in detailed balance. Phys Lett 46A(7):443–444

Haken H (1975) Cooperative phenomena in systems far from thermal equilibrium and in nonphysical systems. Rev Mod Phys 47:67–121

Haken H (1977) Synergetics: an introduction. Springer, Heidelberg

Kirchhoff G (1844) Über die Auflösung der Gleichungen auf welche man bei der Untersuchung der linearen Verteilung galvanischer Ströme geführt wird. Ann Phys 72:495

Kubo R, Matsuo K, Kitahara K (1973) Fluctuation and relaxation of marovariables. J Stat Phys 9:51–96

Meschkowski H (1959) Differenzengleichungen. Vandenhoeck & Ruprecht, Göttingen

Perron O (1913) Die Lehre von den Kettenbrüchen. B.G. Teubner, Stuttgart

Risken H (1972) Solutions of the Fokker-Planck equation in detailed balance. Z Phys B 251:231–243

Risken H (1996) The Fokker-Planck equation, Springer series in synergetics, vol 18. Springer, Heidelberg

Rommelfanger H (1986) Differenzengleichungen. Bibliographisches Institut, Mannheim

Schnakenberg J (1976) Network theory of microscopic and macroscopic behavior of master equation systems. Rev Mod Phys 48(4):571–585

van Kampen NG (1965) Fluctuation in nonlinear systems. In: Burgers AE (ed) Fluctuation phenomena in solids. Academic Press, New York

van Kampen NG (1981) Stochastic processes in physics and chemistry. North-Holland, Amsterdam

Weidlich W, Haag G (1983) Concepts and models of a quantitative sociology: the dynamics of interaction populations, Springer series of synergetics, vol 14. Springer, New York

Part II
Applications Natural Sciences

Chapter 5
Some Applications in Physics and Chemistry

Contents

5.1 Chemical Reactions

A basic question in chemistry is the rate of chemical reactions, or in other words, how long it takes for a chemical reaction to attain equilibrium and the concentration of the different reactants at equilibrium.

The law of mass action is a mathematical model that explains and predicts behaviours of solution in dynamic equilibrium. It states that the rate of a chemical reaction is directly proportional to the product of the masses of the reactants. This implies that the ratio between the concentration of reactants and products is constant, for a chemical reaction mixture at equilibrium.

Let us consider the mass action law by applying it to a simple chemical reaction in which reactants A and B react and produce the two products C and D:

$$aA + bB \rightarrow cC + dD \tag{5.1}$$

where a, b, c, d are the coefficients for a balanced chemical reaction.

The law of mass action states that if the chemical reaction is at equilibrium then the ratio will be constant

G. Haag, *Modelling with the Master Equation*, DOI 10.1007/978-3-319-60300-1_5

$$\frac{[C]^c[D]^d}{[A]^a[B]^b} = K, \tag{5.2}$$

where the square brackets represent their concentrations. The units for K depend upon the units used for concentrations. The equilibrium point depends on the physical conditions of the environment (temperature and pressure).

The kinetic of chemical reactions are usually described as deterministic processes by differential equations for number concentrations of the reactants and products.

What are the reasons for formulating chemical kinetics in a probabilistic framework?

The basic point is that a chemical reaction process is in fact statistical in nature. The concentration or number of molecules in the system is an integer-valued random variable. Therefore, the process should be described by the probability density of this random variable (McQuarrie 1967; Nitzan et al. 1974; Haag and Hänggi 1979, 1980). The mean value of this distribution will then be the observed concentration and the variance will be an indicator of the inherent statistical fluctuations around its mean. For many simple reactions, the mean will be essentially the same as for the deterministic solution. In case of low concentrations, it can be supposed that the molecules of the reactants and the products behave statistically independent of each other and the statistical distribution of the chemical process can be described by a multidimensional Poisson distribution. However, there are a great number of important exceptions for which the deterministic approach is not adequate. In such a case, stochastic models have to be used.

5.1.1 Unimolecular Chemical Reaction

As a first example, let us investigate an exactly solvable one-dimensional chemical reaction scheme, namely the decay of a molecule A to a molecule B

$$A \xrightarrow{k} B. \tag{5.3}$$

It is assumed that the reaction mixture is a well-stirred spatially homogeneous dilution. Therefore, spatial aspects do not have to be considered.

Let the random variable n be the number of A molecules in the system at time t. Because of this reaction scheme (Fig. 5.1), detailed balance is not fulfilled.

The change of probability of state n namely $dP(n)/dt$ is due to two opposing effects: an inflow of probability into state n via $k(n+1)P(n+1)$ and an outflow of probability from state $n \to n-1$ via $knP(n)$. The balance equation is the Master equation (5.6).

The transition rates of this one-dimensional Master equation read

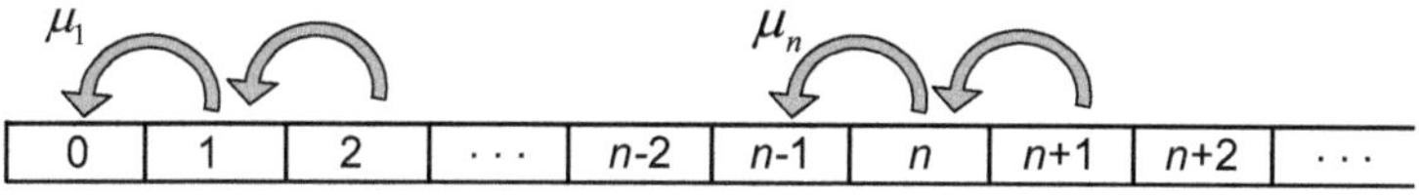

Fig. 5.1 Possible transitions of the one-dimensional chemical reaction Master equation (5.6)

$$\lambda_n = k \text{ and } \mu_n = 0 \tag{5.4}$$

$$\nu_n = 0 \text{ and } \omega_n = 0. \tag{5.5}$$

By use of the combinatorial mass-action kinetics, we obtain the following Master equation

$$\frac{dP(n)}{dt} = k(n+1)P(n+1) - knP(n) \tag{5.6}$$

for $n = 0, 1, 2, \ldots$.

As solution tool we introduce the generating function

$$G(z,t) = \sum_{n=0}^{\infty} P(n,t) z^n \text{ with } |z| \leq 1. \tag{5.7}$$

The initial state at time $t = 0$ may be characterized by $n_o = n(t = 0)$ A-molecules. This requires a δ- like initial probability

$$P(n,0) = 1 \quad \text{for } n = n_0 \tag{5.8}$$

$$P(n,0) = 0 \quad \text{otherwise.} \tag{5.9}$$

Now, we transform (5.6) into a partial differential equation

$$\begin{aligned} \frac{\partial G(z,t)}{\partial t} = \sum\nolimits_{n=0}^{\infty} \frac{dP(n,t)}{dt} z^n &= \sum\nolimits_{n=0}^{\infty} \left[\frac{k}{z}(n+1)P(n+1,t)z^{n+1} - knP(n,t)z^n \right] \\ &= \sum\nolimits_{n=0}^{\infty} \left[\frac{k}{z} - k \right] nP(n,t)z^n \\ &= k(1-z) \sum\nolimits_{n=0}^{\infty} \frac{\partial}{\partial z} z^n P(n,t) \\ &= k(1-z) \frac{\partial G(z,t)}{\partial z} \end{aligned} \tag{5.10}$$

which reads

$$\frac{\partial G(z,t)}{\partial t} = k(1-z)\frac{\partial G(z,t)}{\partial z}. \tag{5.11}$$

It should be mentioned that even though the system has a finite number of states, the sum runs from zero to infinity. This introduces no difficulty since $P(n,t)$ must vanish for all times when $n \geq n_0$, where n_0 is the total number of particles in the system.

To solve (5.11), we apply the method of separation of variables

$$G(z,t) = f(z)g(t) \tag{5.12}$$

and obtain

$$\frac{\frac{\partial g(t)}{\partial t}}{g(t)} = k(1-z)\frac{\frac{\partial f(z)}{\partial z}}{f(z)} = const = -mk \tag{5.13}$$

with a constant m. This finally leads to the solution

$$G(z,t) = \sum_m f_m(z)g_m(t) \tag{5.14}$$

with

$$g_m(t) = g_m^0 e^{-mkt} \text{ and } f_m(z) = f_m^0(z-1)^m \tag{5.15}$$

and the abbreviation $a_m = f_m^0 g_m^0$ to the generating function

$$G(z,t) = \sum_m a_m(z-1)^m e^{-kmt}. \tag{5.16}$$

The initial condition leads to the requirement

$$G(z,0) = \sum_m a_m(z-1)^m = z^{n_0} \tag{5.17}$$

with solution

$$a_m = \binom{n_0}{m} 1^{n_0-m} = \binom{n_0}{m} \tag{5.18}$$

and after insertion in (5.16) to the final form of the generating function.

$$G(z,t)=\sum_{m=0}^{n_0}\binom{n_0}{m}\left((z-1)\,e^{-kt}\right)^m=\left(1+(z-1)e^{-kt}\right)^{n_0}. \tag{5.19}$$

By noting the relations (Sect. 2.1)

$$\left(z\frac{\partial}{\partial z}\right)^N G(z,t)\bigg|_{z=1}=\sum_{m=0}^{\infty}n^N P(n,t)1^N=\left\langle n(t)^N\right\rangle, \tag{5.20}$$

we obtain for the mean value

$$E(n(t))=\left(\frac{\partial G}{\partial z}\right)_{z=1}=\langle n(t)\rangle \tag{5.21}$$

and for the variance

$$\sigma^2(t)=\frac{\partial^2 G}{\partial z^2}\bigg|_{z=1}+\frac{\partial G}{\partial z}\bigg|_{z=1}-\left(\frac{\partial G}{\partial z}\bigg|_{z=1}\right)^2. \tag{5.22}$$

Applying (5.21) and (5.22) on the generating function (5.19), we obtain the equations of motion for the mean value and the variance

$$\langle n(t)\rangle=n_0e^{-kt} \tag{5.23}$$

$$\sigma^2(t)=n_0e^{-kt}\left(1-e^{-kt}\right). \tag{5.24}$$

Figure 5.2 shows the evolution of the mean value (5.23) and the variance (5.24) for an initial value of $n_o=100$ A-molecules. The chosen reaction constant in this example reads $k=0.02$.

We note that the mean value of the stochastic representation coincides with the deterministic result. This shows that the two representations are consistent in its mean. However, this is true for unimolecular reactions, but not for all chemical reactions.

Since the generating function (5.19) enables us to calculate all moments of the distribution, the probability distribution function representing the fluctuations in this chemical reaction can be calculated (Bartholomay 1958) (see Chap. 2)

$$P(n,t)=\binom{n_0}{n}e^{-nkt}(1-e^{-kt})^{n_0-n}. \tag{5.25}$$

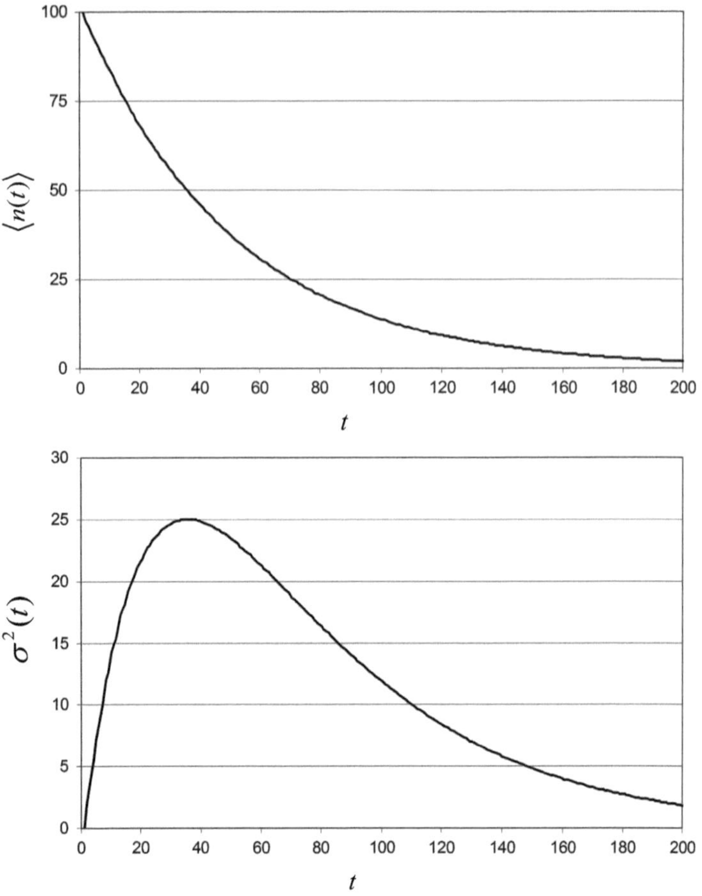

Fig. 5.2 Mean value and variance of the unimolecular chemical reaction $A \rightarrow X$. The variance reaches its maximum $\sigma^2_{\max} = n_0/4$ at time $t_{\max} = (1/2)\ln 2$

5.1.2 Linear Chemical Reaction

As a second example, let us investigate a simple one-dimensional chemical reaction scheme with next neighbour transitions only

$$A \xrightarrow{k_1} X \tag{5.26}$$

$$X \xrightarrow{k_2} A \tag{5.27}$$

Because of the reaction scheme (Fig. 5.3), detailed balance is fulfilled in this application.

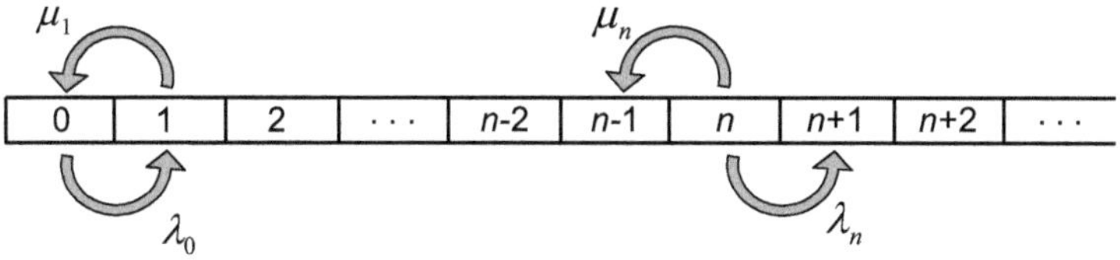

Fig. 5.3 Possible next-neighbour transitions of the one-dimensional chemical reaction Master equation

In this example, it is also assumed that the reaction mixture is a well stirred spatially homogeneous dilution. Therefore, spatial aspects do not have to be considered.

The transition rates of the one-dimensional birth-death Master equation read

$$\lambda_n = k_1 N_A \text{ and } \mu_n = k_2 n \tag{5.28}$$

$$\nu_n = 0 \text{ and } \omega_n = 0. \tag{5.29}$$

By use of the combinatorial mass-action kinetics, we obtain the following Master equation

$$\begin{aligned} \frac{dP(0)}{dt} &= k_2 P(1) - k_1 N_A P(0) \\ \frac{dP(n)}{dt} &= k_1 N_A P(n-1) + k_2(n+1)P(n+1) - (k_1 N_A + k_2 n)P(n) \end{aligned} \tag{5.30}$$

for $n = 1, 2, \ldots.$

In (5.30), N_A denotes the number of A-molecules which are held constant. With the notation

$$a = \frac{k_1 N_A}{k_2} = X_0, \tag{5.31}$$

where X_0 is the deterministic stationary state of the chemical process, we recast the transition factor for the stationary solution in the form

$$\xi_{st}(n) = \frac{\lambda_{n-1}}{\mu_n} = \frac{X_0}{n}. \tag{5.32}$$

The stationary probability distribution is easily calculated to be a Poissonian

$$P_{st}(n) = P_{st}(0) \prod_{k=1}^{n} \xi_{st}(k) = P_{st}(0) \frac{X_0^n}{n!} = \frac{X_0^n}{n!} e^{-X_0} \tag{5.33}$$

with mean value

$$\langle n \rangle = X_0 \tag{5.34}$$

and variance

$$\sigma^2 = X_0. \tag{5.35}$$

The scaled Master equation reads

$$\begin{aligned} \frac{dP(0)}{d\tau} &= P(1) - aP(0) \\ \frac{dP(n)}{d\tau} &= aP(n-1) + (n+1)P(n+1) - (a+n)P(n) \end{aligned} \tag{5.36}$$

with the scaled time $\tau = k_2 t$. As a solution tool, we apply the shift-operator technique introduced in Sect. 3.4.2. Then we reformulate the Master equation accordingly:

$$\frac{dP(n.\tau)}{d\tau} = (E-1)nP(n,\tau) + \left(E^{-1}-1\right)aP(n,\tau) \tag{5.37}$$

with

$$E^{\pm 1}f(n) = f(n \pm 1) \tag{5.38}$$

and

$$nE^{\pm 1}f(n) = E^{\pm 1}(n \mp 1)f(n) \tag{5.39}$$

$$n^2E^{\pm 1}f(n) = E^{\pm 1}\left(n^2 \mp 2n + 1\right)f(n). \tag{5.40}$$

Using (5.37) with (5.38)–(5.40) yields

$$\begin{aligned} \frac{d\langle n \rangle}{d\tau} &= \sum_n n\frac{dP(n,\tau)}{d\tau} \\ &= \sum_n n(E-1)nP(n,\tau) + \sum_n n\left(E^{-1}-1\right)aP(n,\tau) \\ &= \sum_n E(n-1)nP(n,\tau) - \sum_n n^2P(n,\tau) + \sum_n E^{-1}(n+1)aP(n,\tau) - \sum_n anP(n,\tau) \\ &= \langle (n-1)n \rangle - \left\langle n^2 \right\rangle + \langle (n+1)a \rangle - \langle an \rangle \\ &= a - \langle n \rangle \end{aligned} \tag{5.41}$$

or all together

$$\frac{d}{dt}\langle n\rangle = a - \langle n\rangle \quad \text{with solution} \quad \langle n\rangle = (n_0 - a)e^{-\tau} + a \tag{5.42}$$

and

$$\begin{aligned}\frac{d\langle n^2\rangle}{d\tau} &= \sum_n n^2 \frac{dP(n,\tau)}{d\tau} \\ &= \sum_n n^2(E-1)nP(n,\tau) + \sum_n n^2\left(E^{-1}-1\right)aP(n,\tau) \\ &= a(2\langle n\rangle + 1) - 2\langle n^2\rangle + \langle n\rangle \end{aligned} \tag{5.43}$$

for the equation of motion for the second moment of the distribution

$$\frac{d}{d\tau}\langle n^2\rangle + 2\langle n^2\rangle = a(2\langle n\rangle + 1) + \langle n\rangle. \tag{5.44}$$

The evolution of the first and the second moment of the probability distribution are obtained as solutions of the differential Eqs. (5.42) and (5.44). The dynamics of the variance finally is given by (5.45)

$$\frac{d\sigma^2(\tau)}{d\tau} = \langle n^2\rangle - \langle n\rangle^2. \tag{5.45}$$

5.1.3 Autocatalytic Non-linear Chemical Reaction

We consider the autocatalytic non-linear chemical reaction scheme originated by Nicolis (1972) as a third example. In the 1980s, many papers concentrated on this reaction since there was a dispute concerning the correct form of the mean square concentration fluctuations (Saito 1974; Mazo 1975; Glansdorff and Prigogine 1971). The generating function technique was applied as solution method of the Master equation. We will directly solve the Master equation, using the transition-factor method and will compare our method with the generating function technique at the end of this chapter.

The autocatalytic non-linear chemical reaction reads:

$$A \xrightarrow{k_1} X \tag{5.46}$$

$$2X \xrightarrow{k_2} A. \tag{5.47}$$

It is assumed that the reaction mixture is a well-stirred spatially homogeneous dilution. Therefore, spatial aspects do not have to be considered.

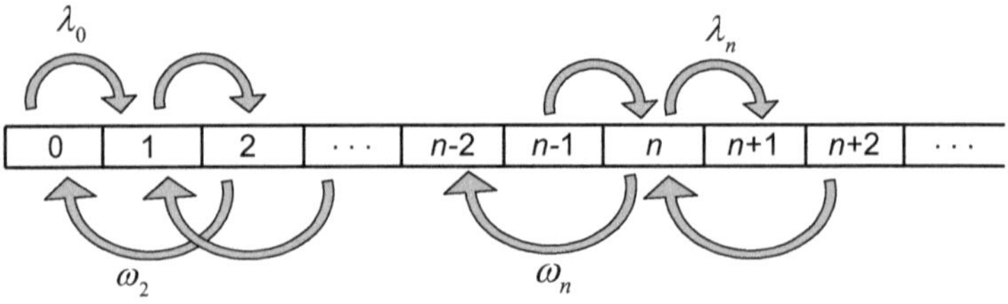

Fig. 5.4 Flow of probability for finding n molecules of species X

Because of the reaction scheme (Fig. 5.4), detailed balance is not fulfilled.

We will proceed in three steps: Firstly, the continued fraction representation is used to develop an approximate solution based on the T-factor method. Secondly, a reduced difference equation is derived based out of the continued fraction solution. The reduced difference equation for this chemical reaction is then solved exactly via the solution method of Laplace (Meschkowski 1959). Thirdly, a comparison of the obtained results and of the different solution methods is presented.

The transition rates of the one-dimensional birth-death Master equation of Sect. 4.3 read

$$\lambda_n = k_1 N_A \text{ and } \mu_n = 0 \tag{5.48}$$

$$\nu_n = 0 \text{ and } \omega_n = k_2 n(n-1), \tag{5.49}$$

where N_A denotes the number of A-molecules which are held constant.

By use of the combinatorial mass-action kinetics, we obtain the following Master equation

$$\begin{aligned} \frac{dP(0)}{dt} &= 2k_2P(2) - k_1N_AP(0) \\ \frac{dP(n)}{dt} &= k_1N_AP(n-1) + k_2(n+1)(n+2)P(n+2) - (k_1N_A + k_2n(n-1))P(n) \end{aligned} \tag{5.50}$$

for $n = 1, 2, \ldots$.

With the notation

$$a = \frac{k_1 N_A}{k_2} = 2X_0^2, \tag{5.51}$$

where X_0 is the deterministic stationary state of the chemical process, we recast (5.50) for the stationary solution in the form

$$0 = 2P_{st}(2) - aP_{st}(0)$$

$$0 = aP(n-1) + (n+1)(n+2)P_{st}(n+2) - (a + n(n-1))P_{st}(n)$$

According (4.115), the stationary transition factor $\xi(n)$ is given by

$$\xi(n) = \frac{a}{n(n-1) + n(n+1)\xi(n+1)} \quad \text{for } n = 1, 2, \ldots. \tag{5.53}$$

The continued fraction solution of the Master equation (5.52) can be obtained numerically according to Sect. 4.3.

Approximate Stationary Solution
The approximate solution using T-factor analysis, is based on the observation that the transition factor $\xi(n)$ varies generally on a slower scale than the probabilities themselves, see again Sect. 4.3. Assuming

$$\xi(n) \approx \xi(n+1) \tag{5.54}$$

and inserting it in (5.53), we obtain

$$\xi(n) = \frac{a}{n(n-1) + n(n+1)\xi(n)}, \tag{5.55}$$

a quadratic equation with the only positive solution

$$\xi(n) \approx \frac{1}{2}\left(\sqrt{1 + 4\frac{a}{n^2}} - 1\right) \quad \text{for } n \gg 1. \tag{5.56}$$

The maxima n_w of the probability distribution is obtained by setting $\xi(n_w) \approx 1$, yields

$$n_w \approx \sqrt{\frac{a}{2}} = X_0. \tag{5.57}$$

The most probable value of the distribution function (5.57) corresponds to the deterministic stationary state X_0 in this approximation.

The variance of the distribution can be easily computed by applying (4.105)

$$\sigma^2 = -\left[\left(\frac{\partial \xi(n)}{\partial n}\right)_{n_w}\right]^{-1} = \frac{3}{4}X_0. \tag{5.58}$$

The probability distribution is not a Poisson distribution, as the variance is smaller as its mean value. The approximate value of the mean square concentration fluctuations (5.58) agree with the concentration fluctuations computed by Nicolis

(1972) and Mazo (1975) using the method of generating function. However, the applied method of transition factors in combination with the development of a corresponding continued fraction allowed an approximate easy calculation of all interesting moments.

In the next step, we will solve the stationary Master equation via the development of a reduced difference equation and via the application of the method of Laplace (Meschkowski 1959).

Exact Stationary Solution Via Reduced Difference Equation
According to (4.115), the stationary transition factor $\xi(n)$ is given by

$$\xi(n) = \frac{a}{n(n-1) + n(n+1)\xi(n+1)} \quad \text{for } n = 1, 2, \ldots \tag{5.59}$$

or after some minor transformation

$$\xi_n = \frac{1}{n} \cdot \frac{a}{(n-1)+} \frac{a}{n+} \cdots \equiv \frac{1}{n} \cdot \frac{u(n+1)}{u(n)}, \tag{5.60}$$

where $u(n)$ obeys the linear difference Eq. (5.61) of second order ($n \to z$). Our transformation has not only reduced the order of the difference equation from three (Master equation) to two (reduced difference equation) but also its complexity from a nonlinear case (polynomial of second order) to a linear case (polynomial of first order):

$$u(z+2) + zu(z+1) - au(z) = 0 \ \text{ with } \ a > 0. \tag{5.61}$$

This simplification via transformation allows us to apply the Laplace-transformation (Meschkowski 1959) for $u(z)$

$$u(z) = \int_q^p t^{z-1} f(t)dt. \tag{5.62}$$

Inserting (5.62) in (5.61) yields

$$\int_q^p t^{z-1} \left[f(t)\left(t^2 - a\right) - t\frac{d(f(t)t)}{dt} \right] dt = -K(t)|_q^p \tag{5.63}$$

where

$$K(t) = t^{z+1} f(t). \tag{5.64}$$

Now the integration limits $\{q, p\}$ have to be fixed by imposing

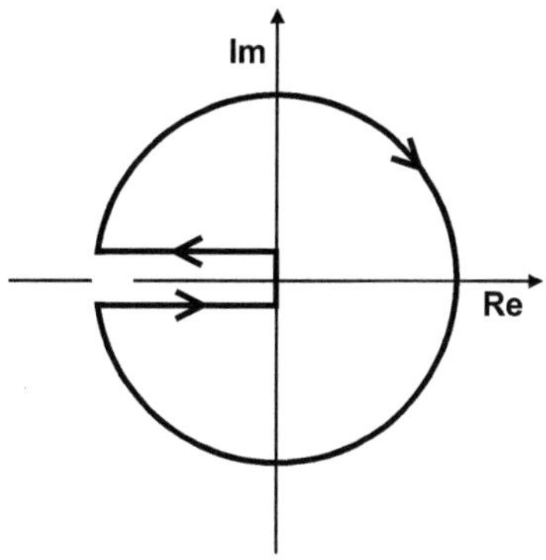

Fig. 5.5 Path of integration Γ

$$K(q) = K(p) = 0. \tag{5.65}$$

If this is the case, the left-hand side of (5.63) yields a first order differential equation for $f(t)$

$$f(t)(t^2 - a) - t\frac{d(f(t)t)}{dt} = 0 \tag{5.66}$$

with the solution

$$f(t) = \frac{1}{t}\exp\left(t + \frac{a}{t}\right). \tag{5.67}$$

For $t = 0$, the factor $\exp\left(t + \frac{a}{t}\right)$ possesses an essential singularity. However, we should note that

$$\lim_{t\to 0} K(t) = \lim_{t\to 0} t^z \exp\left(t + \frac{a}{t}\right) = 0. \tag{5.68}$$

For t negative and real, we can choose the point $z = 0$ as an integration limit.

Denoting the integration path by Γ (see Fig. 5.5), the solution $u(z)$ can be written as

$$u(z) = \oint_\Gamma t^{z-2} \exp\left(t + \frac{a}{t}\right) dt. \tag{5.69}$$

Let us now introduce the transformation $t = i\sqrt{a}s$ and use the relation for the generating function of the Bessel function (Abramowitz and Stegun 1972)

$$\exp\left\{i\sqrt{a}\left(s-\frac{1}{s}\right)\right\}=\sum_{n=0}^{\infty}J_n(2i\sqrt{a})s^n. \tag{5.70}$$

Then we obtain for $u(n)$ the result

$$u(n)=(-2\pi)i^n a^{\frac{n-1}{2}}J_{-(n-1)}(2i\sqrt{a}) \tag{5.71}$$

With the relations $J_{-n}(z)=(-1)^n J_n(z)$ and $I_{-n}(z)=I_n(z)=i^{-n}J_n(iz)$, where $I_n(z)$ denotes the first modified Bessel function, we finally recast (5.71) in the form

$$u(n)=-2\pi i\; a^{\frac{n-1}{2}}I_{(n-1)}(2\sqrt{a}). \tag{5.72}$$

Using (5.72) and (5.60), the stationary transition factor $\xi(n)$ is given by

$$\xi(n)=\frac{\sqrt{a}}{n}\cdot\frac{I_{n-1}(2\sqrt{a})}{I_{n-2}(2\sqrt{a})},\quad \text{for } n=1,2,\ldots. \tag{5.73}$$

We now introduce $a=2X_0^2$ (5.51) and compute the exact stationary probability distribution via (4.33)

$$P_{st}(n)=P_{st}(0)\frac{2^{\frac{n}{2}}X_0^n}{n!}\cdot\frac{I_{n-1}(2\sqrt{2}X_0)}{I_1(2\sqrt{2}X_0)}\quad \text{for } n=0,1,2,\ldots, \tag{5.74}$$

where $P_{st}(0)$ is determined by the normalization condition

$$\frac{1}{P_{st}(0)}=\sum_{n=0}^{\infty}\frac{2^{\frac{n}{2}}X_0^n}{n!}\cdot\frac{I_{n-1}(2\sqrt{2}X_0)}{I_1(2\sqrt{2}X_0)}. \tag{5.75}$$

Noting the relation for the Bessel functions (e.g. Madelung 1964; Abramowitz and Stegun 1972),

$$I_j(4X_0)=2^{-\frac{j}{2}}\sum_{i=0}^{\infty}\frac{2^{\frac{i}{2}}X_0^i}{i!}\cdot I_{i-j}\left(2\sqrt{2}X_0\right), \tag{5.76}$$

we can explicitly calculate the sum in (5.75) to obtain

$$P_{st}(0)=\frac{I_1(2\sqrt{2}X_0)}{\sqrt{2}I_1(4X_0)}. \tag{5.77}$$

The exact normalized stationary probability distribution of the non-linear autocatalytic chemical reaction problem is then given by

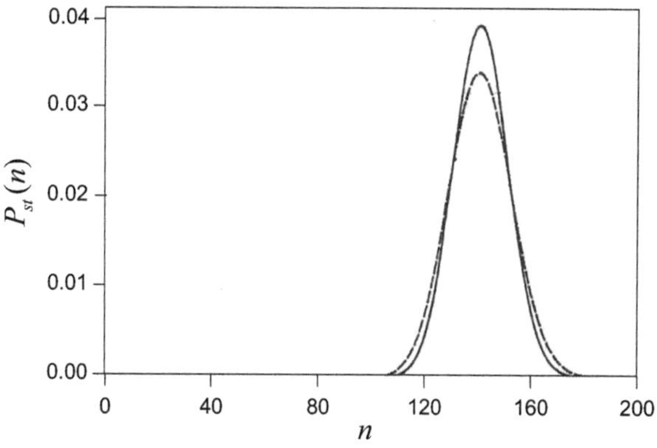

Fig. 5.6 Comparison of the stationary distribution (5.78) with a Poisson distribution to the same mean value $X_0 = 140$ (*dashed line* Poisson distribution)

$$P_{st}(n) = \frac{2^{\frac{n-1}{2}} X_0^n}{n!} \cdot \frac{I_{n-1}\left(2\sqrt{2}X_0\right)}{I_1(4X_0)} \quad \text{for } n = 0, 1, 2, \ldots . \tag{5.78}$$

Obviously, the distribution is not a Poissonian although the transition rates in (5.46), (5.47) obey combinatorial mass-action law kinetics. In Fig. 5.6 a comparison of the exact stationary probability distribution (5.78) with a Poisson distribution to the same mean value X_0 is shown. It is interesting that the nonlinear chemical reaction scheme (5.46), (5.47) leads to a shrinking of the variance by 25%.

Using the probability distribution (5.78), all factorial moments

$$\left\langle n^j \right\rangle_f = \langle n(n-1)\ldots(n-j+1)\rangle \tag{5.79}$$

can be directly calculated

$$\left\langle n^j \right\rangle_f = X_0^j \cdot \frac{I_{j-1}(4X_0)}{I_1(4X_0)} . \tag{5.80}$$

Therefore, the exact mean value of this chemical reaction is given by

$$\langle n \rangle = X_0 \cdot \frac{I_0(4X_0)}{I_1(4X_0)} = X_0\left(1 + \frac{1}{8X_0} + O\left(X_0^{-2}\right)\right), \tag{5.81}$$

and the exact mean concentration fluctuation σ^2 of the distribution reads

$$\begin{aligned}\sigma^2 &= X_0 \cdot \frac{I_0(4X_0)}{I_1(4X_0)}\left(1 - X_0 \cdot \frac{I_0(4X_0)}{I_1(4X_0)}\right) + X_0^2 \\ &= \frac{3}{4}X_0 + \frac{1}{16} + O\left(X_0^{-1}\right)\end{aligned} \tag{5.82}$$

For $X_0 >> 1$, the exact solutions for the mean value and the variance coincide with the approximate solution (5.57), (5.58), obtained via the much simpler calculation procedure (5.56).

The obtained results (5.81), (5.82) are identical with the results obtained by Mazo (1975) and Nicolis (1972) using generating function technique. For completeness, let us consider briefly the solution procedure based on generating function technique, as described in the paper of Mazo (1975).

Exact Stationary Moments Using Generation Function Technique
The technique of generating function (Sect. 2.2.4) leads to the following second order differential equation

$$(1+s)\frac{d^2 f(s)}{ds^2} - 2X_0^2 f(s) = 0. \tag{5.83}$$

with two boundary conditions

$$f(1) = 1 \text{ and } f(-1) = 0. \tag{5.84}$$

Physically, this means that there are equal probabilities for having an even or odd number of X molecules in the system. This boundary conditions follows from the analysis of the birth and death rates of the model. Therefore, it is not an added assumption.

After changing the variables $z = 1/2(1+s), f(s) = g(z)$, the function $g(z)$ satisfies the differential equation

$$\frac{d^2 g(z)}{dz^2} - \frac{4X_0^2}{z} g(z) = 0, \tag{5.85}$$

with $g(0) = 0$ and $g(1) = 1$.

The solution to (5.85), satisfying the boundary conditions, is

$$g(z) = z^{1/2} \frac{I_1\left(4X_0 z^{1/2}\right)}{I_1(4X_0)}, \tag{5.86}$$

where I_1 is the Bessel function of imaginary arguments. Using known properties of the Bessel functions, we get

$$\langle X \rangle = \left(\frac{df(s)}{ds} \right)_{s=1} = \frac{1}{2} \left(\frac{dg(z)}{dz} \right)_{z=1} = X_0 \frac{I_0(4X_0)}{I_1(4X_0)}. \tag{5.87}$$

This result (5.87) agrees completely with (5.81). Furthermore, from (5.86), it follows for the variance

$$\sigma^2 = \frac{3}{4} X_0 + O(1), \tag{5.88}$$

which confirms the result of Nicolis (1972).

It is assumed that the reaction mixture of the considered chemical reaction is well stirred, leading to a spatially homogeneous dilution. Therefore, spatial aspects do not have to be considered yet. Therefore, the next chapter is dealing with the modelling of a chemical diffusion reaction system. As a further aspect, internal excitations of the molecules and transitions between different excitation states are taken into consideration to understand their influence on the observed reaction scheme.

5.1.4 *Linear Chemical Diffusion Reaction System with Internal Transitions*

As an exactly solvable example we now investigate the linear diffusion reaction system $X \leftrightarrow Y$. The molecules X, Y are in contact to reservoirs, enabling the molecules to perform transitions between different internal energy states.

We divide the reaction volume into many cells, labelled by a multi-index $l = (l_1, l_2, l_3)$. The number of species $x_i(\vec{l} \rightarrow)$, $i = 1, 2, \ldots, n$ and $y_j\left(\vec{l}\right)$, $j = 1, 2, \ldots, m$, in cell $\vec{l}$, at time t are related to species which are in the states of internal energy ε_i, and η_j, respectively.

Now the following processes are considered:

(a) By conversion, the species in a given cell $x_i\left(\vec{l}\right)$ transform into $x_j\left(\vec{l}\right)$, and $y_j\left(\vec{l}\right)$ into $y_k\left(\vec{l}\right)$, with rate constants α_{ji}, and β_{kj}, respectively. This could happen via an absorption or emission of phonons or photons due to collisions with other molecules.

(b) Furthermore, we assume that the species are in contact with reservoirs in order to keep the particle numbers $x_i\left(\vec{l}\right)$, $y_j\left(\vec{l}\right)$, constant.

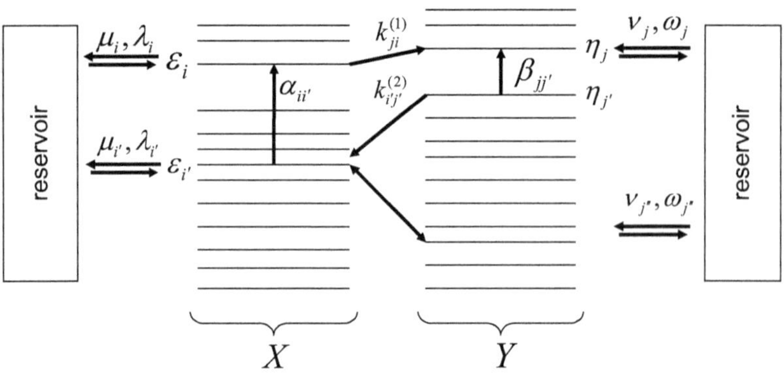

Fig. 5.7 Reaction scheme for species $X \leftrightarrow Y$ within one spatial cell l

(c) Chemical reactions occur, $x_i\left(\vec{l}\right) \rightarrow y_j\left(\vec{l}\right)$ and $y_j\left(\vec{l}\right) \rightarrow x_i\left(\vec{l}\right)$ with forward and backward rate constants $k_{ji}^{(1)}$ and $k_{ji}^{(2)}$, where those rate constants may depend on the internal energy states of the molecules.
(d) Finally, we take into account diffusion processes between neighbouring cells.

The reaction scheme within one spatial cell is shown in Fig. 5.7.

The multivariate probability distribution of the considered inhomogeneous chemical reaction will be denoted by

$$P(\vec{x}, \vec{y}) = P\left(\vec{x}\left(\vec{l}\right), \vec{y}\left(\vec{l}\right)\right), \tag{5.89}$$

where

$$\vec{x}\left(\vec{l}\right) = \left(x_1\left(\vec{l}\right), x_2\left(\vec{l}\right), \ldots, x_n\left(\vec{l}\right)\right) \tag{5.90}$$

$$\vec{y}\left(\vec{l}\right) = \left(y_1\left(\vec{l}\right), y_2\left(\vec{l}\right), \ldots, y_m\left(\vec{l}\right)\right). \tag{5.91}$$

It is apparent, that the Master equation for this spatio-temporal chemical reaction system is rather complicated since there are many different possibilities how a specific configuration X, Y may change in the course of time. Beside the chemical reaction scheme there exist transitions of the X, Y molecules between different internal energy states, respectively. In addition, diffusion processes exist and may lead to spatio-temporal variations in the concentrations of the molecules.

The diffusion-reaction Master equation of birth-death-type for this reaction scheme reads (Haag 1977)

$$\begin{aligned}
\frac{dP(\vec{x},\vec{y})}{dt} =& \sum_{i,j}\Big(k_{ij}^{(1)}\Big(y_j\big(\vec{l}\big)+1\Big)P\Big(x_i\big(\vec{l}\big)-1,y_j\big(\vec{l}\big)+1\Big)-k_{ij}^{(1)}y_j\big(\vec{l}\big)P\Big(x_i\big(\vec{l}\big),y_j\big(\vec{l}\big)\Big)\\
&+\sum_{i,j}\Big(k_{ji}^{(2)}\Big(x_i\big(\vec{l}\big)+1\Big)P\Big(x_i\big(\vec{l}\big)+1,y_j\big(\vec{l}\big)-1\Big)-k_{ji}^{(2)}x_i\big(\vec{l}\big)P\Big(x_i\big(\vec{l}\big),y_j\big(\vec{l}\big)\Big)\Big)\\
&+\sum_{i,j}\alpha_{ji}\Big(\Big(x_i\big(\vec{l}\big)+1\Big)P\Big(x_i\big(\vec{l}\big)+1,x_j\big(\vec{l}\big)-1\Big)-x_i\big(\vec{l}\big)P\Big(x_i\big(\vec{l}\big),x_j\big(\vec{l}\big)\Big)\Big)\\
&+\sum_{i}\mu_i\Big(\Big(x_i\big(\vec{l}\big)+1\Big)P\Big(x_i\big(\vec{l}\big)+1\Big)-x_i\big(\vec{l}\big)P\Big(x_i\big(\vec{l}\big)\Big)\Big)\\
&+\sum_{i}\lambda_i\Big(P\Big(x_i\big(\vec{l}\big)-1\Big)-P\Big(x_i\big(\vec{l}\big)\Big)\Big)\\
&+\sum_{i,j}\beta_{ji}\Big(\Big(y_i\big(\vec{l}\big)+1\Big)P\Big(y_i\big(\vec{l}\big)+1,y_j\big(\vec{l}\big)-1\Big)-y_i\big(\vec{l}\big)P\Big(y_i\big(\vec{l}\big),y_j\big(\vec{l}\big)\Big)\Big)\\
&+\sum_{j}\nu_j\Big(\Big(y_j\big(\vec{l}\big)+1\Big)P\Big(y_j\big(\vec{l}\big)+1\Big)-y_j\big(\vec{l}\big)P\Big(y_j\big(\vec{l}\big)\Big)\Big)\\
&+\sum_{j}\omega_j\Big(P\Big(y_j\big(\vec{l}\big)-1\Big)-P\Big(y_j\big(\vec{l}\big)\Big)\Big)\\
&+\sum_{i,\vec{a}}D_1\Big(\Big(x_i\big(\vec{l}+\vec{a}\big)+1\Big)P\Big(x_i\big(\vec{l}\big)-1,x_i\big(\vec{l}+\vec{a}\big)+1\Big)-x_i\big(\vec{l}\big)P\Big(x_i\big(\vec{l}\big),x_i\big(\vec{l}+\vec{a}\big)\Big)\Big)\\
&+\sum_{j,\vec{a}}D_2\Big(y_j\big(\vec{l}+\vec{a}\big)+1\Big)P\Big(y_j\big(\vec{l}\big)-1,y_j\big(\vec{l}+\vec{a}\big)+1\Big)-y_j\big(\vec{l}\big)P\Big(y_j\big(\vec{l}\big),y_j\big(\vec{l}+\vec{a}\big)\Big)\Big)\Big).
\end{aligned} \tag{5.92}$$

All variables which are not explicitly written in $P(\ldots)$ or in $g(\ldots|\ldots)$ have the same values as on the left-hand side. In the diffusion term of the Master equation we have to sum up over all neighbouring cells $\vec{l}+\vec{a}$ of the cell $\vec{a}$ under consideration and in addition over all cells $\vec{l}$ of the reaction system.

Starting with (5.92), it is easy to calculate the transition factors and the flows

$$J(\vec{x},\vec{y}) = \frac{\frac{dP(\vec{x},\vec{y})}{dt}}{P(\vec{x},\vec{y})} = \frac{dP(\vec{x},\vec{y})/dt}{P(\vec{x},\vec{y})} = \frac{d}{dt}\ln P(\vec{x},\vec{y}), \tag{5.93}$$

where $J(\vec{x},\vec{y})$ can be decomposed into different terms

$$J(\vec{x},\vec{y}) = J_{reac} + \sum_i J_{\text{int}}^{(i)} + \sum_i J_{diff}^{(i)} + \sum_i J_{res}^{(i)} \tag{5.94}$$

with the single terms

$$\begin{aligned}
J_{reac} =& \sum_{i,j}k_{ij}^{(1)}\Big(\Big(y_j\big(\vec{l}\big)+1\Big)g\Big(x_i\big(\vec{l}\big)-1,y_j\big(\vec{l}\big)+1|x_i\big(\vec{l}\big),y_j\big(\vec{l}\big)\Big)-y_j\big(\vec{l}\big)\Big)\\
&+\sum_{i,j}k_{ji}^{(2)}\Big(\Big(x_i\big(\vec{l}\big)+1\Big)g\Big(x_i\big(\vec{l}\big)+1,y_j\big(\vec{l}\big)-1|x_i\big(\vec{l}\big),y_j\big(\vec{l}\big)\Big)-x_i\big(\vec{l}\big)\Big)
\end{aligned} \tag{5.95}$$

$$J_{\text{int}}^{(1)} = \sum_{\substack{i,j\\ i\neq j}} \alpha_{ji}\left(\left(x_i\left(\vec{l}\right)+1\right)g\left(x_i\left(\vec{l}\right)+1, x_j\left(\vec{l}\right)-1|x_i\left(\vec{l}\right), y_j\left(\vec{l}\right)\right)-x_i\left(\vec{l}\right)\right) \tag{5.96}$$

$$J_{\text{int}}^{(2)} = \sum_{\substack{i,j\\ i\neq j}} \beta_{ji}\left(\left(y_i\left(\vec{l}\right)+1\right)g\left(y_i\left(\vec{l}\right)+1, y_j\left(\vec{l}\right)-1|y_i\left(\vec{l}\right), y_j\left(\vec{l}\right)\right)-y_i\left(\vec{l}\right)\right) \tag{5.97}$$

$$J_{diff}^{(1)} = \sum\nolimits_{i,\vec{a}} D_1\left(\left(x_i\left(\vec{l}+\vec{a}\right)+1\right)g\left(x_i\left(\vec{l}\right)-1, x_i\left(\vec{l}+\vec{a}\right)+1|x_i\left(\vec{l}\right), x_i\left(\vec{l}+\vec{a}\right)\right)-x_i\left(\vec{l}\right)\right) \tag{5.98}$$

$$J_{diff}^{(2)} = \sum\nolimits_{j,\vec{a}} D_2\left(\left(y_j\left(\vec{l}+\vec{a}\right)+1\right)g\left(y_j\left(\vec{l}\right)-1, y_j\left(\vec{l}+\vec{a}\right)+1|y_j\left(\vec{l}\right), y_j\left(\vec{l}+\vec{a}\right)\right)-y_j\left(\vec{l}\right)\right) \tag{5.99}$$

$$\begin{aligned} J_{res}^{(1)} &= \sum_i \mu_i\left(\left(x_i\left(\vec{l}\right)+1\right)g\left(x_i\left(\vec{l}\right)+1|x_i\left(\vec{l}\right)\right)-x_i\left(\vec{l}\right)\right) \\ &\quad+\sum_i \lambda_i\left(g\left(x_i\left(\vec{l}\right)-1|x_i\left(\vec{l}\right)\right)-1\right) \end{aligned} \tag{5.100}$$

$$\begin{aligned} J_{res}^{(2)} &= \sum_j \nu_j\left(\left(y_j\left(\vec{l}\right)+1\right)g\left(y_j\left(\vec{l}\right)+1|y_j\left(\vec{l}\right)\right)-y_j\left(\vec{l}\right)\right) \\ &\quad+\sum_j \omega_j\left(g\left(y_j\left(\vec{l}\right)-1|y_j\left(\vec{l}\right)\right)-1\right). \end{aligned} \tag{5.101}$$

The terms J_{reac}, $J_{\text{int}}^{(i)}$, $J_{diff}^{(i)}$ and $J_{res}^{(i)}$ are related to the chemical reaction of the species $X \leftrightarrow Y$, the internal transitions between different energy states of the species X, Y, $\varepsilon_i \leftrightarrow \varepsilon_{i'}$, $\eta_j \leftrightarrow \eta_{j'}$ the diffusion of the molecules X, Y between different spatial cells $\vec{l} \leftrightarrow \vec{l}+\vec{a}$, and the coupling to external reservoirs, enabling energy exchange necessary for the internal transitions (see also Gardiner et al. 1976).

In the next step, we introduce transition factors acting between neighbouring states in the multidimensional configuration space

$$g\left(x_i\left(\vec{l}\right), y_j\left(\vec{l}\right) | x_i\left(\vec{l}\right) - 1, y_j\left(\vec{l}\right)\right) \tag{5.102}$$

$$g\left(x_i\left(\vec{l}\right), y_j\left(\vec{l}\right) | x_i\left(\vec{l}\right), y_j\left(\vec{l}\right) - 1\right). \tag{5.103}$$

The equations of motion for the transition factors read

$$\frac{dg\left(x_i\left(\vec{l}\right), y_j\left(\vec{l}\right) | x_i\left(\vec{l}\right) - 1, y_j\left(\vec{l}\right)\right)/dt}{g\left(x_i\left(\vec{l}\right), y_j\left(\vec{l}\right) | x_i\left(\vec{l}\right) - 1, y_j\left(\vec{l}\right)\right)} = J\left(x_i\left(\vec{l}\right), y_j\left(\vec{l}\right)\right) - J\left(x_i\left(\vec{l}\right) - 1, y_j\left(\vec{l}\right)\right) \tag{5.104}$$

and

$$\frac{dg\left(x_i\left(\vec{l}\right), y_j\left(\vec{l}\right) | x_i\left(\vec{l}\right), y_j\left(\vec{l}\right) - 1\right)/dt}{g\left(x_i\left(\vec{l}\right), y_j\left(\vec{l}\right) | x_i\left(\vec{l}\right), y_j\left(\vec{l}\right) - 1\right)} = J\left(x_i\left(\vec{l}\right), y_j\left(\vec{l}\right)\right) - J\left(x_i\left(\vec{l}\right), y_j\left(\vec{l}\right) - 1\right) \tag{5.105}$$

for all $i, j, \vec{l}$. The different spatial cells are coupled because of the diffusion term J_{diff}. Now we introduce the ansatz for the transition factors

$$g\left(x_i\left(\vec{l}\right), y_j\left(\vec{l}\right) | x_i\left(\vec{l}\right) - 1, y_j\left(\vec{l}\right)\right) = \frac{a_i\left(\vec{l}\right)}{x_i\left(\vec{l}\right)} \quad \text{for } a_i\left(\vec{l}\right) > 0 \text{ and for all } i, j, \vec{l} \tag{5.106}$$

$$g\left(x_i\left(\vec{l}\right), y_j\left(\vec{l}\right) | x_i\left(\vec{l}\right), y_j\left(\vec{l}\right) - 1\right) = \frac{b_j\left(\vec{l}\right)}{y_j\left(\vec{l}\right)} \quad \text{for } b_j\left(\vec{l}\right) > 0 \text{ and for all } i, j, \vec{l}. \tag{5.107}$$

All further transition factors can be obtained by combinations of these two transition factors.

According to Chap. 4, the repeated multiplication of these transition factors leads to the multidimensional probability distribution

$$P(\vec{x},\vec{y}) = \prod_{\vec{l}}\prod_{i}\frac{a_i\left(\vec{l}\right)^{x_i(\vec{l})}}{x_i\left(\vec{l}\right)!}\cdot e^{-a_i(\vec{l})}\prod_{j}\frac{b_j\left(\vec{l}\right)^{y_j(\vec{l})}}{y_j\left(\vec{l}\right)!}\cdot e^{-b_j(\vec{l})}$$

$$= \prod_{\vec{l}}\prod_{i}\frac{\left\langle x_i\left(\vec{l}\right)\right\rangle^{x_i(\vec{l})}}{x_i\left(\vec{l}\right)!}\cdot e^{-\langle x_i(\vec{l})\rangle}\prod_{j}\frac{\left\langle y_j\left(\vec{l}\right)\right\rangle^{y_j(\vec{l})}}{y_j\left(\vec{l}\right)!}\cdot e^{-\langle y_j(\vec{l})\rangle} \tag{5.108}$$

The ansatz used in (5.106), (5.107) corresponds to the multidimensional Poisson distribution (5.108). Hence, the introduced functions $a_i\left(\vec{l}\right), b_j\left(\vec{l}\right)$ correspond to its mean values

$$a_i\left(\vec{l}\right) = \left\langle x_i\left(\vec{l}\right)\right\rangle. \tag{5.109}$$

$$b_j\left(\vec{l}\right) = \left\langle y_j\left(\vec{l}\right)\right\rangle. \tag{5.110}$$

Now, we have to test and prove the compatibility of the ansatz (5.106), (5.107) with the Master equation, in other words with the equation of motion (5.92). For this aim, we insert (5.106), (5.107) with (5.109), (5.110) in (5.95)–(5.101) and obtain

$$J_{reac} = \sum_{j}\left\{k_{ij}{}^{(1)}\frac{\left\langle y_j\left(\vec{l}\right)\right\rangle}{\left\langle x_i\left(\vec{l}\right)\right\rangle} - k_{ji}^{(2)}\right\}x_i\left(\vec{l}\right)$$

$$+\sum_{i}\left\{k_{ji}{}^{(2)}\frac{\left\langle x_i\left(\vec{l}\right)\right\rangle}{\left\langle y_j\left(\vec{l}\right)\right\rangle} - k_{ij}^{(1)}\right\}y_j\left(\vec{l}\right) \tag{5.111}$$

$$J_{\text{int}}^{(1)} = \sum_{j,j\neq i}\left\{\alpha_{ij}\frac{\left\langle x_j\left(\vec{l}\right)\right\rangle}{\left\langle x_i\left(\vec{l}\right)\right\rangle} - \alpha_{ji}\right\}x_i\left(\vec{l}\right) \tag{5.112}$$

$$J_{\text{int}}^{(2)} = \sum_{i,j\neq i}\left\{\beta_{ji}\frac{\left\langle y_i\left(\vec{l}\right)\right\rangle}{\left\langle y_j\left(\vec{l}\right)\right\rangle} - \beta_{ij}\right\}y_j\left(\vec{l}\right) \tag{5.113}$$

$$J_{diff}^{(1)} = \sum_{\vec{a}}\left\{D_1\frac{\left\langle x_i\left(\vec{l}+\vec{a}\right)\right\rangle}{\left\langle x_i\left(\vec{l}\right)\right\rangle} - 1\right\}x_i\left(\vec{l}\right) \tag{5.114}$$

$$J_{diff}^{(2)} = \sum_{\vec{a}} \left\{ D_2 \frac{\langle y_j(\vec{l}+\vec{a}) \rangle}{\langle y_j(\vec{l}) \rangle} - 1 \right\} y_j(\vec{l}) \tag{5.115}$$

$$J_{res}^{(1)} = -\left\{ \mu_i - \frac{\lambda_i}{\langle x_i(\vec{l}) \rangle} \right\} x_i(\vec{l}) \tag{5.116}$$

$$J_{res}^{(2)} = -\left\{ \nu_j - \frac{\omega_j}{\langle y_j(\vec{l}) \rangle} \right\} y_j(\vec{l}), \tag{5.117}$$

where we have omitted all terms independent of $x_i(\vec{l}), y_j(\vec{l})$, in (5.111)–(5.117) because these terms cancel in the equation of motion (5.104), (5.105), if we take the difference on the right-hand side. Insertion of (5.111)–(5.117) in (5.104) and (5.105) leads to the exact coupled set of equations for the mean values (Haag 1977, 1978):

$$\begin{aligned} \frac{d\langle x_i(\vec{l}) \rangle}{dt} &= -\mu_i \langle x_i(\vec{l}) \rangle + \lambda_i + \sum_j \left(k_{ij}^{(1)} \langle y_j(\vec{l}) \rangle - k_{ji}^{(2)} \langle x_i(\vec{l}) \rangle \right) \\ &+ \sum_{i', i' \neq i} \left(\alpha_{ii'} \langle x_{i'}(\vec{l}) \rangle - \alpha_{i'i} \langle x_i(\vec{l}) \rangle \right) + D_1 \nabla_{\vec{l}}^2 \langle x_i(\vec{l}) \rangle \end{aligned} \tag{5.118}$$

$$\begin{aligned} \frac{d\langle y_j(\vec{l}) \rangle}{dt} &= -\nu_j \langle y_j(\vec{l}) \rangle + \omega_j + \sum_i \left(k_{ji}^{(2)} \langle x_i(\vec{l}) \rangle - k_{ij}^{(1)} \langle y_j(\vec{l}) \rangle \right) \\ &+ \sum_{j', j' \neq j} \left(\beta_{jj'} \langle y_{j'}(\vec{l}) \rangle - \beta_{j'j} \langle y_j(\vec{l}) \rangle \right) + D_2 \nabla_{\vec{l}}^2 \langle y_j(\vec{l}) \rangle \end{aligned} \tag{5.119}$$

where the difference operator

$$\nabla_{\vec{l}}^2 \langle \bullet(\vec{l}) \rangle = \sum_{\vec{a}} \left(\langle \bullet(\vec{l}+\vec{a}) \rangle - \langle \bullet(\vec{l}) \rangle \right) \tag{5.120}$$

tends to the Laplace operator in the continuous limit. If we sum up in each spatial cell over all molecules which are in different states of internal energy of the same species X, Y, we obtain the equations of motion

$$\begin{aligned} \frac{d}{dt} \langle \vec{x}(\vec{l}) \rangle &= -\mu \langle \vec{x}(\vec{l}) \rangle + \lambda + k^{(1)} \langle \vec{y}(\vec{l}) \rangle - k^{(2)} \langle \vec{x}(\vec{l}) \rangle \\ &+ D_1 \nabla_{\vec{l}}^2 \langle \vec{x}(\vec{l}) \rangle \end{aligned} \tag{5.121}$$

and

$$\frac{d}{dt}\left\langle \vec{y}\left(\vec{l}\right)\right\rangle = -\nu\left\langle \vec{y}\left(\vec{l}\right)\right\rangle + \omega - k^{(1)}\left\langle \vec{y}\left(\vec{l}\right)\right\rangle + k^{(2)}\left\langle \vec{x}\left(\vec{l}\right)\right\rangle + D_2\nabla_{\vec{l}}^2\left\langle \vec{y}\left(\vec{l}\right)\right\rangle, \tag{5.122}$$

where we have introduced the averaged number of species independent of its internal energy states

$$\left\langle \vec{x}\left(\vec{l}\right)\right\rangle = \sum_i \left\langle x_i\left(\vec{l}\right)\right\rangle \text{ and } \left\langle \vec{y}\left(\vec{l}\right)\right\rangle = \sum_j \left\langle y_j\left(\vec{l}\right)\right\rangle \tag{5.123}$$

and the averaged rate constants

$$\mu = \frac{\sum_i \mu_i \left\langle x_i\left(\vec{l}\right)\right\rangle}{\sum_i \left\langle x_i\left(\vec{l}\right)\right\rangle} \text{ and } \nu = \frac{\sum_j \nu_j \left\langle y_j\left(\vec{l}\right)\right\rangle}{\sum_j \left\langle y_j\left(\vec{l}\right)\right\rangle} \tag{5.124}$$

$$k^{(1)} = \frac{\sum_{i,j} k_{ij}^{(1)} \left\langle y_j\left(\vec{l}\right)\right\rangle}{\sum_j \left\langle y_j\left(\vec{l}\right)\right\rangle} \text{ and } k^{(2)} = \frac{\sum_{i,j} k_{ji}^{(2)} \left\langle x_i\left(\vec{l}\right)\right\rangle}{\sum_i \left\langle x_i\left(\vec{l}\right)\right\rangle} \tag{5.125}$$

$$\lambda = \sum_i \lambda_i \text{ and } \omega = \sum_j \omega_j. \tag{5.126}$$

The Equations of motion (5.121), (5.122) describe behaviour of the species X, Y, averaged over all internal states, respectively. Compared with the more complex equations of motion containing the dynamics of internal transitions of the species (5.118), (5.119) the Eqs. (5.121), (5.122) now contain time-dependent effective rate constants since the rate constants of the averaged process (5.124)–(5.126) still depend on the internal dynamics, namely on the behaviour of $\left\langle x_i\left(\vec{l}\right)\right\rangle$ and $\left\langle y_j\left(\vec{l}\right)\right\rangle$. From the exact time-dependent solution (5.108) we conclude that the linear chemical reaction-diffusion system will remain in a multidimensional Poisson distribution in each spatial cell, if it has initially been prepared in such a state. The system behaves as an ideal mixture although chemical reactions and inner conversion reactions take place. If it was not initially in such a state, the higher order modes of the Master equation would drive the system towards the Poisson distribution. However, the global distribution, averaged over the spatial cells has non-Poisson fluctuations.

This behaviour of the fluctuations was also found by Nicolis and Prigogine (1971), Kuramoto (1973), Nitzan et al. (1974), and Gardiner et al. (1976) in simpler systems.

In chemistry, often only spatially averaged concentrations are of interest, since the mixtures have to be stirred. In this case a spatial averaging is required

$$\langle \vec{x} \rangle = \frac{1}{V} \sum_{\vec{l}} \left\langle \vec{x}\left(\vec{l}\right)\right\rangle \text{ and } \langle \vec{y} \rangle = \frac{1}{V} \sum_{\vec{l}} \left\langle \vec{y}\left(\vec{l}\right)\right\rangle. \tag{5.127}$$

After this averaging procedure, we obtain the coupled equations of motion for the mean number of molecules in the reaction volume

$$\frac{d}{dt}\langle \vec{x} \rangle = -\mu \langle \vec{x} \rangle + \lambda + k^{(1)} \langle \vec{y} \rangle - k^{(2)} \langle \vec{x} \rangle \tag{5.128}$$

$$\frac{d}{dt}\langle \vec{y} \rangle = -\nu \langle \vec{y} \rangle + \omega - k^{(1)} \langle \vec{y} \rangle + k^{(2)} \langle \vec{x} \rangle. \tag{5.129}$$

Both, the internal transition and diffusion processes of the molecules, occur no longer explicitly in the averaged equations. However, the generalised rates depend via the averaging process on the overall dynamic process. This may explain some deviations between the expected behaviour of chemical reactions computed via the pure phenomenological rate equations and the statistical approach.

5.2 Spin-Dynamics

The investigation of the dynamics of spin-systems is of fundamental interest in statistical non-equilibrium physics. Especially the kinetic Ising model with physically motivated transition rates between different microstates has motivated researchers to apply the whole tool set of mathematical solution methods. The appearance of a ferromagnetic order from an initially disordered state, in other words the occurrence of a phase transition, in their conceptual simplicity is one reason for the interest in this widely applicable model type (van Kampen 1965, 1978; Ruijgrok and Tjon 1973; Hilhorst 1973).

Our system consists of an even number of $2S$ spins. Each spin s_i has two possible positions $s_i = \pm 1$. The number of up-spins $s_\uparrow$ minus the number of down spins $s_\downarrow$ is the magnetization and will be denoted by $2s$.

Then the relations hold for the total number of spins

$$s_\uparrow + s_\downarrow = 2S \tag{5.130}$$

and for the magnetization $m = 2s$

$$s_\uparrow - s_\downarrow = 2s = m, \tag{5.131}$$

where S denotes half of the total number of spins for mathematical convenience (see later).

Equations (5.130) and (5.131) imply

$$s_\uparrow = S + s \tag{5.132}$$

and

$$s_\downarrow = S - s, \tag{5.133}$$

where $-S \le s \le S$.

It is assumed, that pairs of next–neighbour spins interact with each other. The corresponding Hamiltonian reads:

$$H = -J \sum_{(i,j)} s_i s_j, \tag{5.134}$$

where the sum extends over nearest neighbours (i, j) on the lattice. Every anti-parallel pair of spins contributes $+J$ to the total energy, in contrary, every parallel pair of spins reduces the total energy by $-J$. When the coupling constant is positive $J > 0$, the interaction favours ferromagnetic order, characterised by non-zero magnetisation. The order appears spontaneously in the absence of any driving field (symmetry breaking potential) if the temperature T is less than a critical temperature T_C. Above T_C, the arrangement of spins is spatially disordered and the magnetisation is zero.

On the one hand, all equilibrium properties of the Ising model follow directly from the partition function

$$Z = \sum_{\vec{c}} \exp(-\beta H), \tag{5.135}$$

where the sum extends over all possible spin configurations $\vec{c}$ with $\beta = 1/k_B T$, and k_B is the Boltzmann constant. On the other hand, the non-equilibrium properties, the dynamics of the spin system, require additional assumptions about the spin-flip properties. For example, the spins may change one at a time or in correlated blocks, the change of the spins may conserve the magnetization or not (Hilhorst 1973). We apply the so-called mean field theory. The interaction between the spins is supposed to take place via a "mean field". The mean field approximation supposes a special limit of the spin system. The actual environment surrounding each spin is replaced by an average field, which is determined self-consistently. The interaction between the spins is supposed to take place through this overall field (mean field) proportional to s in such a way that $(J/S)s$ where J is the interaction constant. The coupling constant is chosen to scale inversely with the system size, so that the energy is extensive, in other words scales linearly with S.

The transition

$$(s_\uparrow, s_\downarrow) \to (s_\uparrow + 1, s_\downarrow - 1) \text{ or equivalently } s \to s + 1 \tag{5.136}$$

is effected by a spin-switch. Vice versa

$$(s_{\uparrow}, s_{\downarrow}) \rightarrow (s_{\uparrow} - 1, s_{\downarrow} + 1) \text{ or equivalently } s \rightarrow s - 1 \tag{5.137}$$

describes a spin-flip in the other direction. The corresponding configurational transition rates are now assumed to be represented by (next-neighbour transitions only)

$$w((s+1) \leftarrow s) \equiv w_{\uparrow}(s) = s_{\downarrow} p_{\uparrow}(s) = (S - s) p_{\uparrow}(s) \tag{5.138}$$

$$w((s-1) \leftarrow s) \equiv w_{\downarrow}(s) = s_{\uparrow} p_{\downarrow}(s) = (S + s) p_{\downarrow}(s) \tag{5.139}$$

and

$$w(s' \leftarrow s) = 0 \quad \text{for } s' \neq s \pm 1, \tag{5.140}$$

where $p_{\uparrow}(s)$, $p_{\downarrow}(s)$ are the transition rates for a spin-flip. We assume, that the transition rates have the following reasonable structure

$$p_{\uparrow}(s) = \nu \exp f(s) \tag{5.141}$$

$$p_{\downarrow}(s) = \nu \exp(-f(s)), \tag{5.142}$$

where the transition rates per spin-flip depend on the function $f(s)$, which describes the coupling between the spins via the mean field $(K/S)s$ as well as the coupling of the spins with a possible external field δ

$$f(s) = \delta + \frac{K}{S} s. \tag{5.143}$$

The strength of the spin-coupling

$$K = \frac{1}{2} \beta J \tag{5.144}$$

depends on the interaction constant J and the temperature $\beta = 1/k_B T$. The higher the temperature T, the lower the strength of the spin-coupling.

Next, the configurational probability is introduced

$$P(s,t) = P(s_{\uparrow}, s_{\downarrow}, t) \tag{5.145}$$

with the normalization condition

$$\sum_{s=-S}^{S} P(s,t) = 1, \tag{5.146}$$

where the sum extends over all possible spin configurations. Then the Master equation for the spin-system reads (next neighbour transitions only)

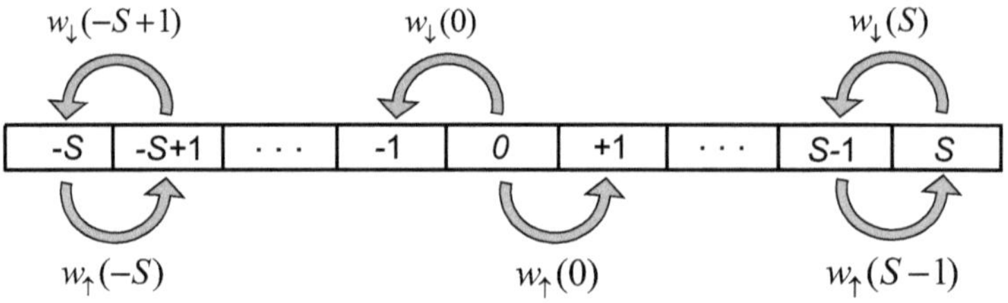

Fig. 5.8 Possible next-neighbour transitions of the one-dimensional Master equation for the spin dynamics

$$\frac{dP(s,t)}{dt} = w_{\downarrow}(s+1)P(s+1,t) + w_{\uparrow}(s-1)P(s-1,t) - \big(w_{\downarrow}(s) + w_{\uparrow}(s)\big)P(s,t) \tag{5.147}$$

for $-S \leq s \leq S$. In Fig. 5.8 the structure of the one-dimensional Master equation for the spin-system is shown.

5.2.1 *Stationary Solution of the Master Equation*

Since (5.147) fulfils the condition of detailed balance

$$w_{\downarrow}(s+1)P_{st}(s+1) = w_{\uparrow}(s)P_{st}(s) \tag{5.148}$$

for $-S \leq s \leq S$, the exact stationary solution of the Master equation (5.147) can be obtained by repeated application of (5.148) as described in Chap. 4

$$P_{st}(s) = P_{st}(0)\prod_{l=1}^{s} \frac{w_{\uparrow}(l-1)}{w_{\downarrow}(l)} \quad \text{for} \quad +1 \leq s \leq S \tag{5.149}$$

$$P_{st}(s) = P_{st}(0)\prod_{l=1}^{s} \frac{w_{\downarrow}(l+1)}{w_{\uparrow}(l)} \quad \text{for} \quad -S \leq s \leq -1, \tag{5.150}$$

where $P_{st}(0)$ is fixed by the normalization condition (5.146). After inserting (5.138)–(5.142) in (5.149) and (5.150) the exact stationary probability distribution of the spin-system can be computed:

$$P_{st}(s) = P_{st}(0)\binom{2S}{S+s}\exp(2F(s)), \tag{5.151}$$

where

$$F(s) = \sum_{l=1}^{s} f(l). \tag{5.152}$$

A further evaluation of the binomial coefficients, using Stirling's formula (Madelung 1964)

$$\ln S! \approx S \ln S - S \tag{5.153}$$

yields with the introduction of the quasi-continuous variable x, describing the subdivision between up- and down-spin regions

$$x = s/S \text{ with } -1 \leq x \leq 1. \tag{5.154}$$

a more mathematically convenient form of the stationary probability distribution

$$P_{st}(Sx) = P_{st}(0)\exp(-SU(x)) \tag{5.155}$$

with the function $U(x)$

$$U(x) = -2F(xS) + ((1+x)\ln(1+x) + (1-x)\ln(1-x)). \tag{5.156}$$

The sum in the term $F(xS)$ can be approximately replaced by the integral

$$F(xS) \approx \int_0^x f(x')dx' = \int_0^x (\delta + Kx')dx' = \delta x + \frac{1}{2}Kx^2. \tag{5.157}$$

The extreme values x_m of the distribution function (5.155) are determined by

$$\left.\frac{\partial U(x)}{\partial x}\right|_{x=x_m} = -2f(x_m) + (\ln(1+x_m) - \ln(1-x_m)) = 0 \tag{5.158}$$

or after introducing the hyperbolic tangent

$$x_m = \tanh f(x_m) = \tanh(\delta + Kx_m). \tag{5.159}$$

where δ describes the possible effect of an external constant magnetic field, interacting with the spin system, and K the interaction of the spins via the mean field.

Let us now consider the special case without an external field, that means for $\delta = 0$. The case $\delta \neq 0$ will be treated later (Sect. 7.2).

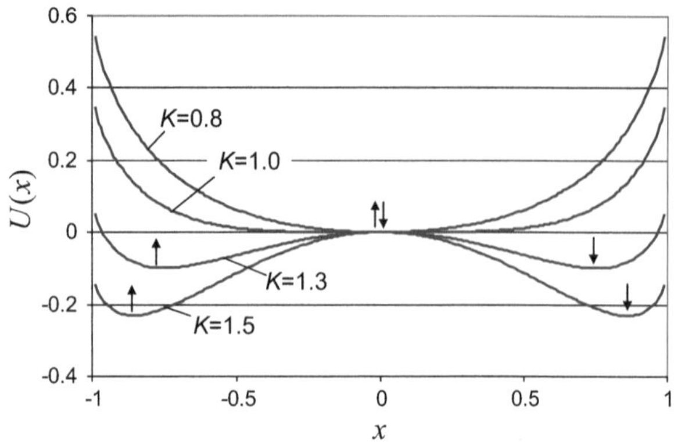

Fig. 5.9 Stationary function $U(x)$ for different values of the spin interaction parameter $K=0.8$, $K=1.0$, $K=1.3$, and $K=1.5$

Special Case Without External Field

In case of $0 \leq K \leq 1$, that means for moderate spin interaction, the spin-distribution has a single maximum at $x_m = 0$. For a large interaction parameter $K > 1$, it has a minimum at $x_m = 0$ and two maxima at $x_{u,d} = \pm x_m$, with $x_m = \tanh K x_m$.

Clearly, $K = 1$ is the Curie point or in other words

$$K = \frac{T_C}{T}. \tag{5.160}$$

In Fig. 5.9, the stationary function $U(x)$ for different values of the spin-spin interaction parameter K is shown. When the temperature approaches the Curie point, starting from a higher temperature $T \rightarrow T_C = 1/2(J/k_B)$ the potential $U(x)$ becomes very flat, leading to an increase of the spin-fluctuations (fluctuation enhancement). At the phase transition point $T = T_C$ the fluctuations become macroscopic and drive the system into the one (up spin) or the other (down spin) ferromagnetic phases (Fig. 5.10). After the phase transition the symmetry of the system will be broken.

5.2.2 *Mean Value and Variance Equations for the Spin Dynamics*

As described in Chaps. 3 and 4, the dynamic mean value and variance equations can be directly derived from the Master equation (5.147), using the definition

$$\langle g(s) \rangle_t = \sum_{s=-S}^{S} g(s) P(s,t) \tag{5.161}$$

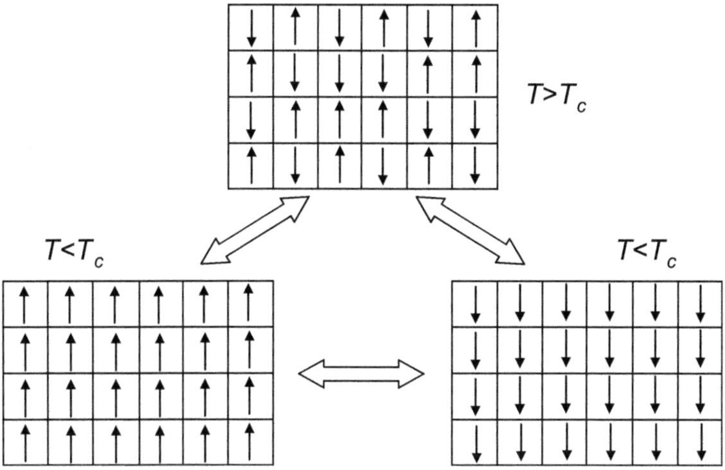

Fig. 5.10 The phase transition from a paramagnetic state to one of two possible ferromagnetic states

and multiplying the right-hand side and the left-hand side with $g(s)$, respectively, and further by summing up over all possible configurations $-S \leq s \leq S$. This leads to the following exact results:

$$\frac{d\langle s\rangle_t}{dt} = \langle w_\uparrow(s)\rangle - \langle w_\downarrow(s)\rangle \tag{5.162}$$

and

$$\frac{d\langle s^2\rangle_t}{dt} = 2\langle s(w_\uparrow(s) - w_\downarrow(s))\rangle + \langle w_\uparrow(s)\rangle + \langle w_\downarrow(s)\rangle. \tag{5.163}$$

For a unimodal probability distributions, one may approximate

$$\langle g(s)\rangle_t \approx g(\langle s\rangle_t) \tag{5.164}$$

and thus arrive at self-contained although approximate equations of motion for the spin system. Introducing the scaled quasi-continuous variable

$$\langle x\rangle_t = \frac{\langle s\rangle_t}{S} \tag{5.165}$$

and the variance

$$\sigma^2(t) = \langle x^2\rangle_t - \langle x\rangle_t^2 \text{ and } \Sigma^2(t) = S\sigma^2(t), \tag{5.166}$$

the equations of motion can be derived from (5.162), (5.163) using (5.164), and (5.165):

$$\frac{d\langle x\rangle_t}{dt} = K\big(\langle x\rangle_t\big) = -\frac{\partial V\big(\langle x\rangle_t\big)}{\partial\langle x\rangle_t} \tag{5.167}$$

$$\frac{d\Sigma^2(t)}{dt} = 2K'\big(\langle x\rangle_t\big)\Sigma^2(t) + Q\big(\langle x\rangle_t\big) \tag{5.168}$$

with the abbreviations

$$\begin{aligned} K\big(\langle x\rangle_t\big) &= S^{-1}w_\uparrow\big(\langle x\rangle_t\big) - S^{-1}w_\downarrow\big(\langle x\rangle_t\big) \\ &= 2\nu\big(\sinh\big(f\big(\langle x\rangle_t\big) - \langle x\rangle_t\cosh\big(f\big(\langle x\rangle_t\big)\big)\big) \end{aligned} \tag{5.169}$$

$$\begin{aligned} Q\big(\langle x\rangle_t\big) &= S^{-1}w_\uparrow\big(\langle x\rangle_t\big) + S^{-1}w_\downarrow\big(\langle x\rangle_t\big) \\ &= 2\nu\big(\cosh\big(f\big(\langle x\rangle_t\big) - \langle x\rangle_t\sinh\big(f\big(\langle x\rangle_t\big)\big)\big) \end{aligned} \tag{5.170}$$

and the potential function $V(\langle x\rangle_t)$ and using $f(\langle x\rangle_t) = K\langle x\rangle_t$ for $\delta = 0$

$$V\big(\langle x\rangle_t\big) = -\frac{2}{K}\left(1+\frac{1}{K}\right)\cosh\big(K\langle x\rangle_t\big) + \frac{2\langle x\rangle_t}{K}\sinh\big(K\langle x\rangle_t\big). \tag{5.171}$$

The movement of $\langle x\rangle_t$ can be visualized as a damped particle moving in a potential V with such a large friction that its velocity only depends on the gradient of the potential. The potential V is responsible for the dynamics of the spin system. However, it is worth emphasising that this potential function V which governs the macroscopic dynamics is not the same as the function U connected with the equilibrium distribution despite the fact, that both functions have their extremal values at the same positions: a single minimum at $\langle x_0\rangle = 0$ when $K < 1$, and a maximum at $\langle x_0\rangle = 0$ and two minima at $\langle x_{u,d}\rangle = \pm x_m$ when $K > 1$. Consequently, for $K < 1$ all solutions of the macroscopic equations tend to $\langle x_0\rangle = 0$, and are therefore globally stable. For $K > 1$, all solutions of (5.167) will approach on of the two minima $\langle x_{u,d}\rangle = \pm x_m$, depending on the initial conditions $\langle x\rangle_{t=0}$ of the spin system (Weidlich and Haag 1983).

5.2.3 *Calculation of the Escape Rate*

The two minima of $V(\langle x\rangle)$ for $K > 1$ correspond to non-zero magnetization in the up or down direction and will be denoted by V_u and V_d. They are stationary solutions of the macroscopic equation (5.167). Distributions located in V_u or V_d are locally stable, but not globally. Every other solution tends to the one or the other state, depending on its initial value. This is a bistable situation. However, fluctuations dominate this process in the long-run. Suppose that the parameter K is well above

1, so that the potential $V(\langle x \rangle)$ exhibits a well-developed potential barrier between both (locally) stable stationary solutions. Then, there is a tiny probability per unit of time for a very large fluctuation which carries the system across the potential barrier, so that the magnetization flips e.g. from up to down. Therefore, on a very long time-scale the single peak located at V_u loses probability to the other potential minimum until there are two peaks of equal height at both minima. The ultimate probability distribution is the one described by $P_{st}(s)$. We will now compute this escape rate.

Suppose, one starts at $t = 0$ with an arbitrary distribution $P(s, 0)$. This distribution evolves and develops two peaks, located at x_u and x_d. They approach their local equilibrium, consisting of two Gaussian peaks at V_u and V_d in the linear-noise approximation.

The probability is now divided into two parts π_u and π_d, one belonging to each peak. Of course, the total probability is:

$$\pi_u + \pi_d = 1. \tag{5.172}$$

Subsequently over a very long time scale τ, these peaks exchange probability:

$$\frac{d\pi_u}{dt} = -\frac{d\pi_d}{dt} = -\frac{1}{\tau}(\pi_u - \pi_d). \tag{5.173}$$

Following a method that was developed by Kramers (1940) and explicitly calculated by van Kampen (1977, 1981) for special barriers, the escape time τ can be calculated for the diffusion dominated case. It is assumed that after the local equilibrium has been reached, the distribution $P(s_{u,d}, t)$ will vary very slowly. Hence, one may approximately take $\frac{dP(s,t)}{dt} \approx 0$ and conclude for the almost stationary solution (detailed balance is not exactly fulfilled)

$$w_\downarrow(s)P^{as}(s) - w_\uparrow(s-1)P^{as}(s-1) = -J, \tag{5.174}$$

where J, which is the net probability flux from $s - 1$ to s is constant. Now we divide (5.174) by the stationary factors (detailed balance)

$$w_\downarrow(s)P^{st}(s) - w_\uparrow(s-1)P^{st}(s-1) = 0 \tag{5.175}$$

and obtain

$$\frac{P^{as}(s)}{P^{st}(s)} - \frac{P^{as}(s-1)}{P^{st}(s-1)} = -\frac{J}{w_\downarrow(s)P^{st}}(s). \tag{5.176}$$

Summing up this identity from $s = s_d + 1 = -d + 1$ to $s = s_u = u$, we obtain

$$\frac{P_u^{as}}{P_u^{st}} - \frac{P_d^{as}}{P_d^{st}} = -J \sum_{s=-d+1}^{u} \frac{1}{w_\downarrow(s)P^{st}}(s). \tag{5.177}$$

Since the probability of both peaks is concentrated around s_u and s_d for the whole time evolution, it seems reasonable to assume:

$$\frac{P(n_{u,d},t)}{P^{st}(n_{u,d})} = \frac{\pi_{u,d}(t)}{\pi_{u,d}^{st}}. \tag{5.178}$$

Taking into account, that the local peaks do not change in shape but only in height, and that the probability of both stationary parts $\pi_u^{st} = \pi_d^{st} = 1/2$, the left-hand side of (5.178) is

$$\frac{\pi_u}{1/2} - \frac{\pi_d}{1/2} = 2(\pi_u - \pi_d). \tag{5.179}$$

The flow of probability between the two minima is described by J, and given by

$$J = \frac{d\pi_u}{dt} = -\frac{d\pi_d}{dt}. \tag{5.180}$$

Hence, Eqs. (5.177) with (5.178) and (5.179) finally provides an estimation of the escape rate τ (van Kampen 1978)

$$\tau = \frac{1}{2} \sum_{s=-d+1}^{u} \frac{1}{w_\downarrow(s)P^{st}}(s). \tag{5.181}$$

It is remarkable that (5.181) expresses the escape rate τ in terms that are known and do not require the Master equation to be solved. Obviously, the value of the sum is mainly determined by the terms around $s = 0$, where $P^{st}(s)$ is very small. Thus one may put $w_\downarrow(s) \approx w_\downarrow(0) = S$ in (5.181) and finds, taking into account the normalization of the stationary solution $P^{st}(s)$, in the quasi-continuous case

$$\begin{aligned}
\tau &= \frac{1}{2S} \int_{-x_u}^{x_u} e^{SU(x')} S d x' \int_{-1}^{1} e^{-SU(x)} S dx \\
&= S \int_{-1}^{1} e^{SU(0)-1/2|U''(0)|x'^2} dx' \int_{-1}^{1} e^{-SU(x_u)+1/2|U''(x_u)|x^2} dx \\
&\propto \frac{2}{\sqrt{|U''(0)U''(x_u)|}} \exp S(U(0) - U(x_u)).
\end{aligned} \tag{5.182}$$

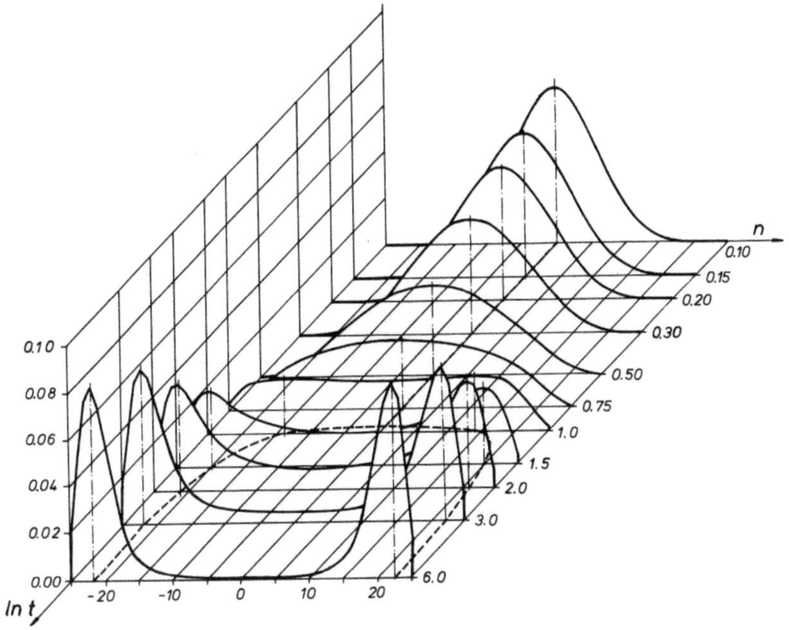

Fig. 5.11 Evolution of a symmetrical initial distribution $P(s,0)$ towards its stationary distribution $P^{st}(s)$, by a direct simulation of the Master equation (5.147) for a spin system with $S=25$, and $K=1.5$ (Source: Weidlich and Haag 1983)

The exponential in (5.182) is the dominant factor, and it shows that the escape rate $\Gamma \propto e^{-cS}$ scales with the total number of spins. Furthermore, the factor c is the difference of the maximum and minimum values of U and not of the potential V. If we know at $t=0$ that the spin system is magnetized in the up-direction, then after a time span of order τ we can no longer be sure that it has not flipped to the down direction. To ensure a large escape time τ the interaction parameter K should be well above its critical value and the number of spins S should be large enough.

In Fig. 5.11, the evolution of a symmetrical initial probability distribution $P(s,0)$ is shown. In this example, the Master equation (5.147) is treated as a homogeneous linear system of equations for the different values of $P(s,t)$ and solved numerically. The initial fluctuation enhancement and the subsequent drift dominated development of the bimodal distribution towards its stationary distribution $P^{st}(s)$ is obvious.

In Fig. 5.12, the initial distribution $P(s,0)$ is asymmetric. The evolution of the spin system leads to a drift dominated redistribution ending up in a bimodal distribution with peaks of different height, after fluctuation enhancement. The final development into the symmetrical stationary distribution $P^{st}(s)$ is fluctuation dominated and extremely slow compared with the drift dominated part.

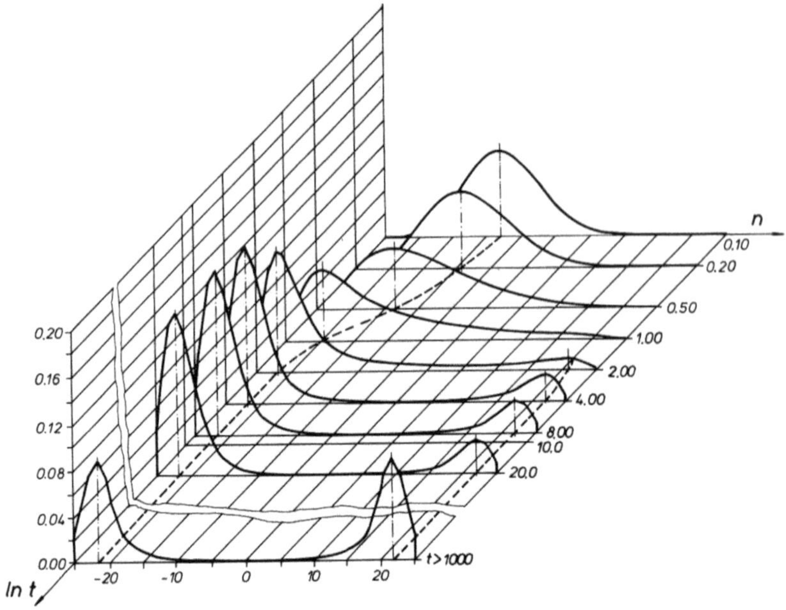

Fig. 5.12 Evolution of an asymmetrical initial distribution $P(s,0)$ into its stationary distribution $P^{st}(s)$, by a direct simulation of the Master equation (5.147) for a spin system with $S=25$, and $K=1.5$ (Source: Weidlich and Haag 1983)

5.3 Quantum Statistics of the Laser Light

We consider N two-level atoms in quasi-resonant interaction with one electromagnetic field mode. The atoms and the field mode are coupled to external reservoirs providing for pump- and loss processes. The density matrix for this system serves as a simple but fundamental model including the relevant photon statistics for Laser action. We skip the derivation of the Laser Master equation starting from first principles. For a detailed quantum statistical analysis and derivation of the Pauli Master equation for the Laser, we refer to the fundamental work of Weidlich et al. (1967a, b), Haken (1970), and Agarwal (1973).

In Fig. 5.13, the principle set up of a Laser is shown. As an example let us consider a solid-state Laser consisting of a rod of active material in which specific atoms are embedded. Mirrors are fixed at the end of the rod.

Each atom may be excited from outside via a specific pump process, e.g. by an optical excitation of the atoms via absorption of the pump light (see Fig. 5.14). Obviously, the Laser is an open system, far away from thermal equilibrium.

The atoms in their excited state typically decay within a short life time of about 10^{-8} seconds (spontaneous emission). However, there also exists a process called stimulated emission. This means that when a wave track reaches an atom in its excited state, the atom is able to pass in its ground state and to emit a photon in phase with the incoming photon. In other words, an acceleration of the emitted light

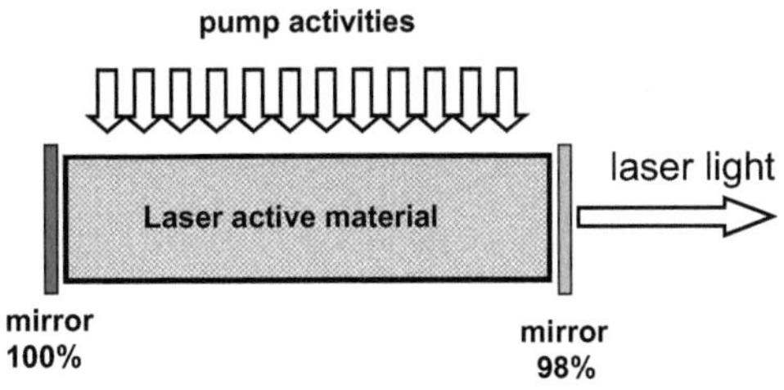

Fig. 5.13 Principle structure of a Laser system

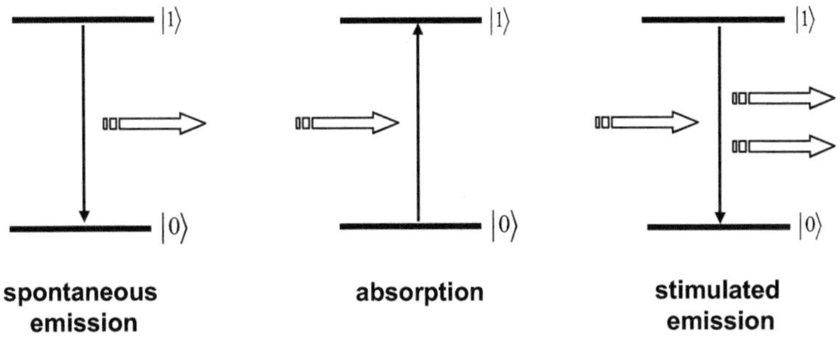

Fig. 5.14 Absorption, spontaneous and stimulated emission

intensity (output power) occurs. The mirrors act for the selection of wave tracks along the axis of the rod. Those wave tracks running in the axial direction are reflected several times between the mirrors, while all other wave tracks leave the Laser device very quickly.

Let us now consider what happens when we start increasing the pumping energy of the system. If the pump energy is below a certain threshold value, the Laser acts like a typical lamp and the atoms emit light wave tracks independent of each other. At the so-called Laser threshold a phase transition occurs, at a certain critical pump power. The atoms of the Laser active material seem to be slaved; all atoms behave in a coordinated way and emit wave tracks in phase. They emit single giant wave tracks with a dramatically increased light intensity (output power).

Above the so-called Laser threshold, the pump energy entering the system is partly converted into Laser light with its unique properties (see Fig. 5.15).

5.3.1 The Pauli Master Equation of the LASER

The Pauli Master equation for the probability in configuration space to find m atoms and n photons excited $P(m, n)$ reads (Weidlich et al. 1967a, b; Haken 1970)

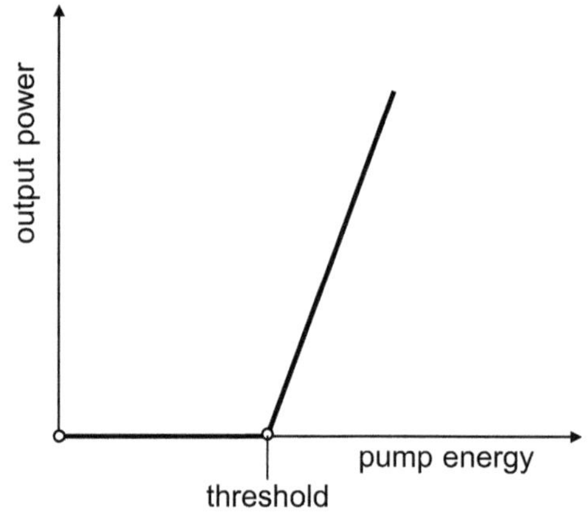

Fig. 5.15 Output power versus input power of a Laser below and above its threshold

$$\begin{aligned}\frac{dP(m,n)}{dt} &= 2\gamma_{\uparrow}(N-m+1)P(m-1,n) - 2\gamma_{\uparrow}(N-m)P(m,n)\\ &+ 2\gamma_{\downarrow}(m+1)P(m+1,n) - 2\gamma_{\downarrow}mP(m,n)\\ &+ 2\nu_{\uparrow}nP(m,n-1) - 2\nu_{\uparrow}(n+1)P(m,n)\\ &+ 2\nu_{\downarrow}nP(m,n+1) - 2\nu_{\downarrow}nP(m,n)\\ &+ \alpha n(m+1)P(m+1,n-1) - \alpha n(n-m)P(m,n)\\ &+ \alpha(n+1)(N-m+1)P(m-1,n+1)\\ &- \alpha(n+1)mP(m,n) \end{aligned} \tag{5.183}$$

with $\alpha = 2\pi g^2/\gamma$ where γ is the inhomogeneous atomic level width (i.e. spread of levels). In Fig. 5.16, the transition rates of the Laser are explained. The parameters $\gamma_{\uparrow}$ and $\gamma_{\downarrow}$ describe transitions of the atomic system via pump processes and losses. Atoms of the Laser active material may be exited via external excitation processes or emit photons and relax to its ground state. This has an impact on the number of exited atoms m. In addition, the number of photons n may change due to the coupling with the atomic system—absorption, spontaneous and stimulated emission of photons—and the output of photons via the mirrors. The parameters $\nu_{\uparrow}$ and $\nu_{\downarrow}$ are the transition rates related to the light field (photon system). The parameter α describes the interaction of the atomic system with the light field. In a Laser system, this process is dominating.

The transition rates of the Laser equation (5.183) are based on fundamental physical considerations. The total transition rate $w(m,n;m',n')$ can be separated into three parts: the transition rate of the atomic system $w_A(m,n;m',n')$, the transition rate of the light field $w_F(m,n;m',n')$ and the transition rate describing the interaction between the atomic system and the light field $w_{AF}(m,n;m',n')$:

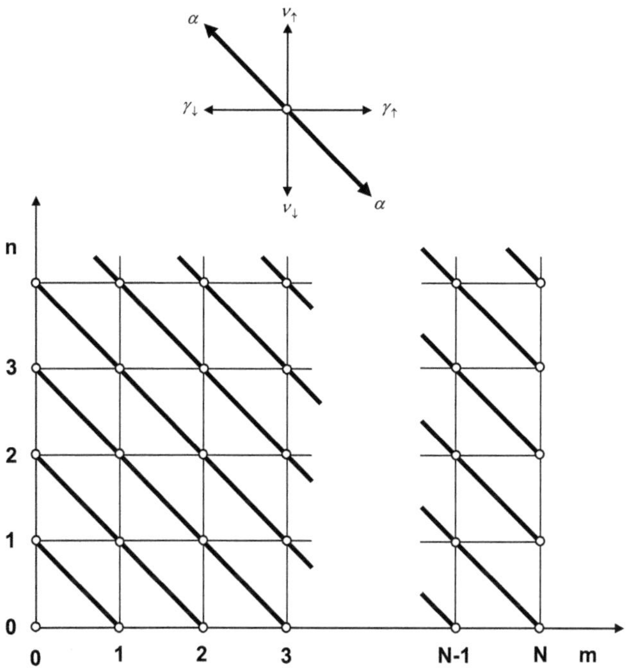

Fig. 5.16 Transition rates of the Laser model in configuration space

$$w(m,n;m',n') = w_A(m,n;m',n') + w_F(m,n;m',n') + w_{AF}(m,n;m',n') \quad (5.184)$$

The transition rates related to the atomic system (excitation or damping of atoms)

$$w_A(m,n;m',n') = 2\gamma_\uparrow(N-m')\delta_{m-1,m'}\delta_{n,n'} + 2\gamma_\downarrow m'\delta_{m+1,m'}\delta_{n,n'} \quad (5.185)$$

and the photonic system (light field) (excitation or damping of photons)

$$w_F(m,n;m',n') = 2\nu_\uparrow(n'+1)\delta_{m,m'}\delta_{n-1,n'} + 2\nu_\downarrow n'\delta_{m,m'}\delta_{n+1,n'} \quad (5.186)$$

and the atom—field interaction

$$w_{AF}(m,n;m',n') = \alpha(n'+1)m'\delta_{m+1,m'}\delta_{n-1,n'} + \alpha n'(n-m')\delta_{m-1,m'}\delta_{n+1,n'} \quad (5.187)$$

fulfil detailed balance, while detailed balance is not valid for the total transition rate $w(m,n;m',n')$. This complicates the mathematical treatment of the Laser Master equation. Therefore, approximation methods have to be used to find an appropriate stationary solution (Haag et al. 1977).

5.3.2 Stationary Solution of the Laser Master Equation

The stationary solution of the Laser Master equation will be calculated via the T-factor method. For this aim we divide (5.183) by $P(m,n)$ and obtain

$$\begin{aligned} J(m,n) = \frac{dP(m,n)/dt}{P(m,n)} &= 2\gamma_{\uparrow}(N-m+1)g(m-1,n|m,n) \\ &- 2\gamma_{\uparrow}(N-m) + 2\gamma_{\downarrow}(m+1)g(m+1,n|m,n) \\ &- 2\gamma_{\downarrow}m + 2\nu_{\uparrow}ng(m,n-1|m,n) - 2\nu_{\uparrow}(n+1) \\ &+ 2\nu_{\downarrow}ng(m,n+1|m,n) - 2\nu_{\downarrow}n \\ &+ \alpha n(m+1)g(m+1,n-1|m,n) \\ &- \alpha n(N-m) \\ &+ \alpha(n+1)(N-m+1)g(m-1,n+1|m,n) \\ &- \alpha(n+1)m = 0 \end{aligned} \tag{5.188}$$

for $m=0,1,\ldots,N$ and $n=0,1,2,\ldots$.

Equation (5.188) provides an inhomogeneous infinite algebraic system of equations for the determination of the stationary T-factors of the Laser system.

The approximation methods developed in Chap. 4 can be immediately applied to the Laser equation (5.188) if we assume in agreement with the experimental situation, that the transition probabilities of the atom—field interaction $w_{AF}(m,n;m',n')$ strongly prevail over those of the atomic transitions related to the pumping activities $w_A(m,n;m',n')$ and to the transitions related to the coupling of the light field $w_F(m,n;m',n')$. Hence, the strong coupling $w_{AF}(m,n;m',n')$ leads to a quick establishment of a quasi stationary state along the lines $(m+n)=k$ as depicted in Fig. 5.16 whereas the transition rates $w_A(m,n;m',n')$ and $w_F(m,n;m',n')$, induce only weak transitions between the different stripes. In this limit of dominating atom-field-interaction, it is plausible to assume that the distribution function only depends on the sum-variable $m+n=k$

$$P(m,n) = P(0,0)\binom{N}{m}\prod_{l=1}^{m+n}\xi(l) \tag{5.189}$$

up to a combinatorial factor

$$\binom{N}{m} = \frac{N!}{(N-m)!m!}, \tag{5.190}$$

taking into account the degeneration of the atomic system. The initial value $P(0,0)$ is fixed by normalization. This assumption (5.189) leads to the specific structure of the transition rates:

$$g(m,n|m-1,n) = \frac{(N-m+1)}{m}\xi(m+n) \tag{5.191}$$

$$g(m,n|m,n-1) = \xi(m+n) \tag{5.192}$$

If we insert (5.191), (5.192) in (5.188) we obtain

$$J(m,n) = \xi(m+n+1)\big(2\gamma_{\downarrow}(N-m) + 2\nu_{\downarrow}(n+1)\big)$$
$$-\big(2\gamma_{\uparrow}(N-m) + 2\nu_{\uparrow}(n+1)\big) + \xi^{-1}(m+n)\big(2\gamma_{\uparrow}m + 2\nu_{\uparrow}n\big) - \big(2\gamma_{\downarrow}m + 2\nu_{\downarrow}n\big) = 0. \tag{5.193}$$

It is worth mentioning that solely based on the assumed structure of the probability distribution (5.189), all terms originating from the atom-field interaction vanish, regardless of the size of the coupling constant α. Furthermore, in (5.193) it is assumed, that the transition factors $\xi = \xi(m+n)$ are a function of the sum variable $m+n=k$. However, since the coefficients in (5.193) show an explicit dependency of m and n, Eq. (5.193) is inconsistent. In order to get to a consistent equation, we have to sum up over all combinations leading to the same value of $k=m+n$. Since there are $\binom{N}{m}$ possibilities, regardless of the arrangement, to find m atoms excited and $(N-m)$ atoms in its ground state, we have to perform a weighted averaging

$$J(k) = \sum_{m=0}^{k} \binom{N}{m} J(m, k-m) \tag{5.194}$$

and thus, we obtain the ξ-transformed Master equation (Haag 1978)

$$w_{\downarrow}(k+1)\xi(k+1) + w_{\uparrow}(k-1)\xi^{-1}(k) - \big(w_{\uparrow}(k) + w_{\downarrow}(k)\big) = 0 \tag{5.195}$$

for $k=0,1,2,\ldots$

with the transformed transition rates

$$w_{\uparrow}(k) = h_N(k+1)\big(\gamma_{\uparrow} + \nu_{\uparrow} f_N(k+1)\big) \tag{5.196}$$

$$w_{\downarrow}(k) = h_N(k)\big(\gamma_{\downarrow} + \nu_{\downarrow} f_N(k)\big) \tag{5.197}$$

and abbreviations

$$h_N(k) = \sum_{m=0}^{k} m \binom{N}{m} \tag{5.198}$$

and

$$f_N(k) = h_N(k)^{-1} \sum_{m=0}^{k} (k-m) \binom{N}{m}. \tag{5.199}$$

The Master equation (5.195) fulfils detailed balance and we can easily compute the stationary transition factors

$$\xi_{st}(k) = \frac{w_\uparrow(k-1)}{w_\downarrow(k)} = \frac{\gamma_\uparrow + \nu_\uparrow f_N(k)}{\gamma_\downarrow + \nu_\downarrow f_N(k)}. \tag{5.200}$$

In the limit of dominating atom-field-interaction, the probability to find m atoms and n photons excited is therefore given by

$$P_{st}(m,n) = P_{st}(0,0) \binom{N}{m} \prod_{k=1}^{m+n} \frac{\gamma_\uparrow + \nu_\uparrow f_N(k)}{\gamma_\downarrow + \nu_\downarrow f_N(k)}. \tag{5.201}$$

By summing up over all atomic states, since we are not interested to know exactly how many atoms are excited, we obtain the distribution function $P(n)$ for the photons (light-field)

$$\begin{aligned} P_{st}(n) &= \textstyle\sum_{m=0}^{N} P_{st}(m,n) = P_{st}(0,0) \sum_{m=0}^{N} \binom{N}{m} \prod_{k=1}^{m+n} \frac{\gamma_\uparrow + \nu_\uparrow f_N(k)}{\gamma_\downarrow + \nu_\downarrow f_N(k)} \\ &\approx P_{st}(0) \textstyle\prod_{k=1}^{N/2+n} \frac{\gamma_\uparrow + \nu_\uparrow f_N(k)}{\gamma_\downarrow + \nu_\downarrow f_N(k)} \end{aligned} \tag{5.202}$$

Since the number of atoms $N \gg 1$, we can use the function $f_N(k)$ in its limiting case $N \to \infty$

$$f_N(k) = \begin{cases} 0 & \text{for } 0 \le k \le N/2 \\ (2k/N - 1) & \text{for } k > N/2 \end{cases}. \tag{5.203}$$

Therefore, we can consider two different shapes of the photon-distribution function

(a) below Laser threshold ($\gamma_\uparrow < \gamma_\downarrow$)
(b) above Laser threshold ($\gamma_\uparrow > \gamma_\downarrow$)

Case (a)
Below the Laser threshold, the function $f_N(k) \approx 0$, and we obtain the typical statistical fluctuation distribution of a fluorescent tube (thermal fluctuations) for the photon distribution

$$P(n) = P(0) \left(\frac{\gamma_\uparrow}{\gamma_\downarrow}\right)^n = \frac{\gamma_\downarrow - \gamma_\uparrow}{\gamma_\downarrow} \left(\frac{\gamma_\uparrow}{\gamma_\downarrow}\right)^n \quad \text{for } (\gamma_\uparrow < \gamma_\downarrow). \tag{5.204}$$

Case (b)
Above threshold, if the pump parameter $\gamma_\uparrow$ exceeds the parameter of spontaneous emissions $\gamma_\downarrow$ the distribution function shows a maximum, obtained via

$$\xi(n_w) = \frac{\gamma_\uparrow + \nu_\uparrow \frac{2n_w}{N}}{\gamma_\downarrow + \nu_\downarrow \frac{2n_w}{N}} = 1 \tag{5.205}$$

$$n_w = \frac{N}{2}\left(\frac{\gamma_\uparrow - \gamma_\downarrow}{\nu_\downarrow - \nu_\uparrow}\right) \geq 0, \tag{5.206}$$

and its variance is given by

$$\sigma^2 = -\left[\frac{\partial \xi(n)}{\partial n}\right]^{-1}_{n_w} = \frac{N}{2}\left(1 + \frac{\gamma_\uparrow}{\nu}(n_{th} + 1) - \frac{\gamma_\downarrow}{\nu} n_{th}\right) \tag{5.207}$$

with $\nu_\uparrow = \nu\, n_{th}$; and $\nu_\downarrow = \nu(n_{th} + 1)$; and $\nu = \nu_\downarrow - \nu_\uparrow$.

$$P_{st}(n) = P_{st}(0)\exp\left(-\frac{(n - n_w)^2}{2\sigma^2}\right) \quad \text{for } n \geq 0. \tag{5.208}$$

This solution (5.202) was first found by Dohm (1972, 1973) by a different derivation method, starting from the same fundamental equation (5.183). As already Dohm stressed, the solution (5.202) becomes exact in the limit of vanishing rate constants $\gamma_\uparrow, \gamma_\downarrow, \nu_\uparrow, \nu_\downarrow$ where their ratios remain finite—in the limit of dominating atom-field-interaction. The form of (5.208) essentially agrees with Risken's (1972) solution of the Laser equation developed via the Fokker–Planck approach.

To illustrate the phase transition for different pumping parameter values $\gamma_\uparrow/\gamma_\downarrow$ the photon distribution and the associated T-factors are shown in Figs. 5.17 and 5.18.

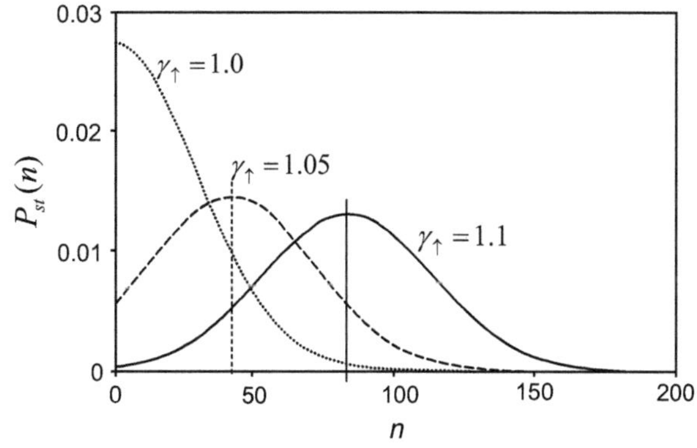

Fig. 5.17 Photon distribution $P_{st}(n)$ for different values of the pump parameter $\gamma_\uparrow = 1.0$, $\gamma_\uparrow = 1.05$, $\gamma_\uparrow = 1.1$, and $\gamma_\downarrow = 1.0$, $\nu_\uparrow = 0.01$, $\nu_\downarrow = 0.1$

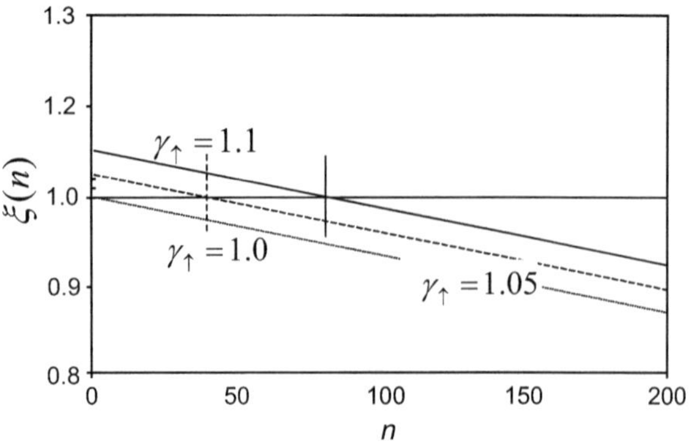

Fig. 5.18 Associated T-factors to the distribution $P_{st}(n)$ for different values of the pump parameter $\gamma_\uparrow = 1.0$, $\gamma_\uparrow = 1.05$, $\gamma_\uparrow = 1.1$, and $\gamma_\downarrow = 1.0$, $\nu_\uparrow = 0.01$, $\nu_\downarrow = 0.1$

References

Abramowitz M, Stegun IA (1972) Handbook of mathematical functions. Dover, New York

Agarwal GS (1973) Master equation methods in quantum optics. Prog Opt 11:1–76

Bartholomay AF (1958) On the linear birt and death processes of biology as Markov chains. Bull Math Biophys 20:97–118

Dohm V (1972) Nichtgleichgewichtsphasenübergang in einem exakt lösbaren Lasermodell. Solid State Commun 11:1273

Dohm V (1973) Thesis, Techn. Hochschule Aachen, Aachen

Gardiner CW, McNeil KJ, Walls DF (1976) Correlations in stochastic theories of chemical reactions. J Stat Phys 14(4):307–331

Glansdorff P, Prigogine I (1971) Thermodynamics of structure, stability and fluctuations. Wiley, New York

Haag G (1977) Transition factor method for discrete master equations and applications to chemical reactions. Z Phys B 29:153–159

Haag G (1978) Neue Lösungsmethoden diskreter Mastergleichungen. Thesis, University of Stuttgart, Stuttgart

Haag G, Hänggi P (1979) Exact solutions of discrete master equations in terms of continued fractions. Z Phys B 34:411–417

Haag G, Hänggi P (1980) Continued fraction solutions of discrete master equations not obeying detailed balaance II. Z Phys B 39:269–279

Haag G, Weidlich W, Alber P (1977) Approximation methods for stationary solutions of discrete master equations. Z Phys B 26:207–215

Haken H (1970) Handbuch der Physik, vol XXV/2c. Springer, Berlin

Hilhorst HJ (1973) Kinetics of the stochastic Ising chain in a class of spin-flip models. Physica 66:497–508

Kramers HA (1940) Brownian motion in a field of force and the diffusion model of chemical reactions. Physica 7:284–304

Kuramoto Y (1973) Some remarks on the theory of fluctuations around nonequilibrium states. Prog Theor Phys 49(5):1782

Madelung E (1964) Die mathematischen Hilfsmittel des Physikers. Springer, Berlin

Mazo RM (1975) On the discrepancy between results of Nicolis and Saito concerning fluctuations in chemical reactions. J Chem Phys 62:4244

McQuarrie DA (1967) Stochastic approach to chemical kinetics. J Appl Probab 4:413–478

Meschkowski H (1959) Differenzengleichungen. Vandenhoeck & Ruprecht, Göttingen

Nicolis G (1972) Fluctuations around nonequilibrium states in open nonlinear systems. J Stat Phys 6:195–222

Nicolis G, Prigogine I (1971) Fluctuations in nonequilibrium systems. Prog Nat Sci 68 (9):2102–2107
Nitzan A, Ortoleva P, Deutch J, Ross J (1974) Fluctuations and transitions at chemical instabilities: the analogy to phase transitions. J Chem Phys 61:1056–1074
Risken H (1972) Solutions of the Fokker-Planck equation in detailed balance. Z Phys 251:231–243
Ruijgrok TW, Tjon JA (1973) Critical slowing down and nonlinear response in an exactly solvable stochastic model. Physica 65:539–565
Saito N (1974) Fluctuations in chemical reactions around the steady states. J Chem Phys 61:3644–3650
van Kampen NG (1965) Fluctuation in nonlinear systems. In: Burgers AE (ed) Fluctuation phenomena in solids. Academic Press, New York
van Kampen NG (1977) Stochastic processes in physics and chemistry. J Stat Phys 17:71
van Kampen NG (1978) An introduction to stochastic processes for physicists. In: Garido L, Seglar P, Shepherd PJ (eds) Stochastic processes in nonequilibrium systems. Proceedings, Sitges 1978. Springer, New York
van Kampen NG (1981) Stochastic processes in physics and chemistry. North-Holland, Amsterdam
Weidlich W, Haag G (1983) Concepts and models of a quantitative sociology: the dynamics of interaction populations, Springer series of synergetics, vol 14. Springer, New York
Weidlich W, Risken H, Haken H (1967a) Quantum mechanical solutions of laser masterequation, Part I. Z Phys 201:396
Weidlich W, Risken H, Haken H (1967b) Quantum mechanical solutions of laser masterequation, Part II. Z Phys 204:223

Part III
Applications Social Sciences

Chapter 6
The Master Equation in Dynamic Decision Theory

Contents

6.1 The Discrete Choice Approach

A central objective in the social sciences concerns the understanding of the decision behaviour of individuals, especially of interacting individuals and of groups of individuals or agents.

The use of disaggregate choice models has been strongly advocated by many researcher such as Golledge and Timmermans (1988), Ben-Akiva and Lerman (1985), Bahrenberg et al. (1984), Pitfield (1984), Johnson and Hensher (1982), Domencich and McFadden (1975), Reggiani (2000), Haag (1989), Akerlof and Kranton (2000) and many others, especially since it is has been possible to capture stochastic and behavioural aspects of decision processes. New insights into the decision making of individuals, group dynamics and environmental impacts may be obtained by focusing on the micro-level of individuals. Therefore, considerable efforts have been devoted to the development of behavioural choice models, capable of considering individual choices from a set of discrete alternatives at a point in time (Train 2003). With very few exceptions the emphasis on such discrete choice models has been strictly cross-sectional even if the choice processes studied were inherently dynamic in nature. Attention in the last decades has been paid to modelling dynamic decision processes.

G. Haag, *Modelling with the Master Equation*, DOI 10.1007/978-3-319-60300-1_6

Choice problems are concerned with individuals or agents who have to select one alternative among a set of alternatives. Utility is a ubiquitous concept in social sciences and especially in economics as an abstract measurement of goal-attainment provided by a decision, e.g. a product or service (Fischer et al. 1988; Heckman 1981; Hensher 1988). It is the idea of introducing an "objective function" for the comparison of alternatives, which can be characterized by a set of attributes. Since utility in this concept cannot be measured directly; we can make inferences about utility based on person's behaviour, if we presume that people act rationally. In ICT (information and communication technologies), a rational agent is defined as one that acts to optimize his goal.

Another weak point of this concept consists in the fact that this utility model does not take into account the degree to which the utility of one alternative exceeds the utility of another. For instance, regardless if the utility difference between two alternatives is big or only marginal, it has no influence on the choice behaviour (Hensher and Wrigley 1984). In reality, uncertainty in utility estimates would give more confidence to the alternative with the much higher utility. Furthermore, this utility concept is solely static. No information is obtained or given how long the process last until a clear decision for one of the alternatives is fixed, if we suppose that choices are made rationally. However, uncertainty remains, since we can not measure all characteristics of the different alternatives, as seen by the individual.

In random utility theory we consider an individual k, $k = 1, 2, \ldots, N$ who has to choose an alternative A_j among a set of L mutually exclusive alternatives A_j, $j = 1, 2, \ldots, L$. It is assumed that the decision maker chooses the option which is considered most desirable to it, given the attributes z_j^k of each alternative A_j as seen by the individual k and his personal tastes s^k. More precisely, it is assumed that a decision maker (individual or agent) k measures the desirability of each alternative A_j by an objective function u_j^k, termed utility, and chooses that alternative which yields the maximum utility. The specification of the functional form of the objective function u_j^k, $j = 1, 2, \ldots, L$ and $k = 1, 2, \ldots, L$ is the starting point for defining a particular choice model. With only a few exceptions (Fischer and Nijkamp 1987; Domencich and McFadden 1975), random utility based choice models assume that u_j^k is linear in the parameters and additive in the variables

$$u_j^k = v_j^k\left(z_j^k, s^k\right) + \varepsilon_j^k \quad \text{for} \quad j = 1, 2, \ldots, L, \tag{6.1}$$

where the term $v_j^k\left(z_j^k, s^k\right)$ is non-stochastic and contains all the attributes of A_j and characteristics which are observed by the decision maker. Therefore, this part is named the deterministic part of the utility function. The term ε_j^k is stochastic (the random part) and reflects all effects related to unobserved attributes of the decision maker and the choice alternatives as well as the idiosyncratic tastes of k.

Thus, the probability p_j^k that the decision maker with utility function $v_j^k\left(z_j^k, s^k\right)$ selects alternative A_j is given by

$$p_j^k = \Pr\left(v_j^k\left(z_j^k, s^k\right) + \varepsilon_j^k \geq v_i^k\left(z_i^k, s^k\right) + \varepsilon_i^k\right), \quad \forall_{i \neq j} \text{ for } j = 1, 2, \ldots, L \text{ and } k = 1, \ldots, N. \tag{6.2}$$

The error terms ε_j^k are unobserved random variables, described by a probability distribution $P(\vec{\varepsilon})$ and by the corresponding cumulative distribution function. In the multidimensional case (6.3), this means

$$\begin{aligned} p_j^k &= \Pr\left(\varepsilon_l^k < v_j^k - v_l^k + \varepsilon_j^k\right)\forall_{l \neq j} \\ &= \int_{\varepsilon_j=-\infty}^{\infty} d\varepsilon_j \int_{\varepsilon_1=-\infty}^{v_j - v_1 + \varepsilon_j} \int_{\varepsilon_2=-\infty}^{v_j - v_2 + \varepsilon_j} \cdots \int_{\varepsilon_L=-\infty}^{v_j - v_L + \varepsilon_j} P(\vec{\varepsilon}) d\varepsilon_L \ldots d\varepsilon_{j+1} d\varepsilon_{j-1} \ldots d\varepsilon_2 d\varepsilon_1 \end{aligned} \tag{6.3}$$

The solution of (6.3) requires multidimensional integration, which can be performed if the distribution function of the error term is known.

6.2 The Probit Model

If it is assumed that the error terms can be described by a multivariate joint normal distribution with a mean vector and a covariance matrix, we obtain the so-called probit model. However, the probit model (Braun and Saam 2015) does not yield a closed form solution. The required multidimensional integration hampers its practical application.

A further simplification of the choice model can be established if the error terms are identically and independently normally distributed (6.4). Then the multidimensional integral reduces to a product of single dimensional integrals

$$\begin{aligned} p_j^k &= \Pr\left(\varepsilon_l^k < v_j^k - v_l^k + \varepsilon_j^k\right)\forall_{l \neq j} \\ &= \int_{\varepsilon_j=-\infty}^{\infty} P\left(\varepsilon_j^k\right) \prod_{l \neq j} \left(\int_{\varepsilon_l=-\infty}^{v_j - v_l + \varepsilon_j} P\left(\varepsilon_l^k\right) d\varepsilon_l^k\right) d\varepsilon_j^k \\ &= \int_{\varepsilon_j=-\infty}^{\infty} P\left(\varepsilon_j^k\right) \left(\prod_{l \neq j} F\left(\varepsilon_l^k \leq v_j^k - v_l^k + \varepsilon_j^k\right)\right) d\varepsilon_j^k \end{aligned} \tag{6.4}$$

However, the simplified form is still cumbersome to apply and is far away from any closed form solution. Therefore, a lot of research in the last decades has focused developing more practical decision models.

6.3 The Multinomial Logit Model (MNL)

In their fundamental work Domencich and McFadden (1975) scored a further simplification step. Error terms ε_j are assumed to be independently and identically Gumbel distributed (see Sect. 2.3.4). There is no theoretical reason to believe that the double exponential Gumbel distribution is a good assumption for the error terms. However, the resulting multinomial logit model (MNL) is rather easy to use: it is explicit and shows almost indistinguishable results in practical studies, compared with the much more complicated probit model.

Suppose that the stochastic terms ε_i are independent of each other and Gumbel distributed. The cumulative Gumbel distribution of a random number ε_i is given by (2.72) (see Sect. 2.3.4):

$$F(\varepsilon_i \leq \varepsilon) = \int_{-\infty}^{\varepsilon_i} P_1(\varepsilon')d\varepsilon' = \exp(-e^{-\varepsilon_i}). \tag{6.5}$$

The corresponding Gumbel density function reads (2.71)

$$P_1(\varepsilon_j) = e^{-\left(\varepsilon_j + e^{-\varepsilon_j}\right)} = e^{-\varepsilon_j}\exp(-e^{-\varepsilon_j}). \tag{6.6}$$

Using (6.6) in (6.4), we have

$$\begin{aligned} p_j^k &= \int_{\varepsilon_j=-\infty}^{\infty} e^{-\varepsilon_j}\exp(-e^{-\varepsilon_j}) \prod_{l\neq j} \exp\left(-e^{-\left(v_j - v_l + \varepsilon_j\right)}\right) d\varepsilon_j \\ &= \int_{\varepsilon_j=-\infty}^{\infty} e^{-\varepsilon_j} \prod_{l} \exp\left(-e^{-\left(v_j - v_l + \varepsilon_j\right)}\right) d\varepsilon_j \\ &= \int_{\varepsilon_j=-\infty}^{\infty} e^{-\varepsilon_j}\exp\left(-\sum_l e^{-\left(v_j - v_l + \varepsilon_j\right)}\right) d\varepsilon_j \\ &= \int_{\varepsilon_j=-\infty}^{\infty} e^{-\varepsilon_j}\exp\left(-e^{-\varepsilon_j}\sum_l e^{-\left(v_j - v_l\right)}\right) d\varepsilon_j. \end{aligned} \tag{6.7}$$

In (6.7), the exponential term (first line) can be brought inside the product, so that the expression can be rewritten (second line), and the product finally stretches over all l and not only the terms $l \neq j$.

We solve the integral by substitution: $t = \exp(-\varepsilon_j)$, $dt = \exp(-\varepsilon_j)d\varepsilon_j$

$$p_j^k = -\int_{t=\infty}^{0} \exp\left(-t\sum_l e^{-(v_j-v_l)}\right)dt = \int_{t=0}^{\infty} \exp\left(-t\sum_l e^{-(v_j-v_l)}\right)dt$$
$$= -\frac{1}{\sum_l e^{-(v_j-v_l)}} \exp\left(-t\sum_l e^{-(v_j-v_l)}\right)\Bigg|_0^{\infty} \tag{6.8}$$
$$= \frac{1}{\sum_l e^{-(v_j-v_l)}} = \frac{e^{v_j}}{\sum_l e^{v_l}}$$

Therefore, with the assumption of the double exponential error term the choice probabilities p_j^k for an agent k assume a very simple formula (Domencich and McFadden 1975)

$$p_j^k = \frac{\exp\left(v_j^k\right)}{\sum_{l=1}^{L} \exp\left(v_l^k\right)} \quad \text{for} \quad j = 1, 2, \ldots, L \text{ and } k = 1, 2, \ldots, N. \tag{6.9}$$

The choice probabilities p_j^k of each individual k of the different alternatives range from zero to one

$$0 \leq p_j^k \leq 1 \quad \text{with} \quad \sum_{j=1}^{L} p_j^k = 1 \tag{6.10}$$

For the MNL it is easy to show, that the ratio of choice probabilities between two alternatives is independent of the presence of other alternatives

$$\frac{p_j^k}{p_l^k} = \frac{\exp\left(v_j^k\left(z_j^k, s^k\right)\right)}{\exp\left(v_l^k\left(z_l^k, s^k\right)\right)}. \tag{6.11}$$

This property of the MNL is a weak point, known as the *red-bus/blue-bus problem*. However, the compact and easy to use formula of the MNL for choice probabilities compared with the probit model as well as some other very interesting properties contribute to its wide use in economics, geography, biology and social sciences. Further applications and extensions of the model are given in Leonardi (1983), Fischer and Aufhauser (1988), and Mueller et al. (2000), among others. However, the multinomial logit model exhibits some further weak points and restrictions:

(a) Stationarity is required. The time span needed by an individual to choose a given alternative is not specified.

(b) Interactions among individuals are excluded, which does not allow treatment of self-reinforcing mechanisms or identity economics effects
(c) The MNL does not take into account the degree to which the utility of one alternative A_j exceeds the utility of another alternative A_l. This statement concerns both, the temporal dimension (how long does it take until a decision is realised) and impact on the choice distribution
(d) The number of alternatives is fixed
(e) Independence of irrelevant alternatives or the red-bus/blue-bus problem.[1] If a new alternative is added to the choice set, it may happen that some individuals who would otherwise have chosen an alternative in the initial choice set would instead choose the new alternative. However, the ratio of choice probabilities between any two alternatives is unaffected in the MNL model by the presence of a third alternative. This means, any new alternative introduced to a choice set will take its choice share proportionally from all other alternatives in the choice set (Samuelson 1985).
(f) No memory effects are included (Markov assumption)

Hence, the multinomial logit model is not suitable for modelling decision problems where collective effects (Arthur 1989; Haag 1989; Braun and Saam 2015) and effects influenced by social identities, as described in the next chapter, are essential.

6.4 Identity Economics

Instead of the economics-dominated utility function Akerlof (1997) and Akerlof and Kranton (2000, 2008) introduced social identities into standard economics by expanding the utility concept. The "new" utility function incorporates both: monetary incentives and the concepts of identity, norms and social categories. This makes it possible to include in decision-making the benefit which is obtained when actions of a person match with the norms and ideals, and the loss of incentives that occurs, if this is not the case. In so far, the social context becomes relevant. In other words, the individual benefit may depend on group processes and the behaviour of other decision makers.

The developed broader definition of identity benefit corresponds to an interactionistic understanding of identity in sociology and anthropology. The individualistic identity of an individual is replaced by an interactionistic identity. The

[1]Let us consider the famous illustration of this problem: We assume a commuter has two options, firstly to take his car or secondly to take a blue bus. Suppose, the choice probability (modal shift) of each is $p_{car} = p_{bluebus} = 0.5$. Now, a new traffic is opened and a red-bus is available, too. All characteristics of both bus lines are the same, beside the non-relevant colour of the two busses. Weak point (e) implies that $p_{car} = p_{bluebus} = p_{redbus} = 0.33$. In reality we would expect that the choice probabilities will likely be closer to $p_{car} = 0.5$, and for the two buses $p_{car} = p_{bluebus} = 0.25$.

introduced new components into the utility function may include externalities and group related effects. Norms and ideals of individuals may change over time; therefore it has to be assumed, that even the identity utility of individuals will become time-dependent.

Hence, the standard utility function v_j^k of an individual k is expanded in a social context by Akerlof and Kranton (2010) and Davis (2004) to include:

(a) Social categories and the assignment of individuals to these categories, i.e. their identity.
(b) Norms and ideals for each category.
(c) The benefits that will be gained if the actions coincide with the norms and ideals of an individual, or the related loss in case of a mismatch

Therefore, in mathematical terms, the standard utility function described in Sect. 6.1 is generalized as follows:

$$v_j^k = u_j^k\left(z_j^k, s^k, N_j^k, I_j^k\right) + \varepsilon_j^k \quad \text{for} \quad j = 1, 2, \ldots, L, \tag{6.12}$$

where z_j^k describes the characteristics, attributes of each alternative A_j as seen by the individual k and s^k his personal tastes, N_j^k the importance of norms and ideals of the different categories, and I_j^k the identity utility or impact of externalities, respectively. The stochastic error term ε_j^k reflects all effects related to unobserved attributes of the decision maker and the choice alternatives as well as the idiosyncratic tastes of k.

The resulting decisions are obtained as usual in economics by maximizing the individual utility functions.

6.5 The Dynamic Decision Model

In their book "Concepts and Models of a Quantitative Sociology", Weidlich and Haag (1983) have developed a theoretical framework for modelling socio-economic systems, using the Master equation formalism. Based on this, Haag (1989) has further developed the Master equation formalism and applied it to decision-making processes of interacting individuals in his book "Dynamic Decision Theory".

What does the Master equation framework afford with respect to the modelling of complex decision-making processes?

The Master equation facilitates the modelling of different members and decision maker of the social system including their interactions in a comprehensive and coherent framework. The human society can be regarded as a multi-component system, whose members, the individuals, adopt different attitudes or kinds of behaviour. The causes of changes in society are assumed to be correlated to the decisions of individuals to change their attitudes. In contrast to physico-chemical systems (see Chap. 5), however, the elementary units—the agents or decision maker—and the interactions within the human society are of rather complex nature (Weidlich 1972). A mixture of fluctuating rational considerations, professional activities and emotional preferences and motivations finally merge into one of relatively few well-demarcated resultant attitudes.

A variety of fluctuating rational, irrational motivations and insecurities and lack of relevant information or misinformation, as well as emotional preferences and motivations are incorporated into the decision-making of individuals (Arthur 1989; Haag and Wunderle 1988).

On the macro-level, in turn, the decision configuration describes the distribution of attitudes of the socio-economic system and may be considered an appropriate set of macro variables for the system under consideration in many applications. As with Akerlof, the identity of the individual (agent), as well as norms and social categories, are criteria for the assignment of individuals to subpopulations.

However, the micro level of details of the complex interplay of rational and emotional, conscious and subconscious, genetic and environmental influences on the decisions of individuals is unknown. Hence, a probabilistic description instead of a deterministic explication of the decision process seems to be adequate. Therefore, the probability that a certain decision configuration is realized will be introduced. The probability distribution over the decision configuration contains the most detailed information about the system. In particular, not only mean values but also higher order moments can be directly computed via the Master equation.

Thus, the Master equation provides a generative mechanism between the micro level characterized by individual decision processes and the macro level of only a few macro variables (order parameters). The generally nonlinear form of the equations of motion expresses the structure of self-consistence, prevalent in all socio-economic systems, namely the cyclic coupling between causes and effects in society.

Through their activities each individual contributes via its cultural, political, religious, social and economic influence to the "collective field" which determines the phase in which the society currently exists. On the other hand, the collective field in turn strongly influences the decision behaviour of individuals by orientating (manipulating) their activities.

The identity economics of Akerlof and Kranton can be seen as an essential part of this collective field (Fig. 6.1).

In mathematical terms, it implies, that mean field considerations are an integral part for the modelling of interactions between individuals via the collective field (Weidlich 1971; Weidlich and Haag 1983). This assumption is well justified in statistical theory, since the application of mean field theory works quite well for long ranging interactions (Haken 1978). Then the "mean field" can be calculated in a self-consistent way. Examples, where mean field considerations are frequently applied are interacting spin systems, the Laser and superconductivity in the natural sciences (see Chap. 5). Since information exchange between individuals is not restricted to next neighbours only, the application of the mean field approach is well justified. In this case, mean field theory leads to a reduction of the complexity

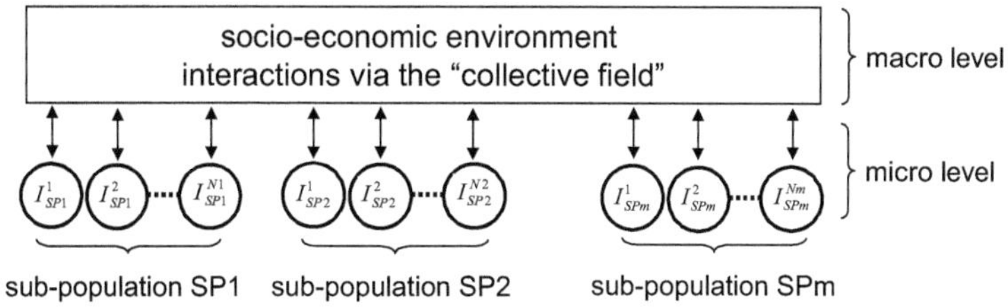

Fig. 6.1 Individuals grouped into sub-populations: each individual is part of the "collective field" of the other individuals

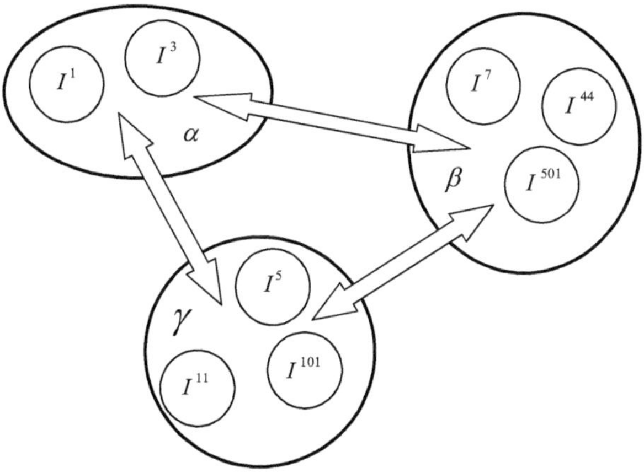

Fig. 6.2 Interactions among individuals belonging to different social groups. The effect of each social group may be modelled via a group specific collective field

of the model, and instead of a large number of small individual components, which interact with each other the effect of interaction is approximated by a single averaged effect, the "collective field".

Figure 6.2, shows exemplary interactions among individuals belonging to different clusters, groups or social categories. The impact of the belonging of an individual to a specific social group and the interaction of this particular agent with members of other social groups may be again modelled via introducing a group specific collective field.

Figure 6.3, shows an example of direct interaction of individuals, characterized as shown via arrows. The strength of each interaction is given by the thickness of the arrows. In this specific case the interaction between each pair of individuals has to be modelled separately.

In the following the "basic model" for the description of decision processes Haag (1989); based on the Master equation formalism is presented, which is even suitable for modelling of collective behavior (Haag and Müller 2015) and comprises effects related to identity economics.

In the stochastic framework, we consider a group of N individuals (agents, decision maker) instead of one individual agent. Therefore, the outcome of this choice model is the probability that a given alternative is chosen by an agent who is

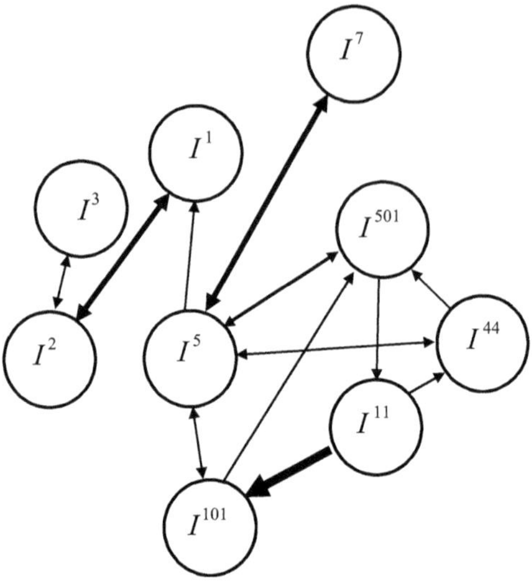

Fig. 6.3 Example of possible interactions among different individuals

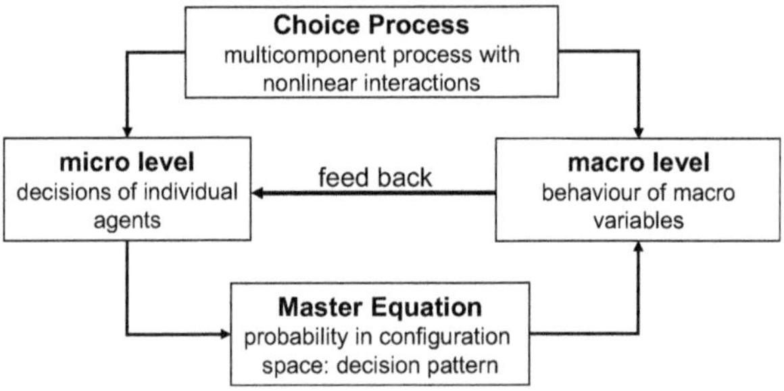

Fig. 6.4 The Master equation within the choice framework

part of the ensemble taking into account the interaction between the different individuals (Haag 1989). In the MNL (see Sect. 6.3), the frequency that a specific agent selects a certain alternative is given under the assumption that all interactions among the individuals can be neglected (diagonal covariance matrix) (Fig. 6.4).

Let us consider a group of N individuals. Each individual has to select one alternative A_j out of a set of $j = 1, 2, \ldots, L$ alternatives. We are interested in the dynamics of the decision configuration

$$\vec{n} = (n_1, n_2, \ldots, n_L), \tag{6.13}$$

where $n_j \geq 0$ is the number of agents who have decided for alternative A_j (see Fig. 6.6). Suppose, changes in the decision configuration are caused by differences in what we call "dynamic utilities" u_j or utilities in the sense of Akerlof and Kranton. Again, the u_j measures the desirability of alternative j. Now the Master equation strategy enters our statistic decision approach (Fig. 6.5).

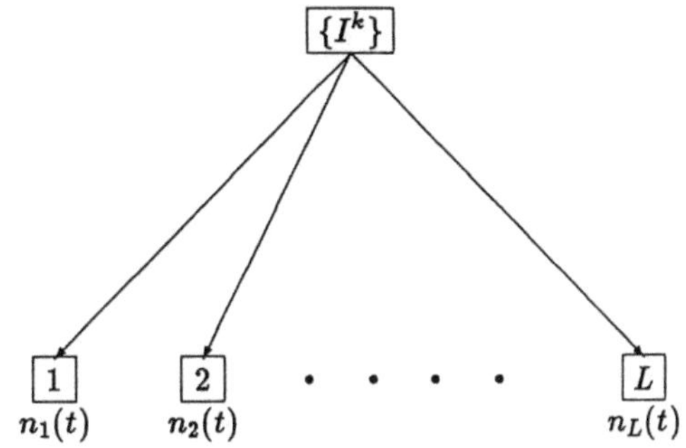

Fig. 6.5 The decision tree for the Master equation choice model for a single individual k

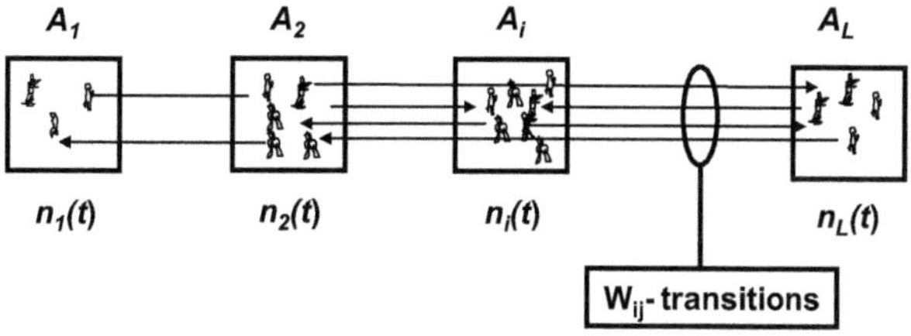

Fig 6.6 The dynamic choice process of agents: possible transitions between different alternatives. The number of changes between two alternatives can be measured in principle

The modelling of the transition rates $w_t\left(\vec{n}+\vec{k},\vec{n}\right)$ is the actual task. What are for example the "driving forces" (key attitudes) that cause an individual to choose another alternative, in other words, to change his decisions?

After the definition of the transition rates all we have to do is to solve the Master equation. In Fig. 6.6, it is illustrated how the total number of agents changing from alternative j to alternative i per time unit, the choice flow w_{ij} is composed by a variety of individual decisions.

In the following, we restrict ourselves to the important case that an agent changes his attitude in other words, selects another alternative. This kind of process is formally equivalent to a simple migratory process (see Chap. 7). We further assume, that the configurational transition rate

$$w_{ij}\left(\vec{n}+\vec{k},\vec{n}\right)=n_jp_{ij}(\vec{n},\vec{\kappa})\quad\text{for}\quad\vec{k}=\left(0,\ldots,+1_i,\ldots,-1_j,\ldots,0\right),\qquad(6.14)$$

is proportional to the number of agents having adopted state j of behaviour multiplied by an "individual" transition rate $p_{ij}(\vec{n},\vec{\kappa})$. This concept assumes a "representative" decision maker who does not exist in reality. However, in a first step, let us subdivide the total ensemble of decision makers into homogeneous sub-populations SPm (Fig. 6.1) with respect to their choice behaviour. The choice behaviour of different interacting sub-populations will be discussed later. The individual transition rate may depend on the decision configuration as well as on a set of trend parameters $\vec{\kappa}$ and explanatory variables. The $p_{ij}(\vec{n},\vec{\kappa})$ here are not individual transition rates in the sense, that different transition rates must be attributed to different agents. Instead we suppose that the different agents of the decision system belonging to the same ensemble operate according to the same probability transition rate depicting the probabilistic decision behaviour of each member of the ensemble. This presumes a homogeneity assumption concerning the choice behaviour of the agents.

If, however, extensive data are available on the decision behaviour of a selected group of agents (micro data), we are able to substitute the assumption (6.14) by direct computation of the configurational transition rates via

$$w_{ij}\left(\vec{n}+\vec{k},\vec{n}\right)=\sum_{l\in\Omega_j}p_{ij}^{(l)}(\vec{n},\vec{\kappa})\quad\text{for}\quad\vec{k}=\left(0,\ldots,+1_i,\ldots,-1_j,\ldots,0\right),\tag{6.15}$$

where we have to sum up over all individual contributions $p_{ij}^{(l)}(\vec{n},\vec{\kappa})$ of all agents $l\in\Omega_j$ changing from state j to state i of behaviour per time unit Δt.

Since (6.15) requires a huge data set of reliable micro data, we will skip this way and continue with the assumption (6.14). The dynamic utilities are related to the transition rates $p_{ij}(\vec{n})$ that an individual who originally preferred alternative j makes a transition to alternative i in the time interval Δt. In our dynamical context, the following analytical form of the transition rates has demonstrated its flexibility in many practical examples:

$$p_{ij}(\vec{n})=v_{ij}\exp\left(u_i(n_i+1)-u_j(n_j)\right)\geq 0,\tag{6.16}$$

where $v_{ij}=v_{ji}>0$ is a symmetric matrix describing the flexibility (willingness) for the members of the ensemble to choose another alternative within a given time interval Δt. In many applications, the flexibility matrix can be reduced to a single parameter v. In the flexibility matrix v_{ij} all effects should be included which will either facilitate or impede a transition from any state j to any state i, independent of any gain of utility.

The individual transition rate (6.16) is chosen to fulfil a set of minimal requirements:

(a) The individual transition rate $p_{ij}(\vec{n})$ is positive by definition. Therefore, $p_{ij}(\vec{n})$ must be a positive function of $u_j(\vec{n})$ and $u_i(\vec{n})$
(b) The individual transition rate $p_{ij}(\vec{n})$ from j to i must be larger than $p_{ji}(\vec{n})$ for the inverse transition $\left(p_{ij}(\vec{n})>p_{ji}(\vec{n})\right)$, if $\left(u_i(\vec{n})>u_j(\vec{n})\right)$
(c) The individual transition rate $p_{ij}(\vec{n})$ must be a monotonically increasing function of the difference $\left(u_i(\vec{n})-u_j(\vec{n})\right)$, since it is compelling that an increasing advantage $\left(u_i(\vec{n})>u_j(\vec{n})\right)$ induces a higher probability to decide for i instead of j

Why an exponential function?

(a) The transition rates $p_{ji}(\vec{n})\geq 0$ have to be positive and each positive function can be transformed and represented via an exponential function without loss of generality, e.g. $g(x)\geq 0$ can be transformed as $g(x)=\exp(\ln g(x))\geq 0$

(b) Weber-Fechner laws (Klinke et al. 2014) describe the human response to a physical stimulus in a quantitative fashion. Weber assumed that the difference between two stimuli is proportional to the magnitude of the stimuli. We employ Fechner's findings and mathematical formalization that subjective sensation is proportional to the logarithm of the stimulus intensity.[2] The inverse action employs an exponential dependence
(c) In the next Sect. 6.5.1 we will demonstrate that by using (6.16), the MNL can be obtained as the stationary solution to our dynamic decision model for non-interacting agents
(d) If the number of transitions w_{ij} between different alternatives per time interval Δt can be counted, the dynamic utilities $u_i(\vec{n})$ can be directly estimated (see Chap. 8).

Note, that the dynamic utilities (6.16) generally depend on the decision configuration (6.11) and therefore include interaction effects among individuals.

Since only differences of the dynamic utilities appear in the transition rates (6.16), the following scaling is convenient, without loss of generality

$$\sum_{i=1}^{L} u_i\left(\vec{n}\right) = 1. \tag{6.17}$$

One plausible and frequently chosen assumption for the dependence of the dynamic utilities on the decision configuration reads

$$u_i(\vec{n}) = \delta_i(t) + \sum_{j=1}^{L} \kappa_{ij} n_j + \sum_{j=1}^{L} \sum_{k=1}^{L} \sigma_{ijk} n_j n_k + O\left(n^3\right), \tag{6.18}$$

where the $u_i(\vec{n})$ are part of a truncated Taylor expansion, and $O(n^3)$ means that terms of order n^3 are omitted. The parameters of (6.18) have the following meaning:

(a) The trend parameter δ_i describes the preference of an agent for the specific alternative A_j, independent of the realized decision behaviour of the other decision makers.
(b) The trend parameter $\kappa_{ii} > 0$ describes agglomeration or synergy effects (self-reinforcing mechanisms)
(c) The trend parameters κ_{ij} for $i \neq j$ represent intra-group interactions
(d) The trend parameters $\sigma_{ijk} = \sigma_{ikj}$ may describe nonlinear effects of higher order such as saturation effects, e.g. for $\sigma_{iii} < 0$

Of course, other modelling assumptions, such as

[2]Empirical research of human perception of loudness and sight justified that the perceived loudness and perceived brightness are proportional to the logarithm of the actual intensity. This is i.e. incorporated in the definition of the scale of loudness, measured in phone, and the physical measure of sound intensity, measured in db (Sacklowski and Röhrl 1973).

$$u_i(\vec{n}) = \delta_i(t) + \sum_{j=1}^{L} \kappa_{ij} \ln n_j \tag{6.19}$$

may also be reasonable, e.g. a convex form, describing decreasing marginal utilities.

Each of the n_j individuals may change from alternative j to alternative i with a given "individual" transition rate, which gives rise to a change of the decision configuration at the macro-level

$$\begin{aligned} \vec{n} &= \left(n_1, \ldots, n_i, \ldots, n_j, \ldots, n_L\right) \rightarrow \vec{n}^{(ij)} \\ &= \left(n_1, \ldots, n_i + 1, \ldots, n_j - 1, \ldots, n_L\right). \end{aligned} \tag{6.20}$$

Hence, the n_j agents contribute to the term

$$w_{ij}\left(\vec{n}^{(ji)}\right) = n_j p_{ij}(\vec{n}) = v_{ij} n_j \exp\left(u_i(n_i + 1) - u_j\left(n_j\right)\right) \tag{6.21}$$

of the configurational transition rate $w_t\left(\vec{n} + \vec{k}, \vec{n}\right)$. Since transitions between all alternatives can take place the total configurational transition rate is the sum over all contributions:

$$\begin{aligned} w_t\left(\vec{n} + \vec{k}, \vec{n}\right) &= \sum_{i,j=1}^{L} w_{ij}\left(\vec{n} + \vec{k}, \vec{n}\right) = \sum_{i,j=1}^{L} n_j p_{ij}(\vec{n}) \quad \text{for} \quad \vec{k} \\ &= \left(0, \ldots, +1_i, \ldots, -1_j, \ldots, 0\right). \end{aligned} \tag{6.22}$$

On the stochastic level, fully consistent with the probabilistic description of individual decision processes, we consider the Master equation for the probability to find a certain decision configuration $P(\vec{n}, t) \geq 0$ realized at time t. The Master equation for this decision process reads

$$\frac{dP(\vec{n}, t)}{dt} = \sum_{i,j=1}^{L} w_{ji}\left(\vec{n}^{(ij)}\right) P\left(\vec{n}^{(ij)}, t\right) - \sum_{i,j=1}^{L} w_{ij}(\vec{n}) P(n, t). \tag{6.23}$$

Since the Master equation (6.23) does not contain any birth- or death events the number of agents (decision maker) does not change, agents can only change from one alternative to another. The Master equation of a simple migratory process has exactly the same structure. In this case the choice decision concerns e.g. the place of living and $P(\vec{n}, t)$ describes the configurational probability to find a certain population distribution within the spatial system.

The time-dependent solution of the Master equation (6.23) contains all the information about the choice processes of a homogeneous group of interacting agents (Chap. 8).

6.5.1 Exact Stationary Solution

The stationary solution of the Master equation corresponds to the choice problem at equilibrium. The construction is facilitated by the fact that the transition rates satisfy the condition of detailed balance

$$w_{ji}\left(\vec{n}^{(ij)}\right)P_{st}\left(\vec{n}^{(ij)}\right) = w_{ij}(\vec{n})P_{st}(n) \quad \text{for all} \quad i,j = 1, \ldots, L, \tag{6.24}$$

which means that the stationary probability flux from $\vec{n}$ to $\vec{n}^{(ij)}$ is equal to the inverse flux from $\vec{n}^{(ij)}$ to $\vec{n}$.

Proof of Detailed Balance

Following the argumentation in Sect. 4.3 we choose a set of smallest closed chains of states. They are sufficient for the proof since arbitrary closed chains can be composed of these smallest ones. A smallest chain Γ connects the following states (see Fig. 6.7):

We choose the following closed loop Γ: $\vec{0} \to \vec{1} \to \vec{2} \to \vec{0}$.

The loop corresponds to members of the population of agents changing their opinion in a cyclic way between $i \to j \to k \to i$, where

$$\begin{aligned} \vec{0} &= \left(n_1, \ldots, n_i, \ldots, n_j, \ldots, n_k, \ldots, n_L\right) \\ \vec{1} &= \left(n_1, \ldots, n_i - 1, \ldots, n_j + 1, \ldots, n_k, \ldots, n_L\right) \\ \vec{2} &= \left(n_1, \ldots, n_i - 1, \ldots, n_j, \ldots, n_k + 1, \ldots, n_L\right) \end{aligned} \tag{6.25}$$

and with the corresponding transitions, which must hold in case of detailed balance

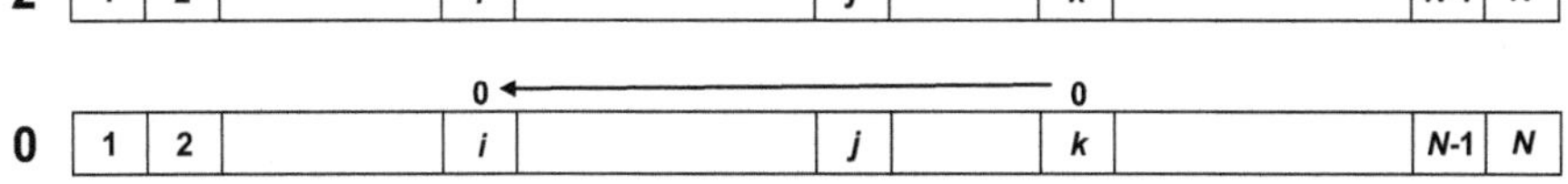

Fig. 6.7 Smallest closed chain in the configuration space (decision space) for the Master Eq. (6.23)

$$
\begin{aligned}
w\left(\vec{1}, \vec{0}\right)P_{st}\left(\vec{0}\right) &= w\left(\vec{0}, \vec{1}\right)P_{st}\left(\vec{1}\right) \\
w\left(\vec{2}, \vec{1}\right)P_{st}\left(\vec{1}\right) &= w\left(\vec{1}, \vec{2}\right)P_{st}\left(\vec{2}\right) \\
w\left(\vec{0}, \vec{2}\right)P_{st}\left(\vec{2}\right) &= w\left(\vec{2}, \vec{0}\right)P_{st}\left(\vec{0}\right)
\end{aligned}
\tag{6.26}
$$

By elimination of the configurational transition rate $P(\vec{n})$, we obtain the condition for the transition rates which has to be fulfilled

$$
\frac{w\left(\vec{0}, \vec{2}\right)}{w\left(\vec{2}, \vec{0}\right)} \cdot \frac{w\left(\vec{2}, \vec{1}\right)}{w\left(\vec{1}, \vec{2}\right)} \cdot \frac{w\left(\vec{1}, \vec{0}\right)}{w\left(\vec{0}, \vec{1}\right)} = 1. \tag{6.27}
$$

Only one term of the right-hand side of (6.28) contributes to each of the transitions in (6.27)

$$
p_{ij}(\vec{n}) = v_{ij}\exp\left(u_i(n_i + 1) - u_j\left(n_j\right)\right) \geq 0. \tag{6.28}
$$

Hence, (6.27) reduces to

$$
\begin{aligned}
\frac{w\left(\vec{0}, \vec{2}\right)}{w\left(\vec{2}, \vec{0}\right)} \cdot \frac{w\left(\vec{2}, \vec{1}\right)}{w\left(\vec{1}, \vec{2}\right)} \cdot \frac{w\left(\vec{1}, \vec{0}\right)}{w\left(\vec{0}, \vec{1}\right)} \Rightarrow & \frac{w_{ik}(n_i - 1, n_k + 1)}{w_{ki}(n_i, n_k)} \frac{w_{kj}\left(n_k, n_j + 1\right)}{w_{jk}\left(n_j, n_k + 1\right)} \\
& \times \frac{w_{ji}(n_i, n_i)}{w_{ij}\left(n_i - 1, n_j + 1\right)} = 1,
\end{aligned}
\tag{6.29}
$$

which can be easily checked. This completes the proof of detailed balance.

Construction of the Exact Stationary Solution

The most important consequence of detailed balance is that the exact stationary solution of the Master equation can be constructed using (6.24). For this aim, we have to choose an arbitrary chain from an initial state via connected neighbouring states in the configuration space to the final state. We shall consider the following chain of states:

$$
\begin{aligned}
&\{N, 0, \ldots, 0\} \to \{N-1, 1, 0, \ldots, 0\} \to \{N-2, 2, 0, \ldots, 0\} \to \cdots \{N-n_2, n_2, 0, \ldots, 0\} \\
&\to \{N-n_2-1, n_2, 1, 0, \ldots, 0\} \to \cdots \{N-n_2-n_3, n_2, n_3, 0, \ldots, 0\} \to \cdots \{n_1, n_2, n_3, \ldots, n_L\},
\end{aligned}
\tag{6.30}
$$

where we have taken into account

$$\sum_{j=1}^{L} n_j = N. \tag{6.31}$$

We start from the chosen reference state $\{N, 0, \ldots, 0\}$ and end up with the general state $\{n_1, n_2, \ldots, n_L\}$. The different states along the chosen chain are connected by non-vanishing transition rates, as described in Sect. 4.3. Hence, we can use this chain to construct the exact configurational probability $P_{st}(n_1, n_2, \ldots, n_L)$ starting from $P_{st}(N, 0, \ldots, 0)$ according to (4.61) by inserting the transition rates (6.16). This leads, for example, to the intermediate result:

$$\begin{aligned} P_{st}(N - n_2, n_2, 0, \ldots, 0) = {} & \frac{N(N-1)\cdots(N-n_2+1)}{n_2!} \\ & \cdot \exp\left\{ 2\sum_{m=1}^{n_2} u_2(m) - 2\sum_{m=N-n_2+1}^{N} u_2(m) \right\} \\ & \cdot P_{st}(N, 0, \ldots, 0) \end{aligned} \tag{6.32}$$

Continuing this procedure along the chosen chain (6.30), we finally obtain the result

$$P_{st}(\vec{n}) = \frac{Z^{-1}\delta\left(\sum_{i=1}^{L} n_i - N\right)}{n_1! n_2! \cdots n_L!} \exp\left[2\sum_{i=1}^{L} F_i(n_i)\right], \tag{6.33}$$

where

$$F_i(n_i) = \sum_{m=1}^{n_i} u_i(m) \quad \text{and} \quad F_i(0) = 0 \tag{6.34}$$

and

$$\delta\left(\sum_{i=1}^{L} n_i - N\right) = \begin{cases} 1 & \text{if } \sum_{i=1}^{L} n_i = N \\ 0 & \text{otherwise} \end{cases}, \tag{6.35}$$

and where the factor Z^{-1} is determined by the normalization condition for the configurational probabilities. It is not a Poisson distribution since the agents of the considered ensemble of decision makers behave not completely independent of each other. We will come back to this point later. The time-dependent solution of the Master equation (6.23) will approach the exact stationary solution (6.33) for $t \to \infty$, keeping everything else constant.

Let us now derive an approximate solution for the most probable decision configuration $\hat{n} = (\hat{n}_1, \hat{n}_2, \ldots, \hat{n}_L)$. It is plausible to assume that the realized distribution, namely the numbers of agents who have decided for alternative $n_1, \ldots, n_j, \ldots, n_L$ corresponds to the most probable distribution.

Using Sterling's formula for factorials for $n >> 1$

$$\ln n! \approx n(\ln n - 1), \tag{6.36}$$

the stationary distribution (6.33) can be written in the form

$$P_{st}(\vec{n}) = Z^{-1}\delta\left(\sum_{i=1}^{L} n_i - N\right)\exp\left[2\sum_{i=1}^{L}\Phi_i(n_i)\right] \tag{6.37}$$

with

$$\Phi_i(n_i) = 2F_i(n_i) - n_i(\ln n_i - 1). \tag{6.38}$$

The maxima and minima $\hat{n} = (\hat{n}_1, \hat{n}_2, \ldots, \hat{n}_L)$ of $P_{st}(\vec{n})$ can now be calculated under the constraint $\sum_{i=1}^{L} n_i = N$ by minimization of

$$\delta\left\{\sum_{i=1}^{L}\Phi_i(n_i) - \lambda\left(\sum_{i=1}^{L} n_i - N\right)\right\} = \sum_{i=1}^{L}\delta n_i\left[2\frac{\partial F_i(n_i)}{\partial n_i} - \ln n_i - \lambda\right] = 0. \tag{6.39}$$

The Lagrangian parameter λ takes into account the constraint (6.35). Since for $n >> 1$ the variable n is a quasi-continuous variable, the sum in (6.39) can be replaced by an integral which amounts to

$$\frac{\partial F_i(n_i)}{\partial n_i} \approx u_i(n_i). \tag{6.40}$$

Inserting (6.40) into (6.39) and solving for $\hat{n}$ yields

$$\hat{n}_j = \frac{N\exp(2u_j(\hat{n}))}{\sum_{i=1}^{L}\exp(2u_i(\hat{n}))} \tag{6.41}$$

for $j = 1, 2, \ldots, L$ and where $\hat{n}_j$ is the most probable number of agents of the ensemble of decision makers who have chosen alternative A_j.

6.5.2 The MNL-Model as a Limiting Case of the Weidlich-Haag Decision Model

The frequency that any individual of the ensemble of decision makers selects alternative A_j is then given by

$$\hat{p}_j = \frac{\exp(2u_j(\hat{p}))}{\sum_{i=1}^{L} \exp(2u_i(\hat{p}))}. \tag{6.42}$$

Comparing the stationary solution (6.42) with the outcome of the multinomial logit model (6.9), one has to identify

$$u_j(\hat{n}) \Leftrightarrow \frac{1}{2} v_j^k. \tag{6.43}$$

Thus both utilities coincide up to an ordinary scaling factor 2. Moreover, in MNL is assumed that the decision maker chooses the option which is considered most desirable, given the attributes z_j^k of each alternative A_j as seen by the individual k and his personal tastes s^k. Contrary to that, the dynamic utilities $u_j(\hat{n})$ are related to the decision behaviour of a homogeneous group of agents with respect to their decision behaviour. However, the whole group of agents is interacting since the decision distribution has an influence on the decision process.

Of course, the homogeneity assumption does not often apply well. In this case a further subdivision of the total population of decision makers into a set of $\alpha = 1, 2, \ldots, A$ sub-populations ($SP_1, \ldots, SP_A$) seems to be appropriate. If, however, each member of the group of decision makers is treated as its own sub-population, the dynamic utilities and the static utility concept can be directly compared. In this specific case, the Master equation and all outcomes are still valid.

Despite the formal similarity of the multinomial logit model, (MNL) (6.9) and the most probable decision configuration according to the dynamic decision model of Weidlich and Haag, (6.42), fundamental differences are observable:

(a) In the Master equation model, interactions between decision makers are taken into account. This means, that (6.42) may exhibit more than one stationary solution and it depends on the initial distribution (initial conditions) which of the several critical points will be approached whether the system exhibits a limit cycle or even chaotic behaviour.
(b) The dynamic utilities are directly related to the transition rates and in so far to the dynamics of the system and not to a static concept. By measuring the number of transitions between different alternatives the dynamic utilities can be measured too (see Sect. 8.7).
(c) The dynamic utilities take into account possible interactions among the considered group of decision makers.

(d) The most probable decision configuration (6.42) represents the solution of the dynamic Master equation choice model in the limiting case $t \to \infty$.
(e) The dynamic utilities (6.16) coincide with the static utilities (6.1) for statistically independent individuals up to an ordinary scaling factor of 2. This justifies the interpretation as dynamic utilities.
(f) For the derivation of the MNL model it is assumed that the error terms are independently and identically Gumbel distributed. The stationary solution of the Master equation model leads to the MNL as a limiting case for non-interacting agents, without any restriction concerning the distribution of uncertainties.
(g) No memory effects are included (Markov assumption)

If the set of transcendental equations (6.41) has only one solution, the corresponding distribution in decision space $P_{st}(\vec{n})$ is unimodal and the equilibrium state is stable. However, if more than one solution exists, the distribution may be multimodal with peaks corresponding to the different possible equilibrium states. It depends on the values of the trend parameters in the dynamic utility function which of these equilibrium states will be realised in the course of time.

Let us now discuss the relation between the stationary solution $P_{st}(\vec{n})$ and the time-dependent solution of the Master equation $P(\vec{n}, t)$. By definition, the stationary solution is the time-independent solution of the Master equation for constant trend parameters. Furthermore, we have already shown (see Chap. 3) that any time-dependent solution approaches the stationary solution if time goes on. However, in general the agents of the system will not be in their equilibrium states. Two different reasons should be distinguished:

1. The trend parameters of the dynamic utilities are time independent, but the distribution $P(\vec{n}, t)$ has not yet reached its equilibrium state. Then the decision configuration is still changing until $P_{st}(\vec{n})$ finally is approached. The corresponding time scale is given by the flexibility matrix $\nu_{ij} = \nu_{ji} > 0$ describing the flexibility (willingness) for the members of the ensemble of agents to choose another alternative within a given time interval Δt
2. In the second case, the trend parameters themselves are slowly time dependent and the transition rates $w_{ij}(\vec{n})$ become time dependent, too, via their dependence on trend parameters. The Master equation is still valid in this more general case, but the probability distribution lacks behind its "stationary" state, related to the current trend parameters.

6.6 Quasi Deterministic Equations of Motion of the Master Equation Choice Model

The process of approaching the stationary solution of the decision distribution will now be considered in detail. The mean value equations are often already sufficient for a comparison of model development with empirical data (Fischer and Nijkamp 1987). We skip the derivation of the approximate mean value equations, since the procedure is straightforward. In analogy to Chap. 3 we obtain:

$$\begin{aligned}\frac{d\langle n_j\rangle}{dt} &= \sum\nolimits_{i=1}^{L}\langle n_i p_{ji}(\vec{n})\rangle - \sum\nolimits_{i=1}^{L}\langle n_j p_{ij}(\vec{n})\rangle\\ &= \sum\nolimits_{i=1}^{L}\langle n_i \nu_{ji}\exp\big(u_j(\vec{n}) - u_i(\vec{n})\big)\rangle - \sum\nolimits_{i=1}^{L}\langle n_j \nu_{ij}\exp\big(u_i(\vec{n}) - u_j(\vec{n})\big)\rangle.\end{aligned} \tag{6.44}$$

This set of in general nonlinear quasi closed mean value equations is further simplified by assuming that the probability distribution is a well-behaved and sharply peaked unimodal distribution. In this special case, it can be assumed that the mean value of a function is comparable with the function of its mean value:

$$\langle f(\vec{n})\rangle_t \approx f(\langle\vec{n}\rangle_t). \tag{6.45}$$

If this assumption is justifiable, quasi-closed mean value and variance equations, as well as approximate equations for higher order moments can be easily derived:

$$\begin{aligned}\frac{d\langle n_j\rangle}{dt} &= \sum\nolimits_{i=1}^{L}\langle n_i\rangle p_{ji}(\langle\vec{n}\rangle) - \sum\nolimits_{i=1}^{L}\langle n_j\rangle p_{ij}(\langle\vec{n}\rangle)\\ &- \sum\nolimits_{i=1}^{L}\nu_{ji}\langle n_i\rangle \mathrm{cxp}\big(u_j(\langle\vec{n}\rangle) - u_i(\langle\vec{n}\rangle)\big) - \sum\nolimits_{i=1}^{L}\nu_{ij}\langle n_j\rangle \exp\big(u_i(\langle\vec{n}\rangle) - u_j(\langle\vec{n}\rangle)\big).\end{aligned} \tag{6.46}$$

If, however, the probability distribution does not have a well-behaved structure, e.g. in case of bistable systems or if small numbers become important the full complexity of the Master equation has to be taken into account.

Clearly, (6.46) represents a set of $L-1$ coupled nonlinear first order differential equations for the most probable decision behaviour of a homogeneous ensemble of agents. It is convenient to proceed to scaled variables (frequencies):

$$x_j(t) = \frac{\langle n_j\rangle}{N} \quad \text{with} \quad \sum_{j=1}^{L} x_j(t) = 1. \tag{6.47}$$

Then we obtain for the dynamic equations for the most probable decision configurations

$$
\begin{aligned}
\frac{dx_j(t)}{dt} &= \sum_{i=1}^{L} p_{ji}(t)x_i(t) - \sum_{i=1}^{L} p_{ij}(t)x_j(t) \\
&= \sum_{i=1}^{L} v_{ji}x_i(t)\exp\big(u_j(\vec{x}(t)) - u_i(\vec{x}(t))\big) - \sum_{i=1}^{L} v_{ji}x_j(t)\exp\big(u_i(\vec{x}(t)) - u_j(\vec{x}(t))\big)
\end{aligned}
\tag{6.48}
$$

for $j = 1, 2, \ldots, L$. The stationary solution of the dynamic decision system can be easily verified by insertion of (6.49) in (6.48):

$$
x_j^{st} = \frac{\exp\big(2u_j(\vec{x}^{st})\big)}{\sum_{i=1}^{L} \exp(2u_i(\vec{x}^{st}))}, \tag{6.49}
$$

where

$$
\lim_{t\to\infty} x_j(t) = x_j^{st}. \tag{6.50}
$$

Comparing (6.49) with (6.42) shows that both, the stationary state of the approximate mean value equation under the assumption of a unimodal, sharply-peaked distribution and the stationary solution of the dynamic equation (6.48) coincide. This confirms the strong relationship between the fully stochastic and the mean value approach.

However, there are still some choice problems that cannot be treated in a satisfactory manner by one of the choice models mentioned in this chapter. Especially, when nested choice processes are considered and the following weaknesses are concerned:

(a) No learning effects or memory effects are considered
(b) The number of alternatives is fixed
(c) The red-bus blue-bus problem

Next let us tackle these points.

6.7 A Master Equation Model of Nested Decision Processes with Memory

The consideration of the history of the choice process apparently violates the Markov assumption. However, the considerable difficulty for the modelling task can be circumvented by introducing a stepwise procedure (Haag and Grützmann 1993).

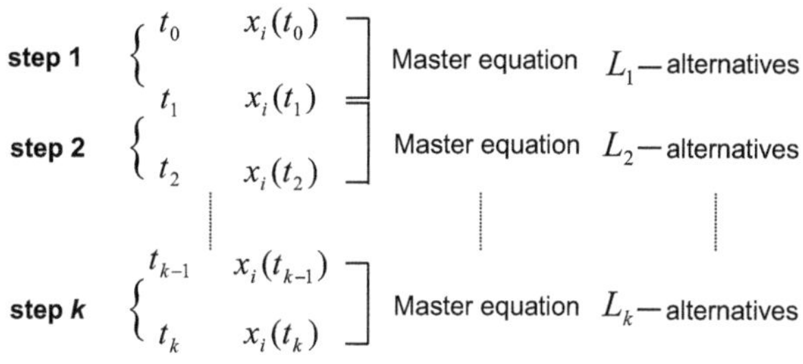

Fig. 6.8 Coupling of different choice sequences

This means that a choice process without previous experience is assumed for the first decision step (first time interval). Therefore, the dynamic Master equation choice model of Sect. 6.5.3 can be used as a basis for constructing the dynamic choice model with memory effects. At the end of the first decision step a specific frequency distribution over the set of alternatives is obtained.

In the second time interval (step 2), the choice behaviour of individuals is influenced by the experience of the different groups of individuals who have decided for various alternatives in the previous decision step. This procedure is repeated, and leads to a choice frequency distribution with memory effects, even in the case where each single decision branch is treated as Markovian (Fig. 6.8). The core point is that the Master equation as well as the mean value equations maintain their validity for each single decision step (decision interval), despite the fact that the whole decision process does not satisfy the Markov assumption.

Let us assume a choice system consisting of $k = 1, 2, \ldots, K$ repeated choice sequences (decision intervals). Moreover, each choice sequence k that pertains to a specific time period $t_{k-1} < t \leq t_k$ may be characterized by a different number of alternatives L_k. Therefore, the inclusion of a new alternative in the choice set or the withdrawal of an old one can be investigated. In order to make this notion operational, $x_i(t_k)$ is introduced for the mean frequency of a decision in favour of alternative i at time t_k and $y_i^{(m)}(t_k)$ for the mean frequency of a decision in favour of alternative i at time t_k if alternative m has been chosen at time t_{k-1}.

Obviously, the frequency $x_i(t_k)$ by which alternative i is selected at time t_k can be computed as weighted sum over all contributions $y_i^{(m)}(t_k)$ to choose alternative i at time t_k given that alternative m has been chosen at time t_{k-1}, multiplied with the frequency by which alternative m has been chosen at time t_{k-1}

$$x_i(t_k) = \sum_{m=1}^{L_{k-1}} y_i^{(m)}(t_k) x_m(t_{k-1}) \quad \text{for} \quad i = 1, 2, \ldots, L_k; \;\; k = 2, \ldots, K. \tag{6.51}$$

Of course, (6.51) provides a coupling of different choice sequences. Since $x_i(t_k)$ and $y_i^{(m)}(t_k)$ are frequencies, the following normalizations are required:

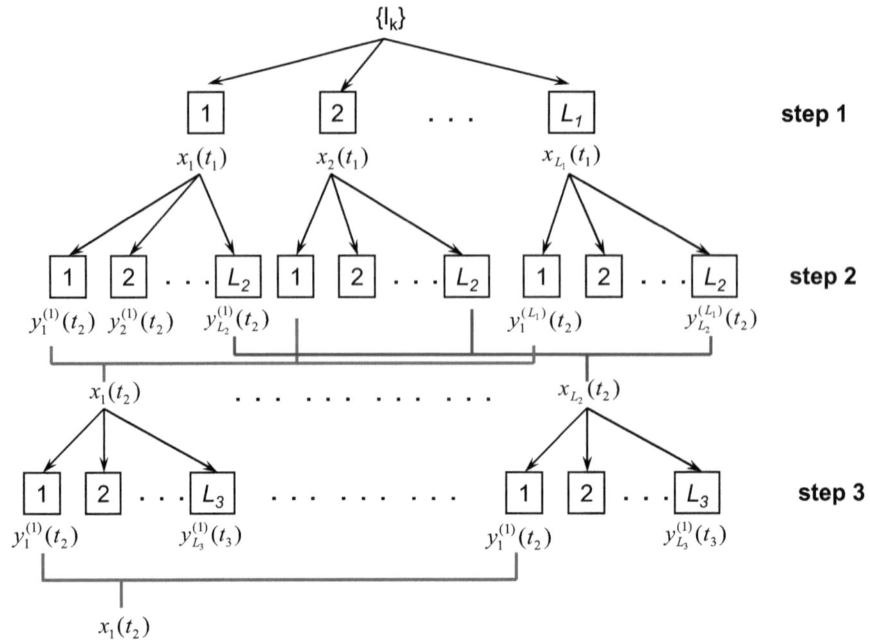

Fig. 6.9 Decision tree of the nested Master equation choice model with memory effects

$$\sum_{i=1}^{L_k} x_i(t_k) = 1 \quad \text{for} \quad k = 1, 2, \ldots, K \tag{6.52}$$

and

$$\sum_{i=1}^{L_k} y_i^{(m)}(t_k) = 1 \quad \text{for} \quad m = 1, 2, \ldots, L_{k-1}; \; k = 2, \ldots, K \tag{6.53}$$

It can be easily proven that the normalization in (6.52) is automatically fulfilled for all $t_k > t_1$

$$\sum_{i=1}^{L_k} x_i(t_k) = \sum_{i=1}^{L_k} \sum_{m=1}^{L_{k-1}} y_i^{(m)}(t_k) x_m(t_{k-1}) = \sum_{m=1}^{L_{k-1}} x_m(t_{k-1}) \sum_{i=1}^{L_k} y_i^{(m)}(t_k) = 1 \tag{6.54}$$

for $k = 2, \ldots, K$.

In Fig. 6.9, the decision tree of the nested choice model with memory effects is presented. This Fig. 6.9 shows how the choice frequencies depend on the history of previous decisions.

Dynamic Equations for Choice Frequencies

The approximate equations of motion for the different choice frequencies $x_i(t_k)$ and $y_i^{(m)}(t_k)$ at time t related to the choice sequence $t_{k-1} < t \leq t_k$ can be computed according (6.48). In other words, it is assumed that each choice sequence can be characterized by a set of stochastic decision processes modelled via the Master equation framework. The resulting approximate mean value equations describe the dynamics of the corresponding choice frequencies.

For the first choice sequence ($k = 1;\ \ i = 1, 2, \ldots, L_1;\ \ t_0 \leq t \leq t_1$), the equations of motion

$$\frac{dx_i(t)}{dt} = \sum_{j=1}^{L} p_{ij}(t)x_j(t) - \sum_{j=1}^{L} p_{ji}(t)x_i(t) \tag{6.55}$$

corresponds to (6.48). This means that memory effects are not taken into account for the first choice sequence. Starting from an initial choice frequency distribution $x_i(t_0)$, $i = 1, 2, \ldots, L_1$ the integration of (6.55) provides the choice frequencies $x_i(t_1)$, $i = 1, 2, \ldots, L_1$ at the end of the sequence t_1.

For the second choice sequence ($k = 2;\ \ i = 1, 2, \ldots, L_2;\ \ m = 1, 2, \ldots, L_1;$ $t_1 < t \leq t_2$), the equations of motion for the frequencies $y_i^{(m)}(t_2)$ read

$$\frac{dy_i^{(m)}(t)}{dt} = \sum_{j=1}^{L_2} p_{ij}^{(m)}(t)y_j^{(m)}(t) - \sum_{j=1}^{L_2} p_{ji}^{(m)}(t)y_i^{(m)}(t). \tag{6.56}$$

The decision frequencies $x_m(t_1)$ for each alternative $m = 1, 2, \ldots, L_1$ at time t_1 provide the initial conditions for the dynamics of the choice frequencies $y_i^{(m)}(t_2)$. Correspondingly, the k-th choice sequence ($k;\ \ i = 1, 2, \ldots, L_k;\ \ m = 1, 2, \ldots,$ $L_{k-1};\ \ t_{k-1} < t \leq t_k$) is described by the set of dynamic equations

$$\frac{dy_i^{(m)}(t)}{dt} = \sum_{j=1}^{L_k} p_{ij}^{(m)}(t)y_j^{(m)}(t) - \sum_{j=1}^{L_k} p_{ji}^{(m)}(t)y_i^{(m)}(t). \tag{6.57}$$

For a given initial choice distribution $x_i(t_0)$ the choice frequencies $x_i(t_1)$ can be computed. This choice distribution at t_1 in turn provides the initial conditions for the second interval $t_1 < t \leq t_2$ and so on. The application of (6.51) and the solution of (6.55) to (6.57) for the frequencies are further used to calculate the frequencies of the alternatives $x_i(t_k)$ at later times $t_k, k = 2, 3, \ldots, K$. However, to solve the dynamic equations one has to specify thc inter- and intra-level interaction.

Inter- and Intra-Level-Interaction

The transition rates $p_{ij}^{(m)}(t)$ for the different choice sequences are now introduced in analogy to (6.21)

$$p_{ij}^{(m)}(t) = v_{ij}(t)\exp\left(u_i^{(m)}(t) - u_j^{(m)}(t)\right) \quad \text{for} \quad i,j = 1,\ldots,L_k; \ m = 1,\ldots,L_{k-1}, \tag{6.58}$$

where $v_{ij}=v_{ji}$ describes the symmetric flexibility matrix, and $u_i^{(m)}(t)$ the corresponding dynamic utility function.

Of course, as already mentioned, there are different reasonable assumptions how to describe the utilities in mathematical terms. Which of these assumptions are most appropriate depends on the particular system under consideration. The utilities introduced $u_i^{(m)}(t)$ take into account the interaction between different choice sequences. In other words, they incorporate the history of previous decisions of the agents/decision maker. Especially synergy effects as well as the favoured preference $\delta_i^{(m)}(t)$ for certain alternatives, e.g. due to a preference of the agents for certain characteristics of the alternatives have to be considered.

One reasonable assumption reads

$$u_i^{(m)}(t) = \delta_i^{(m)}(t) + \kappa^{(m)}(t)x_i(t_{k-1}) \quad \text{for} \quad t_k < t \leq t_{k-1}; \ k = 1,2,\ldots,K, \tag{6.59}$$

where the coupling with the previous choice sequence $k-1$ of the utility function is represented by the term $\kappa^{(m)}x_i(t_{k-1})$, is in this special case. For a positive synergy parameter $\kappa^{(m)} > 0$ the frequency of the alternative i at the previous time interval has a positive influence on the attractiveness of the alternative at later times.

A more comprehensive description of decision processes can be simulated via

$$u_i^{(m)}(t) = \delta_i^{(m)}(t) + \sum_{j=1}^{K-1}\sum_{\alpha}\kappa_{j,\alpha}^{(m)}(t)x_i\left(t_{k-j}\right)^{\alpha} + \sum_{n=1}^{L_k}\sum_{\alpha}\kappa_{n,\alpha}^{(m)}(t)y_n^{(m)}(t_k)^{\alpha}, \tag{6.60}$$

where the coupling with the previous choice sequences, the inter-level interactions, are represented by the terms $\kappa_{j,\alpha}^{(m)}(t)x_i\left(t_{k-j}\right)^{\alpha}$. These terms describe the influence of memory effects related to the different choice sequences. As the memory effects extend over more choice sequences, it is reasonable to assume a decreasing influence of the memory term with time on current choices. The intra-level interactions are described by the terms $\kappa_{n,\alpha}^{(m)}(t)y_n^{(m)}(t_k)^{\alpha}$.

If it seems to be appropriate to assume a logarithmic dependency of the utilities on the choice frequencies $x_i(t_{k-j})$ and $y_i^{(m)}(t_k)$, one has to replace the frequencies by their respective logarithms.

Stationary Solution of the Nested Choice Model

According to (6.55) in connection with (6.58), the stationary solution for the choice sequence $k = 1$ reads

$$\hat{x}_i(t_1) = N_1 \exp(2u_i(t_1)) \quad \text{for} \quad i = 1, 2, \ldots, L_1. \tag{6.61}$$

Moreover, the frequencies for the other choice sequences $k = 2, 3, \ldots, K$ follow as

$$\hat{y}_i^{(m)}(t_k) = M_k^{(m)} \exp\left(2u_i^{(m)}(t_k)\right), \tag{6.62}$$

where the factors N_1 and $M_k^{(m)}$ are determined by the normalization conditions (6.52) and (6.53), leading to

$$N_1 = 1 / \sum_{j=1}^{L_1} \exp(2u_j(t_1)) \tag{6.63}$$

and

$$M_k^{(m)} = 1 / \sum_{j=1}^{L_k} \exp\left(2u_j^{(m)}(t_k)\right) \quad \text{for} \quad k = 2, \ldots, K; \; m = 1, 2, \ldots, L_k. \tag{6.64}$$

Using (6.51), (6.61), and (6.62), the stationary choice frequencies for alternative i at time t_k can be computed in an iterative procedure. For example, the second choice sequence yields

$$\begin{aligned} \hat{x}_i(t_2) &= \sum_{m=1}^{L_1} y_i^{(m)}(t_2) \hat{x}_m(t_1) \\ &= \sum_{m=1}^{L_1} \frac{\exp\left(2u_i^{(m)}(t_2)\right)}{\sum_{j=1}^{L_2} \exp\left(2u_j^{(m)}(t_2)\right)} \hat{x}_m(t_1) \\ &= N_1 \sum_{m=1}^{L_1} \frac{\exp\left(2u_i^{(m)}(t_2)\right)}{\sum_{j=1}^{L_2} \exp\left(2u_j^{(m)}(t_2)\right)} \cdot \exp(2u_m(t_1)) \end{aligned} \quad . \tag{6.65}$$

Therefore, the ratio x_i/x_j depends not only on the utilities u_i and u_j but also on all other utilities, in contrast to the MNL and the Master equation choice model without memory effects. The famous red-bus blue-bus problem does not strike this choice model with memory effects.

For a general choice sequence k we obtain by repeated application of (6.51) with (6.61) and (6.62)

$$\hat{x}_i(t_k) = N_{k-1} \sum_{m=1}^{L_k} \frac{\exp\left(2u_i^{(m)}(t_k)\right)}{\sum_{j=1}^{L_k} \exp\left(2u_j^{(m)}(t_k)\right)} \cdot \exp(2u_m(t_{k-1})), \tag{6.66}$$

where $u_m(t_{k-1})$ defines cumulative utilities, which we will discuss in the next subsection.

Because of the recursive structure of the stationary choice frequencies $\hat{x}_i$ all levels of consideration and all utilities are coupled. The repeated choice process describes a highly-nested system.

The Concept of Cumulated Utilities

In this theory for repeated choice processes, one starts by considering the repeated decisions of a group of agents. The first choice sequence is characterized by a Markov process. At any later stage of the choice process, memory effects are taken into account by making the choice probabilities dependent on previous decisions, which leads to a non-markovian behaviour of the nested process.

Alternatively, it is also possible to describe the dynamics of the choice frequencies at each stage by a markovian choice process (choice process without memory effects) by introducing appropriate cumulative utilities $u_i(t_k)$. Hence, these cumulative utilities depend on the history of the system via memory effects that are represented in the change over time of the system parameters.

In the stationary case, the cumulative utilities $u_i(t_k)$ are defined via

$$\hat{x}_i(t_k) = N_k \exp(2u_i(t_k)). \tag{6.67}$$

If we insert (6.67) in (6.66), we obtain a recursive formula for the cumulative frequencies

$$\exp(2u_i(t_k)) = \frac{N_{k-1}}{N_k} \sum_{m=1}^{L_k} \frac{\exp\left(2u_i^{(m)}(t_k)\right)}{\sum_{j=1}^{L_k} \exp\left(2u_j^{(m)}(t_k)\right)} \cdot \exp(2u_m(t_{k-1})). \tag{6.68}$$

In (6.68), the dependence of the cumulative utilities on their previous values is obvious. If it is assumed that the cumulative utilities depend on the current choice frequencies $x_i(t_k)$ and on the preference parameters $\delta_i(t_k)$ in a certain functional manner, then the recursion relation (6.68) will lead to a renormalization of its parameters.

Therefore, the technique of renormalization group theory (Ma 1982), well known in physics, can be applied. As a result, the parameters, namely the interaction parameters and the preferences become a function of time (time sequences).

The renormalization of the parameters and the change of the trend parameters in the course of time can be seen as the result of a learning process of the agents with respect to changing characteristics and external and internal variations involved in the repeated choice process.

In particular, the repeated choice processes lead to a sequence of parameter values, e.g. in case of (6.59) for the cumulative utility function one obtains for the synergy parameter $\kappa^{(m)}$ the sequence

$$\kappa^{(m)}(t_1), \kappa^{(m)}(t_2), \ldots, \kappa^{(m)}(t_k), \ldots. \tag{6.69}$$

In summary: The adaptations of the markovian Master equation choice model (6.48) to changing data sets (over time) take into account memory effects and other changing interaction parameters. The renormalization procedure leads to changing parameters. This also justifies the application of the choice model (6.48) in case of non-markovian processes.

Simulation of Specific Choice Processes With Memory

In the following section, a few specific examples are presented, to illustrate how the nested Master equation choice model with memory effects may be applied in practical work. We only consider the stationary solution of the repeated choice model. However, when the flexibility parameters v_{ij} of the population of agents do not allow the system to approach its stationary state within a given choice sequence $t_{k-1} < t \leq t_k$ then the choice frequencies $x_i(t)$ must be computed using the approximate equations of motion (6.55) to (6.57). In general, this leads to a smoother behaviour of the $x_i(t)$ compared to their stationary values $\hat{x}_i(t)$.

Figures 6.10, 6.11, 6.12, 6.13, 6.14, and 6.15, show the mean frequency $\hat{x}_i(t_k)$ for different alternatives $i = 1, 2, \ldots, L_k$ during K choice intervals $k = 1, 2, \ldots, K$. The number of repeated choices (choice sequences) used in the examples is $K = 25$. A normal random distribution is assumed as initial distribution of the choice

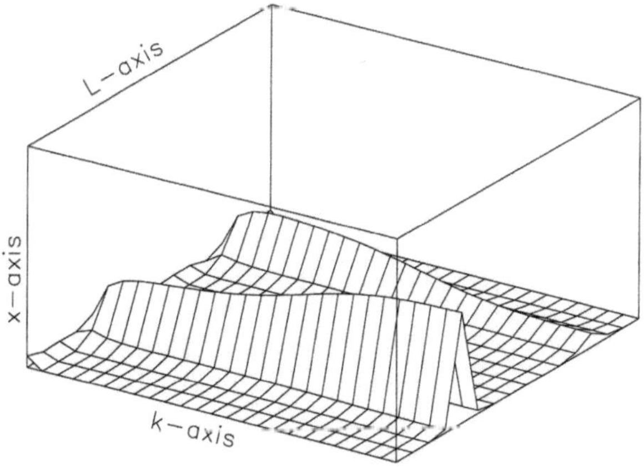

Fig. 6.10 Fixed number of alternatives, $L = 16$, memory effect over one time sequence and $\kappa_1 = 1.5, \kappa_2 = 0.0, \sigma = 0.0, \delta_{16} = 0.0$ (Source: Haag and Grützmann 1993)

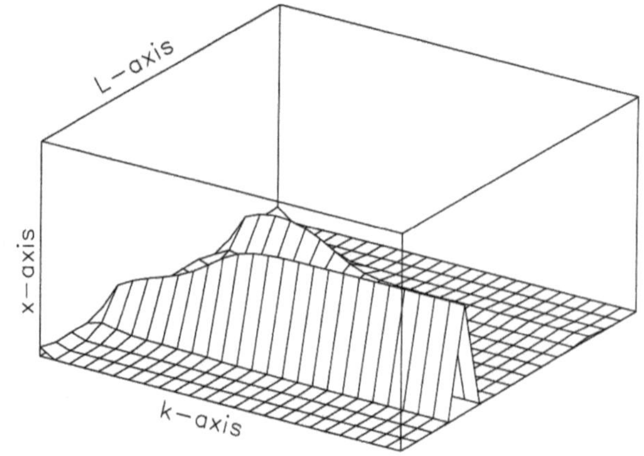

Fig. 6.11 Fixed number of alternatives, $L = 16$, memory effect over two time sequences and $\kappa_1 = 1.5, \kappa_2 = 0.5, \sigma = 0.0, \delta_{16} = 0.0$ (Source: Haag and Grützmann 1993)

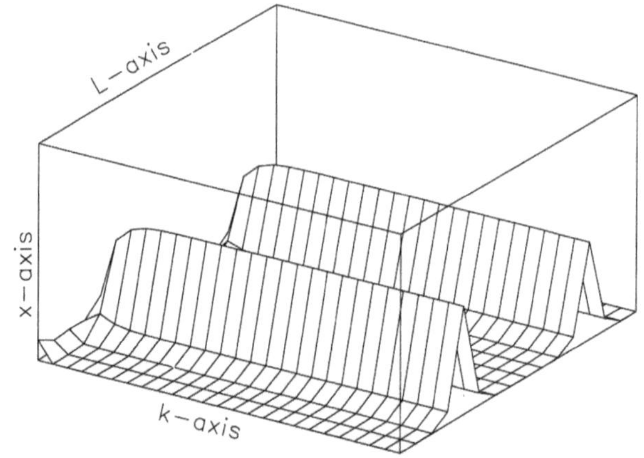

Fig. 6.12 Fixed number of alternatives, $L = 16$, memory effect over one time sequence with saturation effects and $\kappa_1 = 1.5, \kappa_2 = 0.0, \sigma = 0.3, \delta_{16} = 0.0$ (Source: Haag and Grützmann 1993)

frequencies $\hat{x}_i(t_0)$. At later choice sequences (time steps) $t_k > t_0$ the dynamic utility function is assumed to be

$$u_i(t_k) = \delta_i(t_k) + \kappa_1 x_i(t_{k-1}) + \kappa_2 x_i(t_{k-2}) - \sigma x_i(t_{k-1})^2, \tag{6.70}$$

which is rather flexible for the investigation of the influence of memory effects on choice behaviour. A possible dependence of the dynamic utilities on the index (m) is not considered.

The preference parameters $\delta_i(t_k)$ for the different alternatives $i = 1, 2, \ldots, L_k$ are modelled for test purposes via

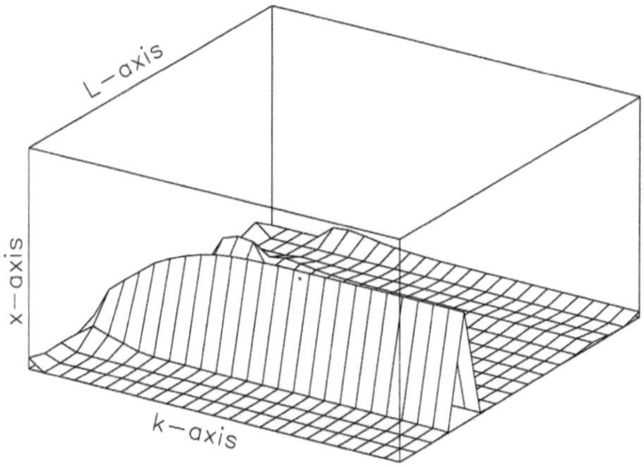

Fig. 6.13 Introduction of a new alternative $L = 15 \rightarrow L = 16$, for $t > 5$, and $\kappa_1 = 2.0, \kappa_2 = 0.0, \sigma = 0.0, \delta_{16} = 2.2$ (Source: Haag and Grützmann 1993)

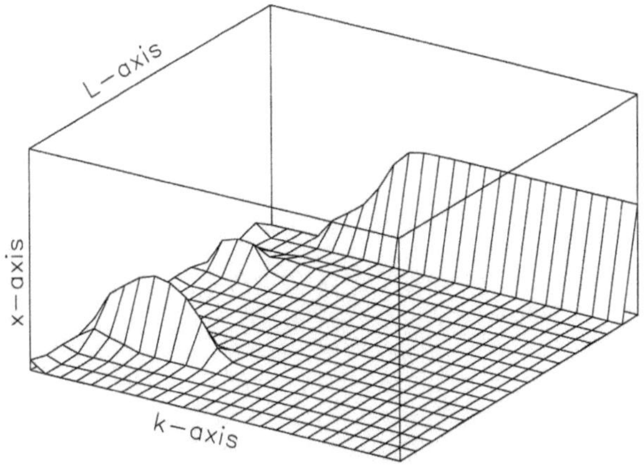

Fig. 6.14 Introduction of a new alternative $L = 15 \rightarrow L = 16$, for $t > 5$, and $\kappa_1 = 2.0, \kappa_2 = 0.0, \sigma = 0.0, \delta_{16} = 2.5$ (Source: Haag and Grützmann 1993)

$$\delta_i(t_k) = -\alpha \cos\left(2\lambda\pi(i-1)/L_k\right), \tag{6.71}$$

where we assume $\alpha = \lambda = 2$, in order to support the dominance of two groups of alternatives.

Figures 6.10, 6.11, and 6.12, illustrate the results of the simulations of the mean frequencies for a fixed number of alternatives $L_k = 16$, for $k = 1, 2, \ldots, 25$.

In Fig. 6.10, the decision process is influenced by the previous choice distribution, where the memory effect extends over one choice sequence. At the beginning, the two preferred groups of alternatives gain a comparable market share. The

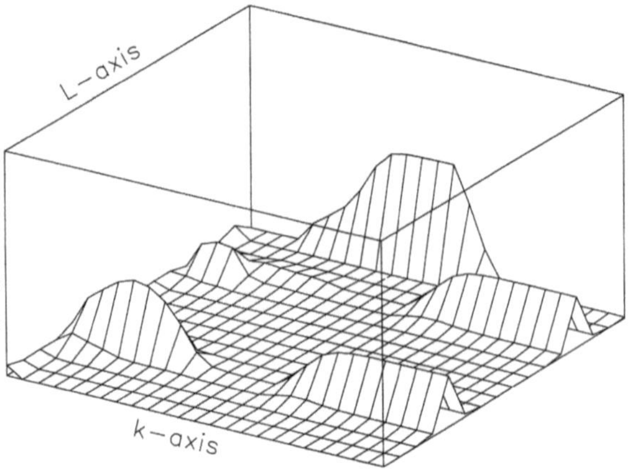

Fig. 6.15 Introduction of a new alternative $L = 15 \rightarrow L = 16$, for $t > 5$, preference parameter $\delta_{16} = 2.5$ for $t_5 \leq t < t_{15}$ and $\delta_{16} = 1.0$ for $t \geq 15$ and $\kappa_1 = 2.0\,, \kappa_2 = 0.0\,, \sigma = 0.0$ (Source: Haag and Grützmann 1993)

memory effect ensures that a strong competition sets in and ultimately leads to the dominance of only one alternative.

Figure 6.11 shows how competition between the two alternatives is amplified when the memory effect extends over two preceding time intervals.

In Fig. 6.12, a saturation effect prevents the exclusive dominance of one alternative, even in case of strong competition. Therefore, the two preferred alternatives gain a comparable market share.

The influence of a newly introduced alternative is investigated in Figs. 6.13, 6.14, and 6.15. It is assumed, that the new alternative is introduced ($L_k = 16$, for $k = 5\,, 6\,, \ldots\,, 25$) for $t_k \geq t_5$.

In Fig. 6.13, the new alternative is given by a rather high value for the preference parameter $\delta_{16} = 2.2$, relative to the preference parameters of the other alternatives. Of course, the new alternative competes with the old ones, but since its market share is zero at the beginning, the support for alternative 16 does not exceed its critical threshold value, and the new alternative loses its market position in the long run.

However, if the value of the preference parameter is set to $\delta_{16} = 2.5$, in other words, if $\delta_{16} \geq \delta_c$ the new alternative gains the competition and dominates in the long run (Fig. 6.14). As pointed out by Ostrusska (1990) as well as Haag and Wunderle (1988) in marketing, the preference parameter for a given product i is related to the marketing activities for that product (e.g. advertisement and prices). The investigation of repeated choice behaviour with memory effects is of crucial importance in product marketing.

In Fig. 6.15, the strong preference for the new alternative $\delta_{16} = 2.5$ is reduced to $\delta_{16} = 1.0$ for later time sequences $t_k \geq t_{15}$. This variation in the preference parameter of alternative 16 results in a recovery of the old alternatives and leads finally to the competitive exclusion of alternative 16.

6.8 The Emergence of Conventions

A main research task in social sciences concerns the application of the Master equation framework to the field of collective behaviour, namely to the question how larger ensembles of actors interact and how this collective behaviour can be modelled.

In a particularly dense description by Elias (1976), it becomes clear that the subject of collective behaviour can and must be linked to the most fundamental socio-scientific issues.

As Elias states:

> Aus der Verflechtung von unzähligen individuellen Interessen und Absichten entsteht etwas, das so wie es ist, von keinem Einzelnen geplant oder beabsichtigt worden ist, und das doch zugleich aus Absichten und Aktionen vieler Einzelner hervorging. Und das ist eigentlich das ganze Geheimnis der gesellschaftlichen Verflechtung, ihrer Zwangsläufigkeit, ihrer Aufbaugesetzlichkeit, ihrer Struktur, ihres Prozesscharakters und ihrer Entwicklung; dies ist das Geheimnis der Soziogenese und der Beziehungsdynamik. (Elias 1976)

Thus, the topic of collective behaviour is nothing less than the "secret of socio-genesis". Of course, phenomena of collective behaviour can be observed in all fields of nature (see the examples of the Laser and ferro-magnetism in Chap. 5). However, the further remarks focus only on those collective processes that take place in human societies, since these are characterized by a number of peculiarities and have been little investigated so far (Diekmann 1997; Dosi et al. 1999; Berninghaus and Schwalbe 1996). Various forms of modelling collective behaviour are discussed in literature (Haag and Grützmann 2002; Braun and Saam 2015). In any case, information and knowledge, norms, external disturbances, innovations or self-reflexive couplings are part of the ingredients of processes of collective behaviour. Without these components, no process of collective behaviour in modern societies would be exist. It is always the decision of actors involved in a network of social forms of coordination, in the respective experiences of the actors with the past, in future expectations as well as in the current economic and social environment.

In this part, we restrict ourselves only to the modelling task. For more general considerations and a detailed review of literature see Haag and Müller (2015).

In a summary one can state that collective behavior requires the following conditions:

Large number of actors: Collective processes are phenomena of large numbers and thus characterized by the irrelevance of personal relationships.

Social coordination modes: The large numbers of autonomous or behavioral acts are always undertaken in the context of existing social coordination forms, to the areas of law, conduct or traditions, and the knowledge available.

Interrelationship between the micro-level and macro-level: Collective behaviour should always be studied and described on at least two different levels, namely the micro-level and the macro-level.

At the macro-level it is about the spatiotemporal representation of collective event sequences and to the necessary process types with which these series of events can be modeled. At the macro-level, all those factors of social coordination are considered, which may be relevant to collective processes, social control systems, institutions, information about the courses of collective processes, which may affect and restrict the behaviour of actors.

Each collective behaviour also has a description on the micro level of actors, namely the subjective perspectives, the cognitive organization of actors. The respective actions of actors must have a necessary and sufficient turn to produce a specific process of collective behavior (Coleman 1987, 1990, 1992, 1993, 1994; Esser 2000; Udehn 2001).

Self-organization: Another characteristic of collective behaviour is characterized by the absence of a central coordination or supervisory body.

Haken (1978) distinguishes organization and self-organization by an example as follows:

Organization: Consider, for example, a group of workers. We then speak of organization or, more exactly, of organized behaviour if each worker acts in a well-defined way on given external orders, i.e., by the boss. It is understood that the thus-regulated behaviour results in a joint action to produce some product.

Self-organization: We would call the same process as being self-organized if there are no external orders given but the workers work together by some kind of mutual understanding, each one doing his job so as to produce a product.

Self-organized processes of collective behavior can also occur without that they are individually intended or desired in this manner. The formation of traffic jams, bubbles in the stock market, financial crisis, or segregation in residential areas are examples of a divergence between individual intentions or objectives and the outcome of collective processes.

A Master Equation Model for Collective Behaviour

In our Master equation model for collective behaviour it is assumed that the probability of an alternative choice depends on the realized frequency distribution of the alternatives in the prior-forth time interval (Haag and Wunderle 1988).

$$u_i(t) = \delta_i(t) + \kappa \cdot x_i(t_{k-1}). \tag{6.72}$$

It is therefore assumed that individuals know how the other individuals have been decided. To what extent this knowledge influences the decision-making process is described by the parameter κ. The parameter κ describes the strength of the dependence of the value of an alternative of the preceding alternative choice or in other words, the effect of the collective behaviour on the subsequent individual decisions. The more individuals have already opted for a specific alternative, the more attractive this alternative is for the remaining decision makers if $\kappa > 0$. The preference for a particular alternative, independent of the previously chosen alternative is described by the preference parameter δ_i.

Assumptions for the Simulations

There are ten competing alternatives i considered. The uncertainties in the evaluation of the alternatives are represented by a noise term $\alpha \cdot ran_i$, the parameter α describes the "degree" of uncertainty. The function ran_i represents the noise distribution (white noice with values between 0 and 1) (Laux 1982; Lancaster 1966).

$$\delta_i(t) = \begin{cases} \delta + \alpha \cdot ran_i & \text{for } i = 1, \ldots, 5 \\ \alpha \cdot ran_i & \text{for } i = 6, \ldots, 10 \end{cases} \tag{6.73}$$

In the next Fig. 6.16, the first 25 decision sequences (time steps) are shown for a system without uncertainties in the decision process ($\alpha = 0$).

Which alternative ultimately "wins" depends in this case to a high extend on the initial constellation. It depends largely upon a relatively high value of the parameter κ which alternative will dominate in the long-run, in other words when the threshold for a "social contagion effect" is reached (see Fig. 6.17).

Based are these simulation results, on an experiment of Salganik et al. (2006) and his team at Columbia University in New York and the Santa Fe Institute in New Mexico with the help of the Internet.

Some 14,000 participants (panelists) could listen to new pop songs on the Internet and were allowed to freely download the songs they liked. download from the net. This resulted in a hit list on the simulated music market.

The individual choices were strongly affected, in the case that each participant was informed how often each song had already been downloaded. This resulted in a different hit list, were rarely purchased songs were even less purchased, and preferred songs became even more popular. Instead of a statistical mixing, the collective behavior (social interaction) led to a polarization.

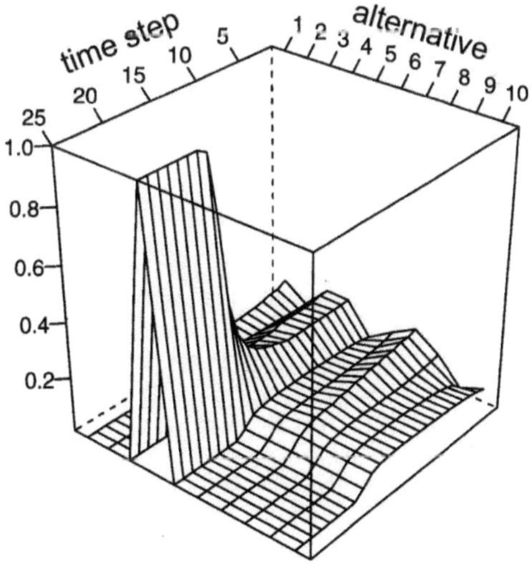

Fig. 6.16 The formation of a dominant alternative in the collective decision-making process with complete information ($\kappa = 4.4$) (Source: Haag and Grützmann 2002)

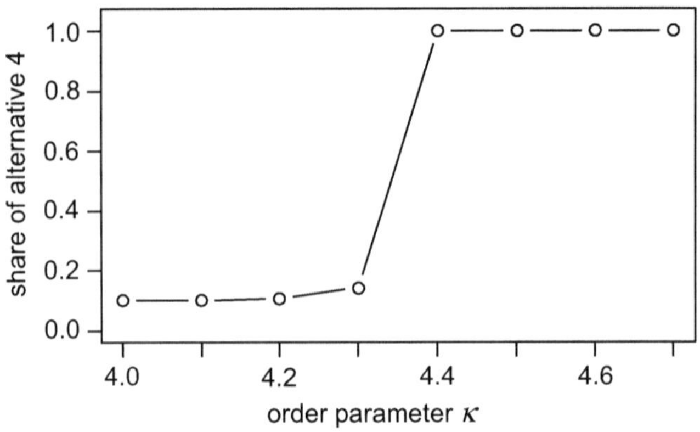

Fig. 6.17 The proportion of the dominating alternative of the strength of "social contagion" (order parameter κ) (Source: Haag and Grützmann 2002)

In an accompanying commentary, the journal states that the social process creating a blockbuster makes it difficult to predict the breakthrough for the one or the other product. Moreover, the greater the social impact, the more unbalanced and unpredictable the collective result is.

This clearly shows that poll ratings affect individual decisions and obviously not only in the simulated music market but also in political elections and purchasing decisions on the stock exchange, just to mention a few but important examples.

References

Akerlof GA (1997) Social distance and social decisions. Econometrica 65(5):1005–1027

Akerlof GA, Kranton RE (2000) Economics and identities. Q J Econ 115(3):715–753

Akerlof GA, Kranton RE (2008) Identity and the economics of organizations. J Econ Perspect 19 (1):9–32

Akerlof G, Kranton R (2010) Identity economics. Princeton University Press, Princeton, NJ

Arthur WB (1989) Competing technologies, increasing returns, and lock-in by historical events. Econ J 99:116–131

Bahrenberg E, Fischer MM, Nijkamp P (eds) (1984) Recent developments in spatial analysis: methodology, measurement, models. Gower Publishing, Aldershot

Ben-Akiva M, Lerman SR (1985) Discrete choice analysis: theory and application to predict travel demand. MIT Press, Cambridge

Berninghaus SK, Schwalbe U (1996) Evolution, interaction and Nash-equilibria. J Econ Behav Organ 29:57–85

Braun N, Saam NJ (eds) (2015) Handbuch Modellbildung und Simulation in den Sozialwissenschaften. Springer, Heidelberg

Coleman JS (1987) Microfoundations and macrosocial behavior. In: Alexander J et al (eds) The micro-macro-link. University of California Press, Berkeley, pp 153–173

Coleman JS (1990) Foundations of social theory. Harvard University Press, Cambridge

Coleman JS (1992) The economic approach to sociology. In: Radnitzky G (ed) Universal economics. Assessing the achievements. Paragon House, New York

Coleman JS (1993) The rational reconstruction of society. Am Sociol Rev 58:1–15
Coleman JS (1994) A rational choice perspective on economic sociology. In: Smelser NJ, Swedberg E (eds) The handbook of economic sociology. Princeton University Press, Princeton, NJ
Davis J (2004) Identity and commitment. Tinbergen Institute Discussion Paper, Amsterdam
Diekmann T (1997) The evolution of conventions with mobile players. Mimeo/National University of Oreland, New York/Maynooth
Domencich T, McFadden D (1975) Travel demand: a behavioural analysis. North Holland, Amsterdam
Dosi G, Marengo L, Bassanini A, Valente M (1999) Norms as emergent properties of adaptive learning. J Evol Econ 9:5–26
Elias N (1976) Über den Prozess der Zivilisation, Bd. 2. Suhrkamp, Frankfurt
Esser H (2000) Soziologie, spezielle Grundlagen, Bd 2. Die Konstruktion der Gesellschaft. Campus, Frankfurt
Fischer MM, Aufhauser E (1988) Housing choice in a regulated market: a nested multinominal logit analysis. Geogr Anal 20:47–69
Fischer MM, Nijkamp P (1987) From static towards dynamic discrete choice modelling. Reg Sci Urban Econ 17:3–27
Fischer MM, Haag G, Sonis M, Weidlich W (1988) Account of different views in dynamic choice processes. IGU-working group on mathematical models, Canberra
Golledge RG, Timmermans H (eds) (1988) Behavioural modelling in geography and planning. Croom Helm, London
Haag G (1989) Dynamic decision theory: applications to urban and regional topics. Kluwer, Dordrecht
Haag G, Grützmann K (1993) A new theory of nested decision processes with memory: the stochastic framework. Pap Reg Sci 72(3):313–335
Haag G, Grützmann K (2002) Die Entstehung von Konventionen als dynamischer Entscheidungsprozess mit Gedächtnis. Studien zur Evolutorischen Ökonomik VI 195:89–109
Haag G, Müller K (2015) Kollektivverhalten. In: Braun N, Saam N (eds) Handuch Modellbildung und Simulation in den Sozialwissenschaften. Springer, Berlin
Haag G, Wunderle P (1988) Quality improvement and self-reinforcing mechanisms in economy. Meth Oper Res 60:709–722
Haken H (1978) Synergetics. An introduction. Springer, Berlin
Heckman JJ (1981) Statistical models for discrete panel data. In: Manski CF, McFadden D (eds) Structural analysis of discrete panel data with economic applications. MIT Press, Cambridge
Hensher DA (1988) Model specification for a dynamic discrete continuous choice automobile demand system. In: Golledge RG, Timmermans H (eds) Behavioural modelling in geography and planning. Croom Helm, London
Hensher DA, Wrigley N (1984) Statistical modelling of discrete choice with panel data. Working paper. 16, Macquarie University, Australia
Johnson LW, Hensher DA (1982) Application of multinomial probit to a two-period panel data set. Transp Res 16(A):457–464
Klinke R, Pape H-C, Kurz A, Silbernagl S (eds) (2014) Physiologie. Thieme Verlag, Stuttgart
Lancaster KA (1966) A new approach to consumer theory. J Polit Econ 74:132
Laux H (1982) Entscheidungstheorie Bd1 und Bd2. Springer, Berlin
Leonardi G (1983) Transient and asymptotic behaviour of a random utility based stochastic search process in continuous space and time. Collaborative Paper, 83, IIASA, Laxenburg
Ma S (1982) Modern theory of critical phenomena. Front Phys. Addison-Wesley
Mueller U, Nauck B, Diekmann A (eds) (2000) Handbuch der Demographie Bd I und Bd II. Springer, Heidelberg
Ostrusska D (1990) Systemdynamik nichtlinearer Marktreaktionsmodelle bei monopolistischer und oligopolistischer Konkurrenz. Thesis, University of Vienna
Pitfield DE (ed) (1984) Discrete choice models in regional science. Pion, London

Reggiani A (ed) (2000) Spatial economic sciences. Springer, Berlin
Sacklowski A, Röhrl E (1973) Einheiten-Lexikon: Begriffe, Größen, Einheiten in Physik und Technik. Deutsche Verlagsanstalt, Stuttgart
Salganik MJ, Sheridan Dodds P, Watts DJ (2006) Experimental study of inequality and unpredictability in an artificial cultural market. Science 10(311):854–856
Samuelson L (1985) On the independence from irrelevant alternatives in probabilistic choice models. J Econ Theory 35:376–389
Train K (2003) Discrete choice models with simulation. Cambridge University Press, Cambridge
Udehn L (2001) Methodological individualism. Background, history and meaning. Routledge, London
Weidlich W (1971) The statistical description of polarization phenomena in society. Br J Math Stat Psychol 24:251–266
Weidlich W (1972) The use of statistical methods in sociology. Collective Phenom 1:51
Weidlich W, Haag G (1983) Concepts and models of a quantitative sociology: the dynamics of interaction populations, Springer series of synergetics, vol 14. Springer, New York

Chapter 7
Applications in Population Dynamics

Contents

7.1 Birth- and Death Processes within a Single Population

7.1.1 Verhulst-Pearl Equation of Population Growth

The mathematician Verhulst (1838) developed a common model of population growth, inspired by "An Essay on the Principle of Population" of Malthus (1798). Verhulst derived his equation to describe the self-limiting growth of a biological population. He assumed that the rate of reproduction is proportional to both, the existing population and the amount of available resources, everything else being equal. The modelling equation is often called the Verhulst-Pearl equation following its rediscovery by Pearl and Reed (1920).

G. Haag, *Modelling with the Master Equation*, DOI 10.1007/978-3-319-60300-1_7

In the following $N(t)$ describes the (continuous) population size at time t. Then the differential equation of Verhulst-Pearl for population growth reads

$$\frac{dN(t)}{dt} = rN(t)\left(1 - \frac{N(t)}{K}\right) \tag{7.1}$$

where r represents the unimpeded growth rate and K the so-called carrying capacity of the population. For a small population size ($N(t) << K$), an exponential growth is expected, if $N(t)$ approaches the carrying capacity K the total growth of $N(t)$ decreases, e.g., due to limited resources, and the population stabilizes.

The solution of (7.1) is given by

$$N(t) = K\left(1 + \frac{K - N_0}{N_0}\exp(-rt)\right)^{-1} \tag{7.2}$$

with the initial population size N_0 at time $t = 0$ and where

$$\lim_{t\to\infty} N(t) = K. \tag{7.3}$$

The carrying capacity K of the population $N(t)$ depicts the highest stable value that the population can reach. Of course, only a few populations follow the simple logistic growth, especially when predators threaten the population. The ecology is not so simple. Among the complications that interfere with the simple logistic model is the fact that predators rarely kill organisms of all ages with equal probability. This unequal mortality affects the age distribution of the population, which in turn affects the specific fertility rate of the population. Likewise, numerous factors influence the mortality of the population. This means that ecologists must also take into consideration the demographic effects.

Empirical observations show, that species in their life history pursued different survival strategies, namely r-strategies (high reproduction rate) or K-strategies (low reproduction rate, but high carrying capacity).

The logistic growth model contains several other very coarse simplifications. One of the most serious is based on the idea that growth rate is highest when the population is on the verge of extinction. For a detailed description, it will be necessary to fix a lower limit M of $N(t)$. Below this limit M the population growth will be negative and the population is doomed to extinction (Fig. 7.1). The easiest way to introduce such a limitation is via adding them as part of a new expression to the logistic equation (Wilson and Bossert 1973).

$$\frac{dN(t)}{dt} = rN(t)\left(1 - \frac{N(t)}{K}\right)\left(\frac{N(t) - M}{N(t)}\right). \tag{7.4}$$

The threshold value of M is of great importance for the evolution of species when $N(t)$ is close to their survival threshold M. The value of M is not so important as long as the population exceeds the value of M by a multiple. However, the determination of the threshold value M for existing populations is quite difficult.

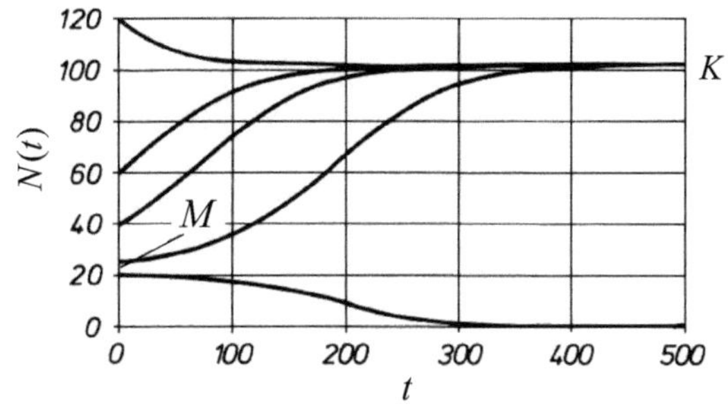

Fig. 7.1 Time path for $N(t)$ according to (7.4). The parameters are related to the spring census gosling count for Canada goose (see Sect. 1.4) $r = 0.00322, K = 102, M = 21$ (Source: Weidlich and Haag 1983)

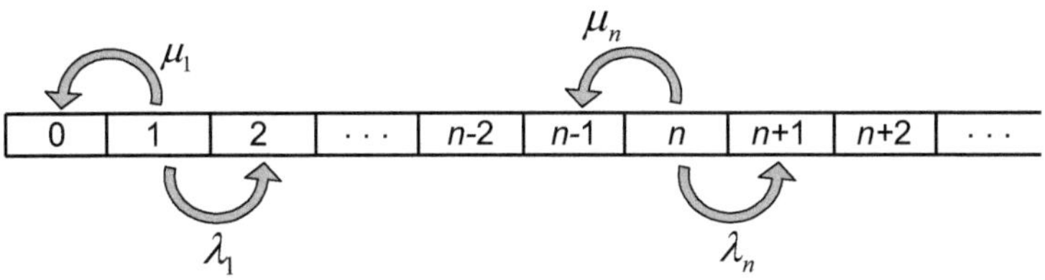

Fig. 7.2 Possible next-neighbour transitions of the one-dimensional population dynamics Master equation. The state zero is an absorbing state

We will see in the following sections that due to the discrete structure of birth- and death processes, especially in case of small population numbers, unexpected increases in the probability of extinction happen.

7.1.2 Stochastic Versus Deterministic Description

Birth- and death processes within a single population may be modelled by a one-dimensional Master equation with next-neighbour transitions, by neglecting migration events and interactions with other species. Figure 7.2 depicts the structure of this Master equation.

Assuming single birth- and death events, in other words next-neighbour transitions only, the Master equation reads

$$\frac{dP(0,t)}{dt} = \mu_1 P(1,t) \tag{7.5}$$

$$\frac{dP(1,t)}{dt} = -(\lambda_1 + \mu_1)P(1,t) + \mu_2 P(2,t) \tag{7.6}$$

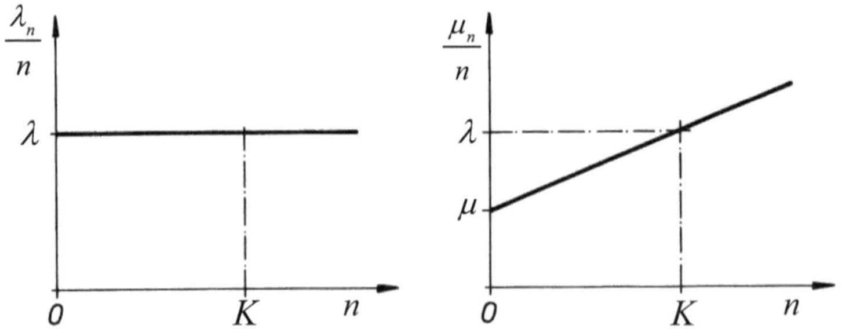

Fig. 7.3 Dependency of the specific birth- and death rate on the size of the population

$$\frac{dP(n,t)}{dt} = \lambda_{n-1}P(n-1,t) - (\lambda_n + \mu_n)P(n,t) + \mu_{n+1}P(n+1,t) \quad \text{for n} = 2,3,\ldots, \tag{7.7}$$

where the birth rate $\lambda_n \equiv w(n+1,n) > 0$, for $n = 0, 1, \ldots$ and the death rate $\mu_n \equiv w(n-1,n) > 0$ for $n = 1, 2, \ldots$ are assumed to have the following n-dependence:

$$\lambda_n = a_1 n > 0 \quad \text{or} \quad \lambda_n = \lambda n > 0 \tag{7.8}$$

$$\mu_n = a_2 n + b_2 n^2 > 0 \quad \text{or} \quad \mu_n = \mu n + \frac{(\lambda - \mu)}{K} n^2 > 0 \tag{7.9}$$

and where the coefficients fulfil $a_1 > 0, a_2 > 0, b_2 > 0$ and $\lambda > 0, \mu > 0, K > 0$. The increase in the death rate per species μ_n/n with growing n could be due to limitations in food resources or increasing risk of infections. An analogous effect, namely a decrease in the specific birth rate λ_n/n for growing n will be neglected as it leads to essentially the same type of results. The birth and death rates per species are depicted in Fig. 7.3.

For next-neighbour transitions only, the condition of detailed balance is fulfilled in a trivial manner and the exact stationary solution of the Master equation (7.5)–(7.7) is obtained

$$P_{st}(n) = P_{st}(0)\prod_{k=1}^{n}\xi(k) = P_{st}(0)\prod_{k=1}^{n}\frac{\lambda_{k-1}}{\mu_k} = \delta_{n0}, \tag{7.10}$$

where

$$\xi(n) = \frac{\lambda_{n-1}}{\mu_n} = \frac{n-1}{n}\frac{a_1}{a_2 + b_2 n} = \left(1 - \frac{1}{n}\right)\frac{a_1}{a_2 + b_2 n}. \tag{7.11}$$

Since $\xi(1) = 0$, the stationary solution (7.10) describes a completely died-out population. In other words, the state zero is an absorbing state. The state zero is stable, since nothing comes out of nothing. The state where everything has died out is the only stable state in population dynamics.

The conventional approximation (ca) assumes $n >> 1$ or $n \rightarrow x$, or in other words, the discrete variable n is replaced by a continuous variable x. This is justified if the population is far from extension,

$$\xi_{ca}(n) = \frac{\lambda_{n-1}}{\mu_n} \approx \frac{a_1}{a_2 + b_2 n}. \tag{7.12}$$

Then the maximum is determined via

$$\xi_{ca}(\bar{n}) \approx 1 \tag{7.13}$$

leading to

$$\bar{n}_{ca} = \frac{(a_1 - a_2)}{b_2} = K \tag{7.14}$$

the "capacity" of the population. Furthermore, with (4.45) and (7.12), the variance of the distribution is obtained

$$\sigma_{ca}^2 = -\left[\left(\frac{\partial \xi(n)}{\partial n}\right)_{n=\bar{n}}\right]^{-1} = \frac{a_1}{b_2} = \frac{\lambda}{(\lambda - \mu)} K. \tag{7.15}$$

This finally leads to the approximate probability distribution of the birth- and death Master equation in population dynamics (4.43) and (4.44)

$$P_{ca}(n) = \frac{1}{\sqrt{2\pi\sigma^2}} \exp\left(-\frac{(n - \bar{n})^2}{2\sigma^2}\right) \quad \text{for} \quad n \geq 0. \tag{7.16}$$

The conventional approximation results in normally distributed fluctuations of the population. However, the nonlinearity parameter in the death rate $b_2 > 0$ has a big influence on the carrying capacity of the population K and the variance of the distribution. All other parameters a_1, a_2 being fixed, smaller values of b_2 increase K and the variance σ^2. This means that the probability to die out, namely to reach the population size zero via a statistical fluctuation becomes unrealistically low, for small values of b_2. We will come back to this point in detail later.

If we consider the distribution function $P(n)$ for small values of n it is obvious that the conventional stationary distribution (7.16) has a completely wrong mathematical structure for neglecting the factor $f(n) > 0$

$$f(n) = \frac{n - 1}{n} \quad \text{for} \quad n = 1, 2, \ldots. \tag{7.17}$$

In Fig. 7.4, the multiplicative influence of $f(n) > 0$ is depicted. The effect of the discrete structure is independent of the birth and death rates.

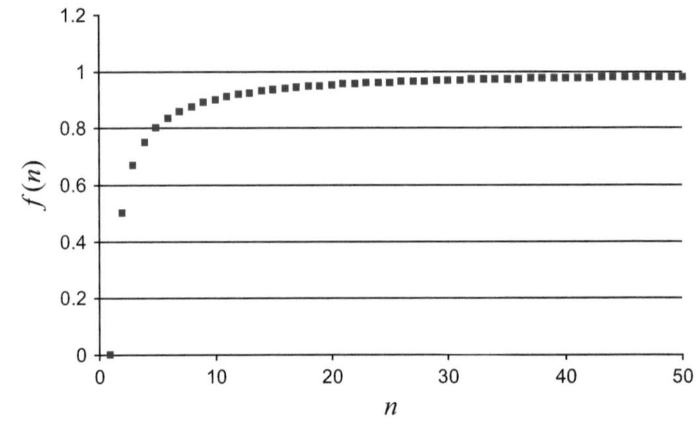

Fig. 7.4 The influence of the discrete structure of $f(n)$

The probability of complete extinction depends decisively on the exact form of (7.11). Therefore, let us exclude only the absorbing state zero and compute the quasi-stationary solution (qs) by using

$$P_{qs}(0) = 0 \tag{7.18}$$

$$P_{qs}(n) = P_{qs}(1)\prod_{k=2}^{n}\xi_{qs}(k) = P_{qs}(1)\prod_{k=2}^{n}\left(1-\frac{1}{k}\right)\frac{a_1}{a_2+b_2k} \quad \text{for } n=2,3,\ldots, \tag{7.19}$$

with the normalization condition

$$\sum_{n=0}^{\infty} P_{qs}(n) = 1. \tag{7.20}$$

The value of $P_{qs}(1)$ is obtained by (7.20) as

$$P_{qs}(1) = \left[1+\sum_{n=0}^{\infty}\prod_{k=2}^{n}\xi_{qs}(k) = 1\right]^{-1}. \tag{7.21}$$

Equation (7.19) describes a meta-stable solution of the Master equation, since the dying-out probability is very small and the transition rates between the different states of the qs-distribution guarantee that the qs-distribution is in a quasi-equilibrium state.

In order to get the maximum of the distribution functions $P_{ca}(n)$ and $P_{qs}(n)$ at the same point, it is advisable to rescale the rate constants. This is of importance, since we want to compare both approximations with actual observations of a wild living giant Canada goose population.

$$a_1^* = \frac{\bar{n}}{\bar{n}-1} a_1, \quad a_2^* = a_2 \quad \text{and} \quad b_2^* = b_2$$

or (7.22)

$$\lambda^* = \frac{K}{K-1}\lambda \quad \mu^* = \mu \quad \text{and} \quad K^* = \frac{K}{K-1}\left(K + \frac{\mu}{\lambda - \mu}\right).$$

The turning values of the distribution $P_{qs}(n)$ are again determined by

$$\xi_{qs}(n) \approx 1, \tag{7.23}$$

which exhibits two positive solutions

$$n_{\pm} = \frac{K}{2}\left(1 \pm \sqrt{1 - \frac{4\ \lambda}{K\lambda - \mu}}\right) = \frac{K}{2}\left(1 \pm \sqrt{1 - \frac{4\sigma^2}{K^2}}\right) \tag{7.24}$$

for

$$K > 2\sigma. \tag{7.25}$$

Expanding the square root in a Taylor series up to first order for $K^2 >> \sigma^2$ yields the two positive solutions

$$n_+ = K^* \approx K - \frac{\lambda}{\lambda - \mu} = K\left(1 - \frac{\sigma^2}{K^2}\right), \tag{7.26}$$

where the first solution n_+ corresponds to the maximum of the population distribution, namely the carrying capacity, and the second solution

$$n_- = M \approx \frac{\lambda}{\lambda - \mu} = \frac{\sigma^2}{K^2}. \tag{7.27}$$

n_- can be identified with the survival threshold M. It is compelling that M is determined by the ratio of the variance σ^2 (statistical fluctuations) and the carrying capacity K (Fig. 7.5).

7.1.3 Mean Value and Variance Equations

Equations of motion for the mean value $\langle n \rangle_t$ and the variance $\sigma^2(t)$ can be obtained directly from the Master equation (7.7) via the methods introduced in Chap. 3:

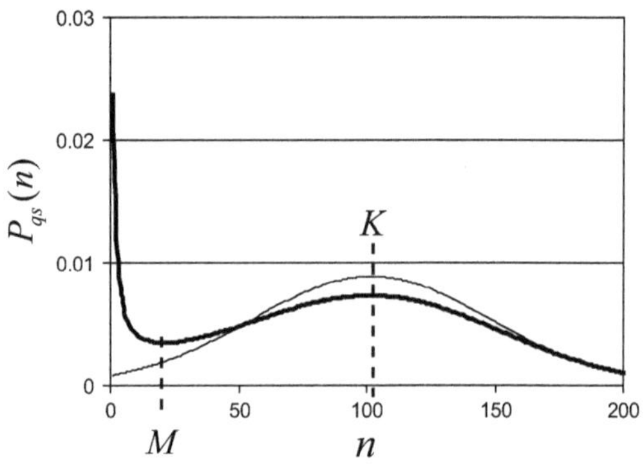

Fig. 7.5 Comparison of the quasi-stationary solution $P_{qs}(n)$ with the distribution function of the conventional approximation $P_{ca}(n)$ (*thin line*). The parameter values are related to the giant Canadian goose population: $a_1 = 0.5587; a_1^* = 0.5642; a_2 = 0.5320; b_2 = 2.622 \cdot 10^{-4}; n_- = 21; n_+ = 102$ (Source: Weidlich and Haag 1983)

$$\frac{d\langle n\rangle_t}{dt} = \langle\lambda_n\rangle_t - \langle\mu_n\rangle_t \tag{7.28}$$

$$\frac{d\sigma^2{}_t}{dt} = 2\Big(\langle n\lambda_n\rangle_t - \langle n\mu_n\rangle_t\Big) + (1 - 2\langle n\rangle_t)\langle\lambda_n\rangle_t + (1 + 2\langle n\rangle_t)\langle\mu_n\rangle_t. \tag{7.29}$$

It is solved with the initial conditions $\langle n\rangle_{t=0} = \langle n\rangle_0$ and $\sigma^2(t=0) = \sigma_0^2$. However, the exact equations (7.28) and (7.29) are not closed. As described in Chap. 3, we may proceed to quasi-closed equations if we introduce the assumptions (7.30)

$$\langle\lambda_n\rangle_t \approx \lambda\langle n\rangle_t \quad \text{and} \quad \langle\mu_n\rangle_t \approx \mu\langle n\rangle_t \tag{7.30}$$

and obtain the two ordinary differential equations

$$\frac{d\langle n\rangle_t}{dt} = r\langle n\rangle_t\left(1 - \frac{\langle n\rangle_t}{K}\right) \tag{7.31}$$

$$\frac{d\sigma^2{}_t}{dt} = (2\lambda - r)\langle n\rangle_t + \frac{r}{K}\langle n\rangle_t^2 + 2r\sigma^2\left(1 - \frac{2\langle n\rangle_t}{K}\right), \tag{7.32}$$

where we have introduced the net growth rate of the population and the carrying capacity K

$$r = (\lambda - \mu). \tag{7.33}$$

Equation (7.31) is the well-known logistic equation of Verhulst and Pearl (7.1) with solution (7.2)

$$\langle n\rangle_t = K\left(1 + \frac{K - \langle n\rangle_0}{\langle n\rangle_0}\exp(-rt)\right)^{-1}. \tag{7.34}$$

As described in the introduction to this chapter, it will be necessary to fix a lower limit M of the population size $\langle n\rangle_t$, where the population growth will be negative and the population will becomes extinct below this limit M. We mentioned before that the value n_-, equation (7.27), can be identified with the survival threshold M of equation (7.3). Therefore, the approximate mean value equations for single population growth may be modified appropriately:

$$\frac{d\langle n\rangle_t}{dt} = r\langle n\rangle_t\left(1 - \frac{\langle n\rangle_t}{K}\right)\left(\frac{\langle n\rangle_t - n_-}{\langle n\rangle_t + n_-}\right). \tag{7.35}$$

The trajectories of equation (7.35) for different initial conditions are shown in Fig. 7.1 for the parameters related to the giant Canada goose population.

7.1.4 Extinction Process and Life Time of Populations

As already mentioned and as Fig. 7.4 clearly shows, the discrete structure of the population distribution has a big influence on the dynamics of population growth when the population is on the verge of extinction. We follow the ideas of Matsuo et al. (1978) and van Kampen (1977, 1978) in calculating the extinction process and in estimating the life time of the quasi-stationary distribution.

Suppose $\pi_1(t)$ is the probability that the population is still alive, and $\pi_0(t)$ describes the probability that the species are extinct at time t. The probabilities $\pi_0(t)$ and $\pi_1(t)$ can be calculated directly using the Master equation (7.3)–(7.7)

$$\pi_1(t) = \sum_{n=1}^{\infty} P_{qs}(n,t) \qquad \text{probability of species being alive} \tag{7.36}$$

$$\pi_0(t) = P_{qs}(0,t) \qquad \text{probability of species being extinct} \tag{7.37}$$

with the normalization condition of the probability distribution

$$\pi_0(t) + \pi_1(t) = 1. \tag{7.38}$$

From the Master equation (7.7), it immediately follows that

$$\frac{d\pi_0(t)}{dt} = -\frac{d\pi_1(t)}{dt} = \mu_1 P(1,t). \tag{7.39}$$

We assume that the population has reached its meta stable state, in other words that the quasi-stationary distribution $P_{qs}(n,t)$ has been already established at time $t-0$ according to (7.18) and (7.19) with $\pi_0(0)=0$ and $\pi_1(0)=1$. We further

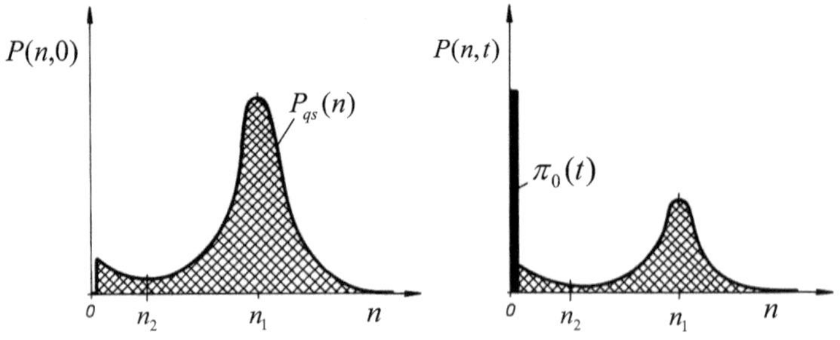

Fig. 7.6 Initial state with weights $\pi_0(0)$ and $\pi_1(0)=1$ and change of the weights due to probability transfer to the absorbing state zero

assume, that the probability flow from the state $P(1,t)$ to the state zero, representing the extinct population is very slow compared to the internal shape keeping probability flows based on internal fluctuations of the quasi-stationary distribution. This implies the assumption

$$P(n,t) \approx \pi_1(t)P_{qs}(n) \qquad \text{for} \quad n = 1, 2, \ldots . \tag{7.40}$$

The initial state and the transfer of probability from the living state to the extinct state are shown in Fig. 7.6

Insertion of (7.40) for $n=1$ into (7.39) leads to a differential equation for the mean life time τ_{qs} of the quasi-stationary distribution

$$\frac{d\pi_1(t)}{dt} = -\mu_1 P_{qs}(1)\pi_1(t) \equiv -\frac{1}{\tau_{qs}}\pi_1(t). \tag{7.41}$$

Therefore, the mean life time of the quasi-stationary distribution τ_{qs} reads

$$\tau_{qs} = \left[\mu_1 P_{qs}(1)\right]^{-1} = \frac{1}{\mu_1}\left[1 + \sum_{n=2}^{\infty}\prod_{k=2}^{n}\xi_{qs}(k)\right], \tag{7.42}$$

where $\xi_{qs}(k)$ is given by (7.11).

Applying (7.41) for the conventional approximation (7.16) yields

$$\tau_{ca} = \left[\mu_1 P_{ca}(1)\right]^{-1} = \frac{\sqrt{2\pi\sigma_{ca}{}^2}}{\mu_1}\exp\left(\frac{\bar{n}^2}{2\sigma_{ca}^2}\right). \tag{7.43}$$

It turns out that the life time calculated by (7.42) may be orders of magnitude smaller than the life time calculated by using the conventional approximation (7.43). The discrete structure of the configuration space cannot be neglected in the dying-out process. This statement becomes obvious when we apply and compare our theory of a quasi-deterministic extinction process and the conventional approximation with actual data. For this purpose, we use data of a natural

Table 7.1 Spring census, gosling count and hunting harvest for Canada goose (Branta Canadensis Maxima) at Waubay National Wildlife Refuge, South Dacota, USA

Year	Total yearlings and adults	Goslings produced	Geese shot in vicinity of refuge
1945	92	58	
1946	70	43	
1947	130	53	
1948	138	43	
1949	186	91	
1950	164	110	154
1951	74	70	45
1952	74	46	44
1953	60	33	9
1954	65	60	26
1955	166	79	121
1956	60	90	36
1957	50	9	35
1958	102	33	44
1959	137	61	73
1960	89	49	48
1961	115	67	22
1962	50	38	29
	101	57	53

population of geese in Canada observed by Hanson (1976). The data set is shown in Table 7.1.

In Table 7.1, the population numbers of this special Canada goose population between the years 1945 and 1962 are depicted. Obviously, there are big fluctuations in the population size caused by big fluctuations in the birth rate but especially the death rate and hunting harvest. The data have already been used by De Angelis in his work "Application of stochastic models to a wildlife population" (1976) for the estimation of the life time of this specific goose population. The system parameters he estimated using continuous variables have led De Angelis to a criticism of the Master equation approach, since the calculated an unrealistically high life time of the goose population of about $\tau \approx 2.6 \times 10^7$ years.

Based on the data presented in Table 7.1, we calculate the system parameters a_1, a_2, b_2 or λ, μ, K, using the estimated values of $\langle n \rangle = 102, \sigma^2 = 2131$, the mean birth rate per year of $\langle \lambda_n \rangle = 57$ geese, and the mean death rate of $\langle \mu_n \rangle = 52.7$ geese per year. This leads to the parameters

$$\begin{aligned} a_1 &= 0.5587 \\ a_1^* &= 0.5642 \\ a_2 &= 0.5320 \\ b_2 &= 2.622 \times 10^{-4} \end{aligned} \tag{7.44}$$

Moreover, we have chosen these parameters (7.44) for a further comparison of the estimated life time using the conventional approach τ_{ca} as well as the quasi-stationary approach τ_{qs}

$$\tau_{qs} \approx 80 \text{ years} \quad \text{and} \quad \tau_{ca} \approx 2300 \text{ years}. \tag{7.45}$$

The discrete structure of the Master equation leads to a considerably shorter life time of the goose population (qs) by a factor of almost 30 compared with the classical approximation (ca). The unrealistically high life time after De Angelis is thus a consequence of the conventional approximation.

7.2 Sudden-Urban Growth

Why were there only six cities over one million inhabitants in 1860 and currently well over 400 is one of the greatest challenges to regional scientists. The tremendous growth of settlements has a big influence on the structure of our societies in the developed and the less developed parts of the world. The political and social consequences of shocks in urban growth or decline are till now not sufficiently analysed and there exists no adequate strategies or recommendations for planners and politicians. Many reasons may have a dominating influence on immigration and emigration flows resulting in a mass movement.

Since the growth rate of the total world population is lower than the rate of the cities population growth, sudden urban growth must feed on migration. This suggests that the migrant sees a clear advantage of living in a city over all other alternatives (Papageorgiou 1980). Furthermore, this attractiveness of cities leads to their sudden growth at a certain threshold of size followed by a more continuous growth afterwards.

In order to explain the phenomenon, Wheaton (1974) introduced, on the basis of microeconomic considerations a relationship between urban opportunities, the size of the city, and the size of the hinterland. According to conventional microeconomic principles, each individual tries to maximize his or her utility under given constraints (e.g. budget constraints). Therefore, if the urban utility exceeds the utility of living in the hinterland, a migration towards the city is expected. Assuming, as in Wheaton's (1974) linear utility functions, and that technological innovations favour urban areas more than the hinterland, only a rather smooth development of the City size can be explained, but not sudden urban growth.

To obtain sudden urban growth Papageorgiou (1980) proposed a third-order polynomial for the urban utility function with respect to city size. Shifts in the absolute level of the utilities now cause smooth changes of the population size of the city until an instability point is reached. A small further upward shift of the utility of the city then leads to a dramatic sudden transition until a new stable equilibrium point is reached.

The assumption of a third-order polynomial of the utility function, and especially that the population reacts in such a sensitive way to population size variations seems to be rather theoretical.

Therefore, the question must be raised whether less complicated assumptions can yield an explanation of sudden urban growth, if simultaneously the equilibrium condition is replaced by a dynamic stochastic consideration (Haag 1989).

7.2.1 A Master Equation Model for Shocks in Urban Evolution

Starting point of our consideration is the empirical fact that the take-off of cities cannot be explained by simple birth- and death processes only. Thus, sudden urban growth feeds on migration.

We consider a city c and its hinterland h. The number of people living in the city is n_c and of the hinterland n_h. Then the total population of the city-hinterland system is

$$n_c + n_h = 2N \tag{7.46}$$

and

$$n_c - n_h = 2n, \tag{7.47}$$

where N denotes half of the population size for mathematical convenience (see later). Neglecting birth and death processes in a first approach, we consider N as constant, while n remains the only relevant variable.

In general, not all residents of the city and the hinterland are potential migrants. Instead, there is an inert part and a mover pool part of the population. We consider the mover pool populations, only.

The transition

$$(n_c, n_h) \rightarrow (n_c + 1, n_h - 1) \quad \text{or equivalently} \quad n \rightarrow n + 1 \tag{7.48}$$

is effected by the migration of one member of the hinterland h to the city c. Vice versa

$$(n_c, n_h) \rightarrow (n_c - 1, n_h + 1) \quad \text{or equivalently} \quad n \rightarrow n - 1 \tag{7.49}$$

describes the transition of a member of the city c to the hinterland h. The corresponding configurational transition rates are now assumed to be represented by

$$w((n+1) \leftarrow n) \equiv w_{\uparrow}(n) = n_h p_{\uparrow}(n) = (N-n)p_{\uparrow}(n) \tag{7.50}$$

$$w((n-1) \leftarrow n) \equiv w_{\downarrow}(n) = n_c p_{\downarrow}(n) = (N+n)p_{\downarrow}(n) \tag{7.51}$$

and

$$w(n' \leftarrow n) = 0 \quad \text{for} \quad n' \neq n \pm 1, \tag{7.52}$$

where $p_{\uparrow}(n), p_{\downarrow}(n)$ are the individual transition rates. We assume that the individual transition rates have the following reasonable form

$$p_{\uparrow}(n) = \nu \exp(u(n_c) - v(n_h)) \tag{7.53}$$

$$p_{\downarrow}(n) = \nu \exp(v(n_h) - u(n_c)), \tag{7.54}$$

where the utility function of the city linearly depends on the size n_c of the city

$$u(n_c) = \alpha_0 + \alpha_1 n_c \tag{7.55}$$

and on the utility of living in the hinterland of n_h

$$v(n_h) = \beta_0 + \beta_1 n_h. \tag{7.56}$$

The utility difference $(u(n_c) - v(n_h))$ can therefore be expressed as a linear function of n

$$f(n) = u(n_c) - v(n_h) = \delta + \widetilde{\kappa} n. \tag{7.57}$$

The coefficients $\delta, \widetilde{\kappa}$ are related to the coefficients $\alpha_0, \alpha_1, \beta_0, \beta_1, N$.
Next, the configurational probability is introduced

$$P(n,t) = P(n_c, n_h, t) \tag{7.58}$$

with the normalization condition

$$\sum_{n=-N}^{N} P(n,t) = 1, \tag{7.59}$$

where the sum extends over all possible configurations. The Master equation for the city—hinterland system reads (next-neighbour transitions only)

$$\frac{dP(n,t)}{dt} = w_{\downarrow}(n+1)P(n+1,t) + w_{\uparrow}(n-1)P(n-1,t) - \big(w_{\downarrow}(n) + w_{\uparrow}(n)\big)P(n,t) \tag{7.60}$$

for $-N \leq n \leq N$. In Fig. 7.7, the structure of the one-dimensional Master equation is shown.

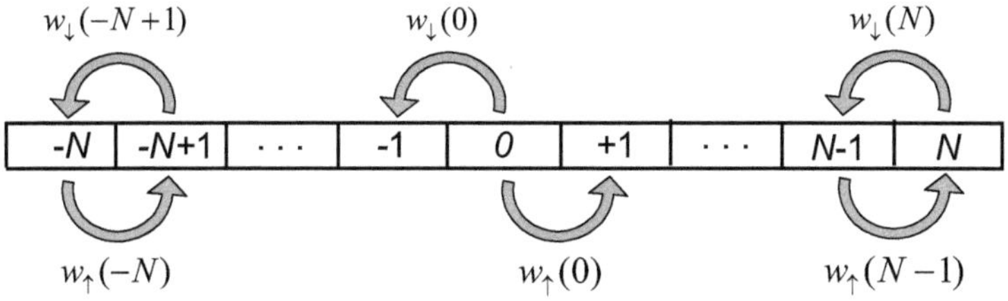

Fig. 7.7 Possible next-neighbour transitions of the one-dimensional Master equation for the city—hinterland system

7.2.2 *Stationary Solution of the City-Hinterland Master Equation*

Since (7.60) fulfills the condition of detailed balance

$$w_\downarrow(n+1)P_{st}(n+1) = w_\uparrow(n)P_{st}(n), \tag{7.61}$$

for $-N \leq n \leq N$, the exact stationary solution of the Master equation (7.60) can be obtained by repeated application of (7.61) as described in Chap. 5

$$P_{st}(n) = P_{st}(0)\prod_{m=1}^{n}\frac{w_\uparrow(m-1)}{w_\downarrow(m)} \quad \text{for} \quad +1 \leq n \leq N \tag{7.62}$$

$$P_{st}(n) = P_{st}(0)\prod_{m=1}^{n}\frac{w_\downarrow(m+1)}{w_\uparrow(m)} \quad \text{for} \quad -N \leq n \leq -1, \tag{7.63}$$

where $P_{st}(0)$ is fixed by the normalization condition (7.59). After inserting (7.53)–(7.56) in (7.62) and (7.63) the explicit stationary solution of the city-hinterland system can be computed:

$$P_{st}(n) = P_{st}(0)\frac{(N!)^2}{(2N)!}\binom{2N}{N+n}\exp(2F(n)), \tag{7.64}$$

where

$$F(n) = \sum_{m=1}^{n} f(m). \tag{7.65}$$

A further evaluation of the binomial coefficients, using Stirling's formula

$$\ln N! \approx N\ln N - N \tag{7.66}$$

yields with the introduction of the quasi-continuous variable

$$x = n/N \quad \text{with} \quad -1 \leq x \leq 1 \tag{7.67}$$

and (2.34) a more convenient form of the stationary probability distribution

$$P_{st}(Nx) = P_{st}(0)\exp(NU(x)) \tag{7.68}$$

with the potential $U(x)$

$$U(x) = 2F(xN) - ((1+x)\ln(1+x) + (1-x)\ln(1-x)). \tag{7.69}$$

The sum in the term $F(xN)$ can be replaced approximately by the integral

$$F(xN) \approx \int_0^x f(x')dx'. \tag{7.70}$$

The extreme values x_m of the distribution function (7.68) are determined by

$$\left.\frac{\partial U(x)}{\partial x}\right|_{x=x_m} = 2f(x_m) - (\ln(1+x_m) - \ln(1-x_m)) = 0, \tag{7.71}$$

or after introducing the hyperbolic tangent

$$x_m = \tanh f(x_m) = \tanh(\delta + \kappa x_m), \tag{7.72}$$

where δ describes the constant term and κ the scaled linear part of the utility difference function (7.57). Equation (7.72) can be solved graphically by depicting the left-hand side (lhs) and the right-hand side (rhs) of (7.72) and discussing the points of intersection (Fig. 7.8).

In case of $0 \leq \kappa \leq 1$, that means for moderate agglomeration trends there exists always only one stationary solution of (7.72) corresponding to a distribution function having one maximum. This maximum is located in the domain $-1 \leq x_m < 0$, or $0 \leq x_m \leq 1$ for $\delta < 0$, or $\delta > 0$, respectively.

However, for a large agglomeration parameter $\kappa > 1$, and if the absolute value of the shift parameter δ exceeds a critical value $|\delta| > \delta_c$, there exists also only one stationary point of (7.72). This case is depicted in the Fig. 7.8a, d.

The critical value of the shift parameter δ_c is determined by the equation

$$\cosh\left(\delta_c - \sqrt{\kappa(\kappa-1)}\right) = \kappa. \tag{7.73}$$

There are three solutions of (7.72) corresponding to two maxima and one minima of the distribution function, if $\kappa > 1$, and if $|\delta| < \delta_c$ according to Fig. 7.8c.

Let us now interpret these results in the sense of sudden urban growth.

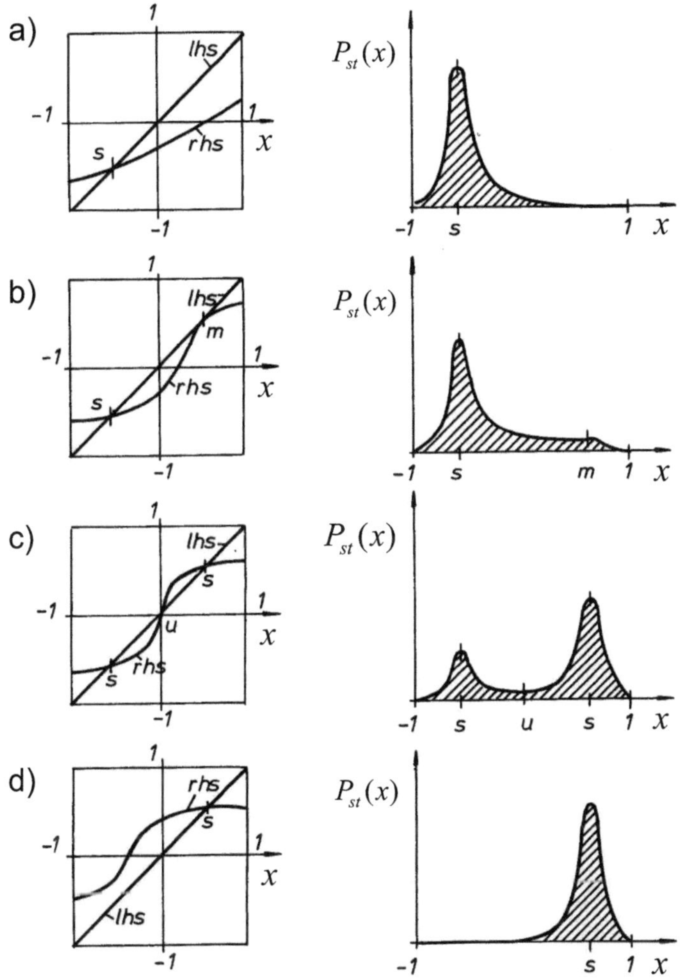

Fig. 7.8 Stationary points and stationary probability distribution for different values of the shift parameter δ, for $\kappa > 1$ (Source: Haag 1989)

We distinguish three phases of development:

(a) If there is no pronounced agglomeration trend, this means, if $0 \leq \kappa < 1$, the maximum of the distribution function $P_{st}(x)$ shifts with the preference parameter $\delta(t)$ slowly, and no sudden growth of the city will happen.
(b) The situation changes dramatically if we start from a preference of the population for the hinterland $\delta < 0$ and assume a slow increase of the preference of the population towards the city. In addition, a large tendency to agglomerate exists $\kappa > 1$. At first, a second maximum of the distribution function appears and this will grow in its height relative to the old one since δ will also change its sign. However, if the trend parameter δ has reached its positive critical value

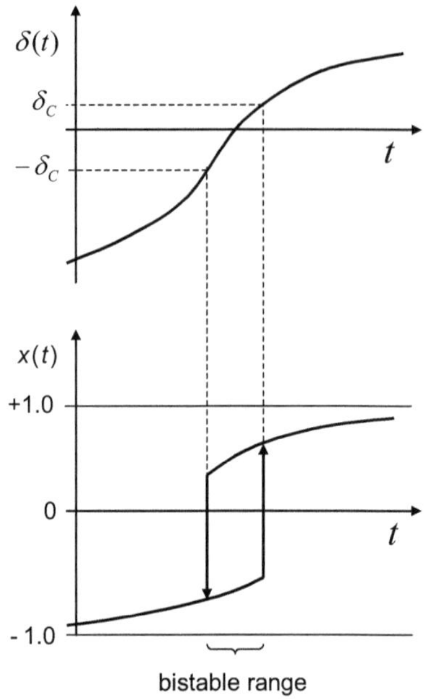

Fig. 7.9 A possible scenario leading to sudden urban growth. Shift of the preference parameter δ and response of the population partition between hinterland and city population ($\kappa > 1$)

$\delta_c > 0$, the population will favour life in the city and the first maximum representing the hinterland will disappear. A sudden increase of the population of the city happens (Fig. 7.9). The slow change of the preference parameter, which is the control parameter, however, leads to a large and sudden shift of the population partition in favour of the city area.

As a main result, we conclude that sudden urban growth may happen even with linear utility functions. However, the trend parameters have to fulfil the following requirements:

(a) The tendency to agglomerate of the population has to be rather large, $\kappa > 1$ since otherwise, only a smooth development of the population share will happen
(b) $d\delta/dt > 0$ in the transition region, in other words, the attractiveness of the city grows at least in a certain interval of time.

The main reason for the sudden increase in the population share of the city is the self-accelerating character of the agglomeration effect: the more people there are in the city, the more people are attracted to migrate to the city. In the transition region, the population distribution is in a metastable state since $P(n, t) \neq P_{st}(n)$ and the expected escape rate (see Chap. 5) will be very slow. Hence, we have to consider the full dynamics of the Master equation or derive appropriate equations of motion for the mean values and variances.

7.2.3 Mean Value and Variance Equations for the City-Hinterland System

As described in Chaps. 3 and 4, the dynamic mean value and variance equations can be derived directly from the Master equation (7.60), using the definition

$$\langle g(n)\rangle_t = \sum_{n=-N}^{N} g(n)P(n,t) \tag{7.74}$$

by multiplying the right-hand side and the left-hand side with $g(n)$, respectively, and summing up over all possible states $-N \leq n \leq N$. This leads to the following exact results:

$$\frac{d\langle n\rangle_t}{dt} = \langle w_\uparrow(n)\rangle - \langle w_\downarrow(n)\rangle \tag{7.75}$$

and

$$\frac{d\langle n^2\rangle_t}{dt} = 2\langle n\left(w_\uparrow(n) - w_\downarrow(n)\right)\rangle + \langle w_\uparrow(n)\rangle + \langle w_\downarrow(n)\rangle. \tag{7.76}$$

For essential unimodal probability distributions, one may approximate

$$\langle g(n)\rangle_t \approx g\left(\langle n\rangle_t\right) \tag{7.77}$$

and thus arrive at self-contained although approximate equations of motion. Introducing the scaled quasi-continuous population share

$$\langle x\rangle_t = \frac{\langle n\rangle_t}{N} \tag{7.78}$$

and the variance

$$\sigma^2(t) = \langle x^2\rangle_t - \langle x\rangle_t^2 \quad \text{and} \quad \Sigma^2(t) = N\sigma^2(t), \tag{7.79}$$

the equations of motion can be derived from (7.75), (7.76) using (7.50), (7.51):

$$\frac{d\langle x\rangle_t}{dt} = K\left(\langle x\rangle_t\right) \tag{7.80}$$

$$\frac{d\Sigma^2(t)}{dt} = 2K'\left(\langle x\rangle_t\right)\Sigma^2(t) + Q\left(\langle x\rangle_t\right) \tag{7.81}$$

with the abbreviations

$$\begin{aligned} K(\langle x\rangle_t) &= N^{-1}w_\uparrow(\langle x\rangle_t) - N^{-1}w_\downarrow(\langle x\rangle_t) \\ &= 2\nu(\sinh(f(\langle x\rangle_t)) - \langle x\rangle_t\cosh(f(\langle x\rangle_t))) \end{aligned} \tag{7.82}$$

$$\begin{aligned} Q(\langle x\rangle_t) &= N^{-1}w_\uparrow(\langle x\rangle_t) + N^{-1}w_\downarrow(\langle x\rangle_t) \\ &= 2\nu(\cosh(f(\langle x\rangle_t)) - \langle x\rangle_t\sinh(f(\langle x\rangle_t))). \end{aligned} \tag{7.83}$$

The mean value equation which describes the migration between the hinterland h and the city c when

$$f(\langle x\rangle_t) = \delta(t) + \kappa(t)\langle x\rangle_t \tag{7.84}$$

is the corresponding utility gain function then reads

$$\frac{d\langle x\rangle_t}{dt} = 2\nu(\sinh(f\langle x\rangle_t) - \langle x\rangle_t\cosh(f\langle x\rangle_t)). \tag{7.85}$$

Equation (7.85) is a nonlinear differential equation for the population share variable $\langle x\rangle_t$. The stationary solution of (7.85) can be read off immediately and agrees with the transcendental equation of the stationary probability distribution $P_{st}(n)$. This demonstrates the consistency of the mean value equations with the stochastic approach.

Assuming that the solution of (7.85) has been found, it is easy to find the explicit solution of the variance equations (7.81)

$$\Sigma^2(t) = \Sigma^2(0)\left[\frac{K(\langle x\rangle_t)}{K(\langle x\rangle_0)}\right]^2 + K^2(\langle x\rangle_t)\int_{x(0)}^{x(t)} \frac{Q(\langle x\rangle_t)}{K^3(\langle x\rangle_t)}dx. \tag{7.86}$$

This solution (7.86) is non-divergent only for $K(\langle x\rangle_0) \neq 0$ and describes the dynamics of the fluctuations of the migration flow around the trajectory $\langle x\rangle_t$.

If a city-hinterland system moves in the critical domain, characterized by (7.73) an enhancement of the fluctuations of the migratory flows occur, anticipating the phase transition from a more rural phase of living to an urban dominated population. Therefore, if we regard shocks in urban evolution as structural phase transitions of a city-hinterland system, the evolution of the fluctuations can be used as a possible indicator of instability (see Fig. 7.10).

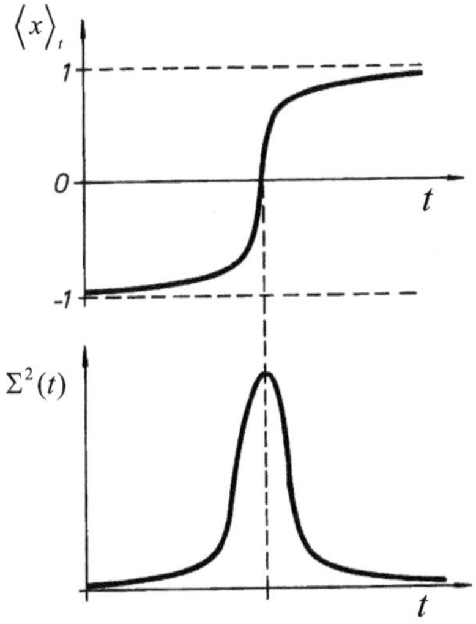

Fig. 7.10 Expected evolution of the population share $\langle x \rangle_t$ and the corresponding dynamics of the fluctuations of the migratory flows $\Sigma^2(t)$

7.3 Predator-Prey-Interaction

The predator-prey interaction is one of the fundamental processes of biology. A very simple but often used and discussed model for the description of predator-prey interaction is the famous Volterra-Lotka model (Lotka 1920; Volterra 1927) whose authors independently of each other developed the theoretical structure of this model.

The Volterra-Lotka equations have a long history of use in economic theory and in geography Dendrinos (1985), Haag and Dendrinos (1983), Dendrinos and Haag (1984), despite their importance in ecology and biology (Haken 2004; Ahmad 1993). Many researchers have been attracted to generalize and extend the underlying system of equations in many respects (May 1973; Ludwig 1974; Goel and Richter-Dyn 1974). In this book, our intention is to compare the classical Volterra-Lotka equations with the Master equation approach.

The classical Volterra-Lotka equations represent a pair of first-order, non-linear, differential equations used to describe the dynamics of biological systems in which two species interact, one as a predator and the other as a prey. We will consider a special case of the Volterra-Lotka model by allowing the prey species to migrate to a habitat where the predator species cannot follow. In this case the unstable cycles of the Volterra-Lotka case will be replaced by a stable limit cycle.

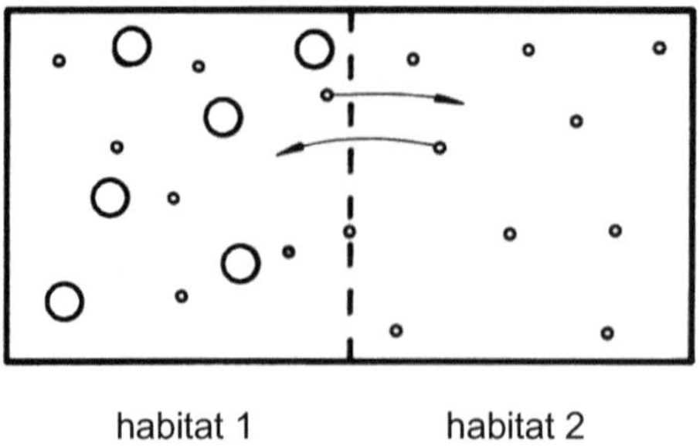

Fig. 7.11 Prey (*small circle*) and predator (*large circle*) population in two habitats 1 and 2. Only the prey population is able to migrate between habitat 1 and habitat 2

7.3.1 The Master Equation for the Volterra-Lotka Model with an Refuge Habitat

Two habitats are assumed, an open habitat 1 and a refuge habitat 2 (Weidlich and Haag 1983). We further assume that the prey species are able to migrate between both habitats while only habitat 1 is accessible for predator species (see Fig. 7.11).

Then the configuration consists of three integers (m_1, m_2, n) the numbers of prey species in both habitats m_1, m_2 and the number of predator species in habitat 1, called n.

In the following, the transition rates for the birth- and death events of the predator and prey species as well as the prey-predator interaction are introduced:

(a) Migration of prey species

$$w_{21}(-1,1,0;m_1,m_2.n) = m_1 p_{21}(m_1,m_2,n) = \widetilde{\nu} m_1 \exp(u(m_1,m_2,n)) \tag{7.87}$$
$$w_{12}(1,-1,0;m_1,m_2.n) = m_2 p_{12}(m_1,m_2,n) = \widetilde{\nu} m_2 \exp(-u(m_1,m_2,n)), \tag{7.88}$$

where $u(m_1, m_2, n)$ describes the propensity of the prey species for one of the two habitats. We assume that the number of predator species influences their propensity, an increasing number of predator species will shift the propensity for living towards habitat 2, where the predators cannot follow. However, it is also plausible to assume that habitat 1 is more preferred by the prey species since food resources are only or better available in habitat 1 than in habitat 2. Often, prey species tend to prefer living in swarms in contrast to predator species that prefer individual hunting.

Therefore, we model the propensity to migrate as following

$$u(m_1,m_2,n) = \delta + \alpha(m_1 - m_2) - \beta n, \tag{7.89}$$

where the parameter $\delta \geq 0$ describes the propensity of the prey species for habitat 1, $\alpha > 0$ the propensity of the prey species for clustering in swarms, and the fear parameter $\beta > 0$ the strength of fear of the prey to be hunted and eaten.

(b) Birth- and death events of the prey population

$$w_{1+}(1,0,0;m_1,m_2,n) = m_1 a_1 \tag{7.90}$$
$$w_{1-}(-1,0,0;m_1,m_2,n) = m_1 b_1 + m_1^2 c_1 \tag{7.91}$$
$$w_{1+}(0,1,0;m_1,m_2,n) = m_2 a_2 \tag{7.92}$$
$$w_{1-}(0,-1,0;m_1,m_2,n) = m_2 b_2 + m_2^2 c_2 \tag{7.93}$$

(c) Birth- and death events of the predator population

$$w_{1+}(0,0,1;m_1,m_2,n) = n a_3 \tag{7.94}$$
$$w_{1-}(0,0,-1;m_1,m_2,n) = n b_3 + n^2 c_3 \tag{7.95}$$

(d) Prey-predator interaction

$$w_{1+}(-1,0,1;m_1,m_2,n) = m_1 n \mu_1. \tag{7.96}$$

We now restrict ourselves to a more simplified predator-prey system. For convenience, the parameters are specialized and a simplified notation for numerical simulations is given:

$$\begin{aligned} a_1 &= a_2 = \kappa_1 & a_3 &= 0 \\ b_1 &= b_2 = 0 & b_3 &= \kappa_2 \\ c_1 &= 0 & c_3 &= 0 \\ c_2 &= \omega & \mu_1 &= \gamma. \end{aligned} \tag{7.97}$$

The total configurational transition rate $w(k_1,k_2,k_3;m_1,m_2,n)$ is the sum of the transition rates concerning the contributions (a)–(d). Therefore, we can cast the Master equation for the probability distribution in the general form

$$\begin{aligned} \frac{dP(m_1,m_2,n)}{dt} = \left(\frac{\partial P}{\partial t}\right)_{Mig}(m_1,m_2,n) + \left(\frac{\partial P}{\partial t}\right)_{BD}(m_1,m_2,n) \\ + \left(\frac{\partial P}{\partial t}\right)_{VL}(m_1,m_2,n), \end{aligned} \tag{7.98}$$

where the three terms on the rhs of the Master equation are explicitly given by

$$\begin{aligned} \left(\tfrac{\partial P}{\partial t}\right)_{Mig}(m_1,m_2,n) = &(m_1+1)p_{21}(m_1+1,m_2-1,n)P(m_1+1,m_2-1,n) \\ &-m_1 p_{21}(m_1,m_2,n)P(m_1,m_2,n) \\ &+(m_2+1)p_{12}(m_1-1,m_2+1,n)P(m_1-1,m_2+1,n) \\ &-m_2 p_{12}(m_1,m_2,n)P(m_1,m_2,n) \end{aligned} \tag{7.99}$$

$$\left(\frac{\partial P}{\partial t}\right)_{BD}(m_1,m_2,n) = \kappa_1(m_1-1)P(m_1-1,m_2,n) - \kappa_1 m_1 P(m_1,m_2,n)$$
$$+\kappa_1(m_2-1)P(m_1,m_2-1,n) - \kappa_1 m_2 P(m_1,m_2,n)$$
$$+\omega(m_2+1)^2 P(m_1,m_2+1,n) - \omega m_2^2 P(m_1,m_2,n)$$
$$+\kappa_2(n+1)P(m_1,m_2,n+1) - \kappa_2 n P(m_1,m_2,n) \qquad (7.100)$$

$$\left(\frac{\partial P}{\partial t}\right)_{VL}(m_1,m_2,n) = \gamma(m_1+1)(n-1)P(m_1+1,m_2,n-1)$$
$$-\gamma m_1 n P(m_1,m_2,n). \qquad (7.101)$$

The only stationary solution of the Master equation is again given by all species are died out, in other words

$$P_{st}(m_1,m_2,n) = \delta_{m1,0}\cdot\delta_{m2,0}\cdot\delta_{n,0}. \qquad (7.102)$$

7.3.2 Mean Value Equations for the Volterra-Lotka Model with Migration

Applying the tool box of Chap. 4 we obtain approximate closed equations of motion for the mean number of prey species in habitat 1, $\tilde{x}\equiv\langle m_1\rangle$, and habitat 2, $\tilde{y}\equiv\langle m_2\rangle$, and the mean number of predator species in habitat 1, $\tilde{z}\equiv\langle n\rangle$

$$\frac{d\tilde{x}}{dt} = -(\tilde{x}p_{21}(\tilde{x},\tilde{y},\tilde{z}) - \tilde{y}p_{12}(\tilde{x},\tilde{y},\tilde{z})) + \kappa_1\tilde{x} - \gamma\tilde{x}\cdot\tilde{z} \qquad (7.103)$$
$$\frac{d\tilde{y}}{dt} = +(\tilde{x}p_{21}(\tilde{x},\tilde{y},\tilde{z}) - \tilde{y}p_{12}(\tilde{x},\tilde{y},\tilde{z})) + \kappa_1\tilde{y} - \omega\tilde{y}^2 \qquad (7.104)$$
$$\frac{d\tilde{z}}{dt} = -\kappa_2\tilde{z} + \gamma\tilde{x}\cdot\tilde{z}. \qquad (7.105)$$

After a transformation to scaled variables and scaled time

$$x = \frac{\kappa_2}{\gamma}\tilde{x} \qquad y = \frac{\kappa_2}{\gamma}\tilde{y} \qquad z = \frac{\kappa_1}{\gamma}\tilde{z} \qquad \tau = \kappa_1 t \qquad (7.106)$$

as well as to scaled trend parameters

$$\nu = \frac{\widetilde{\nu}}{\kappa_1} \qquad \delta = \delta \qquad \alpha = \frac{\gamma}{\kappa_2}\widetilde{\alpha} \qquad \beta = \frac{\gamma}{\kappa_1}\widetilde{\beta} \quad \frac{1}{y_0} = \frac{\omega}{\kappa_1} \cdot \frac{\kappa_2}{\gamma} \qquad a = \frac{\kappa_2}{\kappa_1} \tag{7.107}$$

and the scaled preference function

$$u = \delta + \alpha(x - y) - \beta z, \tag{7.108}$$

scaled mean value equations for the Volterra-Lotka system with migration interaction is obtained

$$\frac{dx}{d\tau} = -\nu\,(x\exp(-u) - y\exp(+u)) + x(1 - z) \tag{7.109}$$

$$\frac{dy}{d\tau} = +\nu\,(x\exp(-u) - y\exp(+u)) + \left(\frac{y_0 - y}{y_0}\right)y \tag{7.110}$$

$$\frac{dz}{d\tau} = -a(1 - x)z. \tag{7.111}$$

These mean value equations are the starting point for the following analysis.

7.3.3 Singular Points and Stability

As far as singular points and their stability are concerned, standard analytical methods will be applied. Evidently, the singular points $P_s(x_s, y_s, z_s)$ of the dynamic system are solutions of the three transcendental equations

$$0 = -\nu(x_s \exp(-u(x_s, y_s, z_s)) - y_s\exp(+u(x_s, y_s, z_s))) + x_s(1 - z_s) \tag{7.112}$$

$$0 = +\nu(x_s \exp(-u(x_s, y_s, z_s)) - y_s\exp(+u(x_s, y_s, z_s))) + \left(\frac{y_0 - y_s}{y_0}\right)y_s \tag{7.113}$$

$$0 = -a(1 - x_s)z_s. \tag{7.114}$$

The linearized equations of motion around the singular points $P_s(x_s, y_s, z_s)$ are then

$$\frac{d}{dt}\begin{pmatrix} \xi \\ \eta \\ \zeta \end{pmatrix} = \begin{pmatrix} a_{11} & a_{12} & a_{13} \\ a_{21} & a_{22} & a_{23} \\ a_{31} & a_{32} & a_{33} \end{pmatrix}\begin{pmatrix} \xi \\ \eta \\ \zeta \end{pmatrix}, \tag{7.115}$$

where

$$\begin{aligned} x(\tau) &= x_s + \xi(\tau) \\ y(\tau) &= y_s + \eta(\tau) \\ z(\tau) &= z_s + \zeta(\tau) \end{aligned} \tag{7.116}$$

and where the matrix elements are given by

$$\begin{aligned} a_{11} &= 1 - z_s + \nu(ax_s - 1)\exp(-u_s) + \nu ay_s \exp(u_s) \\ a_{12} &= -(\nu\, ax_s\exp(-u_s) + \nu(ay_s - 1)\exp(u_s)) \\ a_{13} &= -(x_s + \nu\beta x_s\exp(-u_s) + \nu\beta y_s\exp(u_s)) \\ a_{21} &= -(\nu(ax_s - 1)\exp(-u_s) - \nu ay_s\exp(u_s)) \\ a_{22} &= \frac{(y_0 - y_s)}{y_0} - \frac{y_s}{y_0} + \nu ax_s\exp(-u_s) + \nu(ay_s - 1)\exp(u_s) \\ a_{23} &= \nu\beta x_s\exp(-u_s) + \nu\beta y_s\exp(u_s) \\ a_{31} &= az_s \\ a_{32} &= 0 \\ a_{33} &= -a(1 - x_s) \end{aligned} \tag{7.117}$$

For the performing of the linear stability analysis, we solve (7.115) by using

$$\begin{pmatrix} \xi(\tau) \\ \eta(\tau) \\ \zeta(\tau) \end{pmatrix} = \begin{pmatrix} \xi_0 \\ \eta_0 \\ \zeta_0 \end{pmatrix} \exp(\lambda\tau), \tag{7.118}$$

where the eigen values λ are determined via the secular equation

$$\begin{vmatrix} a_{11} - \lambda & a_{12} & a_{13} \\ a_{21} & a_{22} - \lambda & a_{23} \\ a_{31} & a_{32} & a_{33} - \lambda \end{vmatrix} = 0. \tag{7.119}$$

Stability of a particular singular point $P_s(x_s, y_s, z_s)$ is guaranteed if $\text{Re}(\lambda_j) \leq 0$ for $j = 1, 2, 3$.

In the following, we analyse two special cases in more detail:

(A) The pure Volterra-Lotka model by neglecting migration of the prey species
(B) The full model including migration and saturation of the prey species and taking into account saturation effects in the second habitat.

7.3.4 Case A: The Pure Volterra-Lotka Model

The pure Volterra-Lotka model is obtained by setting the mobility parameter of the prey species to zero $\nu = 0$ leading to vanishing migration effects, in other words the

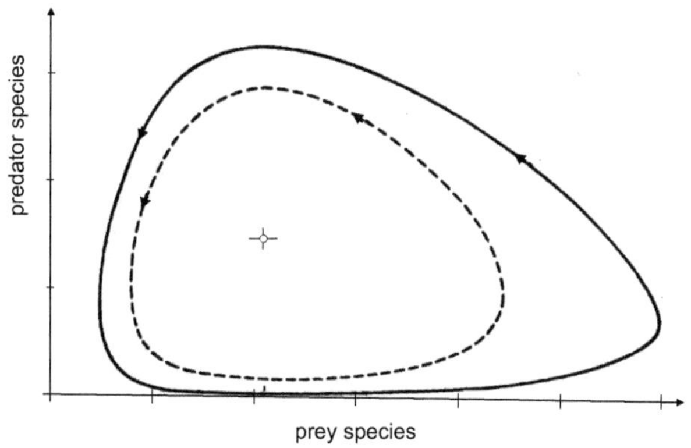

Fig. 7.12 Typical Volterra-Lotka cycles in the x-z-plane. Different cycles appear for different initial conditions

two habitats become decoupled. Therefore, it is enough to consider only habitat 1. From the equations (7.112) and (7.114) two singular points are found

$$P_s^{(0)}(x=0, z=0) \qquad \text{and} \qquad P_s^{(1)}(x=1, z=1). \tag{7.120}$$

The corresponding eigen values of the linearized equations (7.115) obey (7.119) are

$$\begin{aligned} \lambda_1^{(0)} &= 1, \lambda_2^{(0)} = -a \\ \lambda_1^{(1)} &= i\sqrt{a}, \lambda_2^{(1)} = -i\sqrt{a} \end{aligned} \tag{7.121}$$

for the two singular points $P_s^{(0)}(x=0, z=0)$ and $P_s^{(1)}(x=1, z=1)$, respectively. Hence, the point $P_s^{(0)}(x=0, z=0)$, where all prey and predator species are died out is unstable, in contrast to the exact solution of the Master equation, where this would be the only stable point. Moreover, the stationary point $P_s^{(1)}(x=1, z=1)$ represents a "centre" surrounded by closed orbits (see Figs. 7.12 and 7.13). The precise form of the orbits around the centre can be easily determined using the exact equations of motion (7.109)–(7.111). It can also be shown that the orbits exhibit the constant of motion (7.122).

$$[x(\tau) - \ln x(\tau)] + \frac{1}{a}[z(\tau) - \ln z(\tau)] = K. \tag{7.122}$$

The constant of motion K is determined by the initial conditions $x(0), z(0)$ and determines the shape of all orbits in the $x - z$ plane. Small perturbations, in other words, changes in the numbers of prey or predator species, e.g. due to environmental events or hunting change the orbit. The orbits of the Volterra-Lotka model are not stable against perturbations.

Yet we must add that the predator-prey relationship is only one of several factors that may cause population cycles. Other factors may be migration, genetic changes and overpopulation. In the laboratory, it is very difficult to run whole population

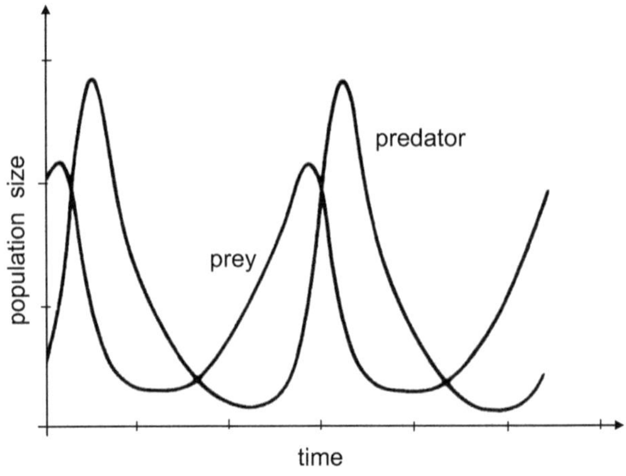

Fig. 7.13 Typical Volterra-Lotka cycles x(t), z(t)

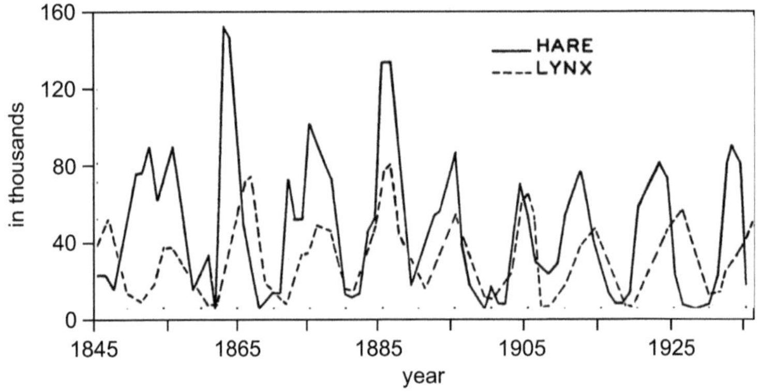

Fig. 7.14 The hare—lynx cycle in Canada. Changes in the abundance of the lynx and the snowshoe hare (after D. A. McLulich: Fluctuations in the numbers of varying hare, Univ. Toronto Press, Toronto 1937)

cycles. So, what is wrong with the Volterra-Lotka equations? They are too simplistic although the underlying assumptions seem to be correct.

Figure 7.14 depicts temporal oscillations in the numbers of lynx and snowshoe hares approximately measured by the number of pelts received by the Hudson Bay Company.

Despite the instability of the Volterra-Lotka cycles against perturbations the observed stability of real natural prey-predator systems (see Fig. 7.14) is sometimes surprising. A wide range of observations of actual predator-prey systems can be explained, in fact if one only takes a few relatively minor changes in the modelling framework.

An important extension of the Volterra-Lotka model consists in giving the preys a refuge, in which they are safe. Hence, there will always exist a certain number of prey organisms that cannot be reached by predators. Such hiding places exist in nature frequently and are probably for a considerable number of balanced predator-prey systems responsible. This may lead to stable cycles (limit cycles) instead of unstable closed curves. Therefore, we will now consider case B.

7.3.5 Case B: Volterra-Lotka Model with Migration

The full Volterra-Lotka model (7.109)–(7.111) with migration facility of the prey organisms and saturation in the refuge habitat will be considered.

Equations (7.112)–(7.114) have the trivial singular point $P_s^{(0)}(x=0, y=0, z=0)$ which turns out to be unstable (as expected)

$$\lambda_1^{(0)} = -a, \lambda_2^{(0)} = 1, \lambda_3^{(0)} = 1 - 2\nu\cosh(\delta). \tag{7.123}$$

For the non-trivial solutions it follows from (7.114) that either yield

B (i) $x_s = 1$ or B (ii) $z_s = 0$.

Case B (i)
It follows from (7.112) and (7.113) by elimination of the migration term the relation

$$z_s = 1 + \left(\frac{y_0 - y_s}{y_0}\right) y_s, \tag{7.124}$$

which can be inserted into (7.113) to yield the transcendental equation for y_s

$$\nu y_s \exp\left(w^{(i)}(y_s)\right) = \left(\frac{y_0 - y_s}{y_0}\right) y_s + \nu\exp\left(-w^{(i)}(y_s)\right) \tag{7.125}$$

with

$$w^{(i)}(y_s) = \rho - \alpha y_s - \beta\left(\frac{y_0 - y_s}{y_0}\right) y_s, \tag{7.126}$$

$$\text{and where } \rho = \delta + \alpha - \beta. \tag{7.127}$$

Equation (7.125) can be solved graphically by plotting both, the lhs and the rhs as functions of y_s. Up to this point, the discussion of (7.125) has been completely general. However, for the further analysis, we suppose the following particular but reasonable trend parameters

$$\beta = \delta > 0 \qquad \text{and} \qquad y_0 = 1. \tag{7.128}$$

This choice implies that the fear parameter of the prey species β equals its preference parameter δ for habitat 1, where the predator species lurk. A plausible explanation for this choice could be, for instance, that food for the prey organisms is better in the more dangerous habitat 1 than in habitat 2 where the predators cannot follow. Furthermore, without loss of generality a specific saturation level $y_0 = 1$ is assumed. For this particular set of trend parameters, the solution of (7.112)–(7.114) is rather simple and reads

$$P_s^{(1)}(x_s = 1, y_s = 1, z_s = 1). \tag{7.129}$$

The stability analysis for this specific stationary point also becomes reasonably transparent, and the secular equation (7.119) assumes the form

$$P(\lambda) = C_0\lambda^3 + C_1\lambda^2 + C_2\lambda + C_3 = 0 \tag{7.130}$$

with the abbreviations

$$\begin{aligned} C_0 &= 1 \\ C_1 &= 1 - 2r \\ C_2 &= a(1 + s) - r \\ C_3 &= a(1 - r + s) \end{aligned} \tag{7.131}$$

$$\begin{aligned} r &= \nu(2\alpha - 1) \\ s &= 2\nu\beta \end{aligned}$$

The necessary and sufficient condition for the asymptotic stability of the singular point $P_s^{(1)}(x_s = 1, y_s = 1, z_s = 1)$ is that all solutions to (7.130) have negative real parts. This condition is fulfilled if the Hurwitz criterion (Haken 1978; Hurwitz 1895) holds:

$$\text{(a)} \quad \frac{C_n}{C_0} > 0 \qquad \text{for} \qquad n = 1, 2, 3 \tag{7.132}$$

$$\text{(b)} \quad \begin{aligned} H_1 &= C_1 > 0 \\ H_2 &= C_1C_2 - C_0C_3 > 0. \\ H_3 &= C_3H_2 > 0 \end{aligned} \tag{7.133}$$

The straightforward calculation of the Hurwitz criterion for the coefficients of (7.130) leads to the simple condition for the clustering parameter

$$\alpha < 1/2 \tag{7.134}$$

for the asymptotic stability of the singular point $P_s^{(1)}(x_s = 1, y_s = 1, z_s = 1)$.

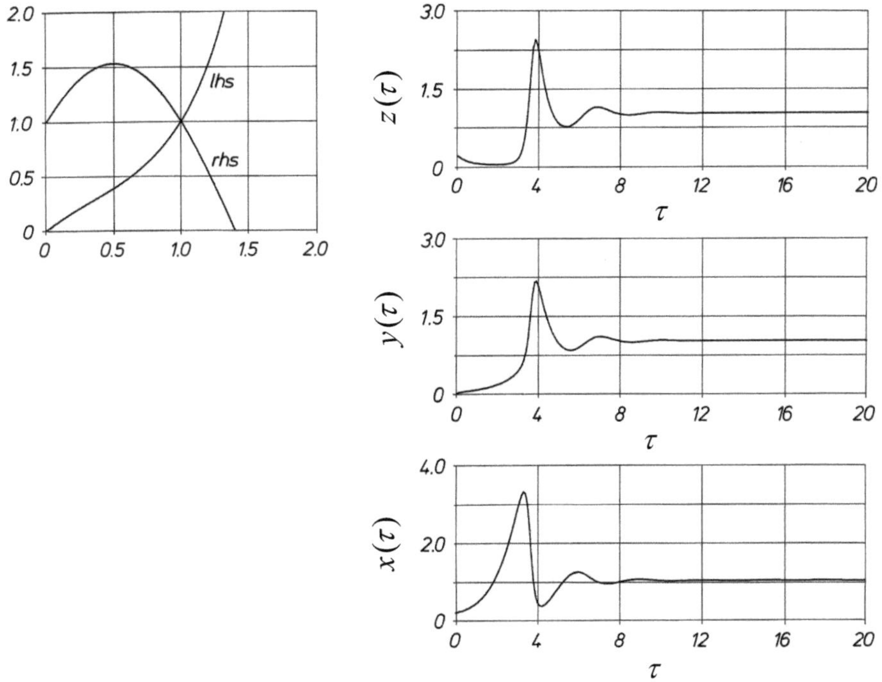

Fig. 7.15 Graphical solution of (7.125) for $\alpha = 0, \beta = 1, \delta = 1, y_0 = 1, \nu = 1, a = 2$ and the trajectories of $x(\tau), y(\tau), z(\tau)$ (Source: Weidlich and Haag 1983)

In the following simulations for three different values of the clustering parameter α, and $\beta = \delta = y_0 = \nu = 1$ and $a = 2$ are presented (Figs. 7.15, 7.16 and 7.17). Beside the graphical solution of (7.125), the time-dependence of the system variables $x(\tau)$, $y(\tau), z(\tau)$ is shown, respectively. Figure 7.15 shows a damped time dependence of the system variables approaching the stable focus $P_s^{(1)}(x_s = 1, y_s = 1, z_s = 1)$ for a clustering parameter of the prey species $\alpha = 0$. The case of a critical slowing down of the oscillatory behaviour of all species is shown in Fig. 7.16 for $\alpha = 0.5$. If the agglomeration parameter α of the prey species exceeds the critical value 0.5, a stable limit cycle is obtained, as shown in Fig. 7.17. Independent of the initial conditions, all the variables $x(\tau), y(\tau), z(\tau)$ approach a regular pulsatory time-dependence. In this case, the singular point $P_s^{(1)}(x_s = 1, y_s = 1, z_s = 1)$ has become unstable for $\alpha = 1.5$.

How can we interpret this result? For a small clustering behaviour of the prey species an equilibrium situation is reached and is characterized by a steady stream of prey species from the refuge (habitat 2) to the open habitat 1, which finally leads to a steady number of prey species in both habitats and a steady number of predator species.

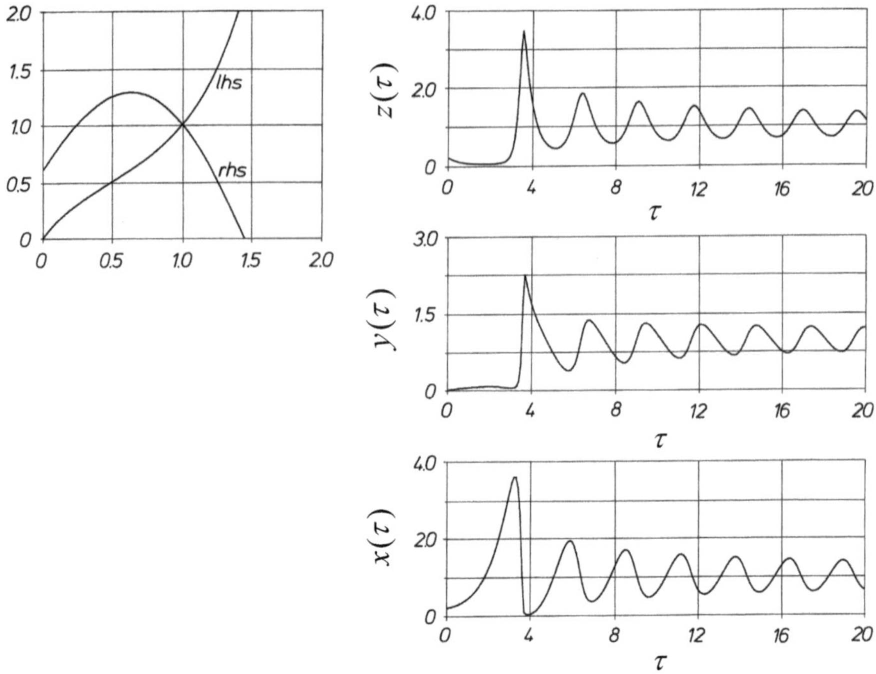

Fig. 7.16 Graphical solution of (7.125) for $\alpha = 0.5\,, \beta = 1\,, \delta = 1\,, y_0 = 1\,, \nu = 1\,, a = 2$ and the trajectories of $x(\tau)\,, y(\tau)\,, z(\tau)$ (Source: Weidlich and Haag 1983)

However, for a large trend parameter α, the prey species tend to swarm formation. The migration of the prey between the two habitats takes place periodically in swarms and thus induces a periodic variation in the predator population as well.

Case B (ii)

The case with $z_s = 0$ is also of interest. Inserting $z_s = 0$ in (7.112) and (7.113) yield the relation

$$x_s = -\left(\frac{y_0 - y_s}{y_0}\right) y_s, \tag{7.135}$$

which can be inserted in (7.113) in order to obtain the transcendental equation (7.136) for y_s

$$\left(\frac{y_0 - y_s}{y_0}\right) \exp\left(-w^{(ii)}(y_s)\right) + \exp\left(w^{(ii)}(y_s)\right) = \frac{1}{\nu}\left(\frac{y_0 - y_s}{y_0}\right) \tag{7.136}$$

with

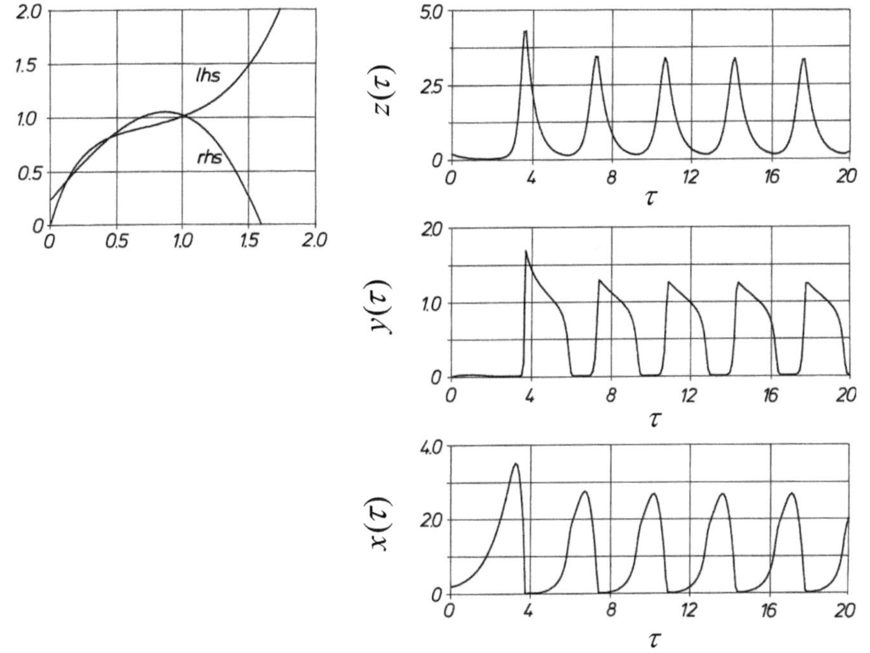

Fig. 7.17 Graphical solution of (7.125) for $\alpha = 1.5, \beta = 1, \delta = 1, y_0 = 1, \nu = 1, a = 2$ and the trajectories of $x(\tau), y(\tau), z(\tau)$ (Source: Weidlich and Haag 1983)

$$w^{(ii)}(y_s) = \delta - \alpha y_s \left(1 + \frac{y_0 - y_s}{y_0}\right). \tag{7.137}$$

The stability analysis then shows that the stationary point given by (7.136), (7.137) and $z_s = 0$ is stable. Figure 7.18 shows the graphical solution of (7.136) and the time-dependence of $x(\tau), y(\tau), z(\tau)$ of the prey and predator population for the set of trend parameters $\delta = 0, \beta = \nu = y_0 = \alpha = 1$ and $a = 2$.

Figure 7.18 shows that the predator population becomes almost extinct, whereas the prey species in the two habitats 1, 2 approach a stable equilibrium state. The reason for the almost extinction of the predator population despite finite values of $x(\tau)$ and $y(\tau)$ of the prey population is because the prey has no preference for habitat 1 ($\delta = 0$) and can therefore freely migrate into the refuge habitat. The number of preys in the dangerous habitat 1 remains small, so that the effective death rate of the predator population $a(1 - x) > 0$ remains positive.

Finally, we consider the special case of migration of the prey but without any saturation restrictions in the habitat 2, by taking the limit $y_0 \to \infty$. Therefore, equations (7.24), (7.125) with (7.126) simplify considerably.

Let us set the singular point $P_0(0, 0, 0)$ aside and consider the case $x_s = 1$ leading to

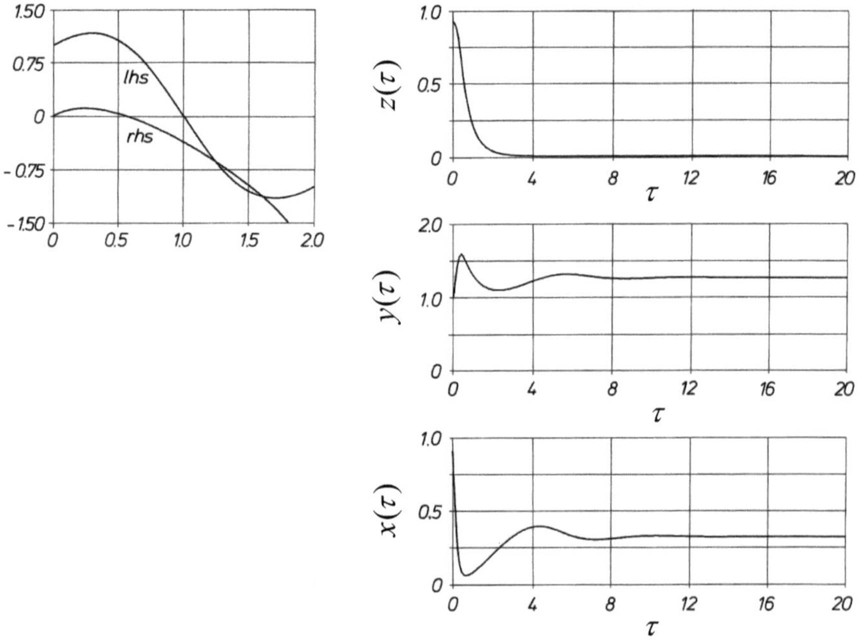

Fig. 7.18 Graphical solution of (7.136) for $\alpha = 1, \beta = 1, \delta = 0, y_0 = 1, \nu = 1, a = 2$ and the trajectories of $x(\tau), y(\tau), z(\tau)$ (Source: Weidlich and Haag 1983)

$$z_s = 1 + y_s \tag{7.138}$$

with

$$\nu y_s \exp(\rho - \sigma y_s) = y_s + \nu \exp(-\rho + \sigma y_s) \tag{7.139}$$

$$\text{and } \rho = \delta + \alpha - \beta \text{ and } \sigma = \alpha + \beta. \tag{7.140}$$

The predator population can only extinct, $z_s = 0$, in the trivial case when the whole prey population also become extinct, thus $x_s = y_s = 0$. The graphical solution of (7.139) shows that 0, 1 or 2 singular points may exist.

In Fig. 7.19, the time-dependence of the system variables $x(\tau), y(\tau), z(\tau)$ is shown for the set of trend parameters $\delta = 2, \beta = \nu = 1$ and $a = 2$ when the trend to swarm formation ($\alpha = 0$) does not exist. In this particular case an explosive multiplication of the predator species in habitat 1 and the prey species in habitat 2, where there exists no saturation limit, may happen. The prey population in habitat 1 exceeds a rather small value due to its moderate preference for habitat 1. Moreover, the prey population follows an exponential growth, whereas the predator population growths much slower and there are always enough prey species for

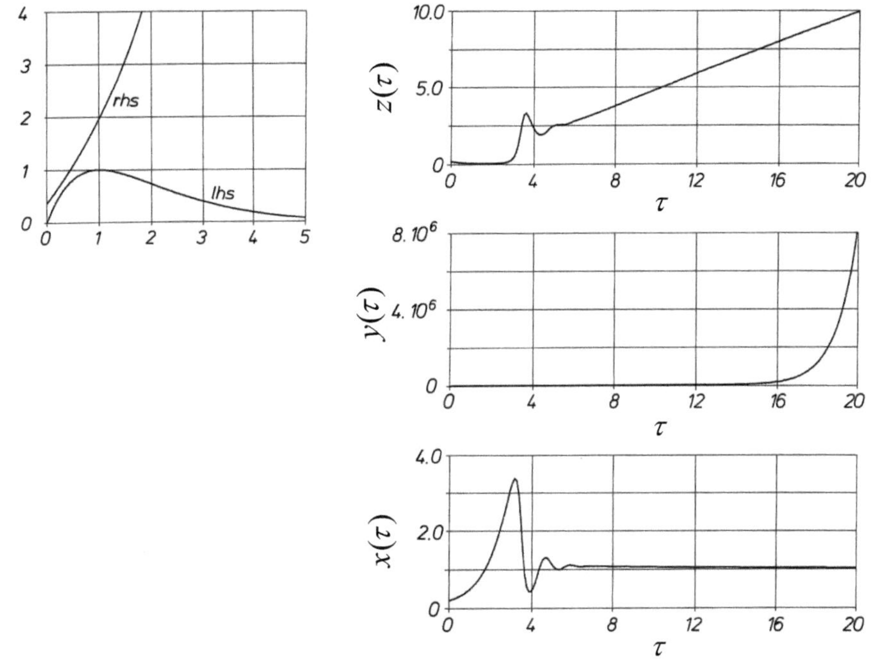

Fig. 7.19 Graphical solution of (7.95) for $\alpha=0\,,\beta=1\,,\delta=2\,,\nu=1\,,a=2$ and the trajectories of $x(\tau)\,,y(\tau)\,,z(\tau)$ (Source: Weidlich and Haag 1983)

the predators to eat. In this situation, the critical point (stationary point) $z_s=0$ becomes unstable and no further stable critical point exists. In this case, an explosion of the predator-prey system is described.

Figure 7.20 also shows damped oscillations despite the assumption of unsaturated birth of the prey. The lack of fear of the prey, $\beta=0$ and their moderate preference for habitat 1 ($\delta=1$) may explain this behaviour.

The complexity of the equations of motion of this extended Volterra-Lotka system with a density-dependent migration term exhibits a huge variety of possible trajectories for both populations, the prey and the predator. This means, that only a small part of the underlying mathematical structure and resulting possible solutions have been discussed in this chapter. It is not excluded that even chaotic solutions exist. This may be a suggestion for the reader to perform further research related to this topic.

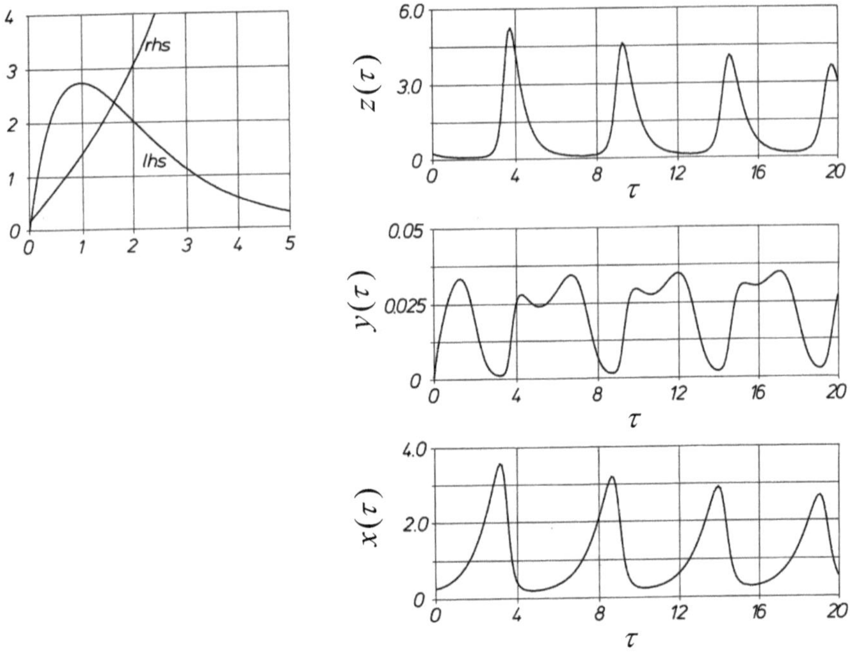

Fig. 7.20 Graphical solution of (7.95) for $\alpha=1, \beta=0, \delta=1, \nu=1, a=2$ and the trajectories of $x(\tau), y(\tau), z(\tau)$ (Source: Weidlich and Haag 1983)

7.4 Deterministic Chaos in Population Dynamics

In statistical physics, chaotic evolution is a well-known phenomenon. The Brownian motion, namely a dust particle suspended in a gas or fluid executes a random motion because of its irregular collisions with molecules of the gas or liquid. It is typical for conventional chaos that a large number of external microscopic degrees of freedom—the colliding molecules in the example mentioned—leads to random forces generating chaos. However, during the last decades another type of chaos has been discovered and denoted as deterministic chaos. It can arise in a set of deterministic differential equations (three or more) or even in systems described by difference equations (one or more). Although in case of deterministic chaos, the trajectories approach or remain within a finite domain of the space of variables, the system does not show any periodicity. Correspondingly, the Fourier analysis of the chaotic orbit yields a continuous broad band spectrum of frequencies, instead of the few selected frequencies characterizing any regular motion (Haken 1983; Campbell 1991; Lorenz 1979).

Deterministic chaos appears in simple models of natural science. The Laser is a good example: starting with rather low pumping rates, the output shows the typical statistical behaviour of a classical lamb. Above a critical threshold (see Chap. 5) we observe a phase transition and a continuous Laser action sets in. For even higher pumping rates, accompanied by another phase transition, the continuous Laser light makes a transition to regular Laser pulses. For very high pumping rates, those Laser pulses become irregular and behave completely chaotic.

It is not surprising that many even more complex systems in social sciences exhibit chaotic behaviour. The question of any possible predictability of economic long-term development is a crucial challenge. Dendrinos (1985) and Mosekilde et al. (1985) have shown that urban migration systems may exhibit strange attractors giving rise to chaotic patterns. Deterministic chaos can also appear in our general migration model even under less restrictive assumptions.

Our society consists of a multitude of interacting individuals. The decision-making behavior of individuals depends on current and future expectations as well as on past experiences and numerous constraints such as budget constraints. To make the system manageable, it is a common practice to cluster groups of individuals or groups of decision makers in so-called subpopulations. This concept assumes a "representative" decision maker who does not exist in reality, but may characterize the corresponding subpopulation. In other words, each subpopulation is represented by a set of roles depicting the mean probabilistic decision behaviour of the members of the subpopulation. This presumes a homogeneity assumption concerning the choice behaviour of agents (see Chap. 6).

We assume that the total population (species, individuals, agents, decision maker, etc.) $N(t)$ consists of $\alpha = 1, 2, \ldots, A$ subpopulations differing in their choice behaviour. In addition, we assume that the decision space is divided into L subspaces (regions, different opinions, different alternatives, …). The population number of subpopulation α belonging to subspace i, $i = 1, 2, \ldots, L$ at time t is $n_{\alpha i}(t)$. For the sake of simplicity we ignore birth and death processes and consider the change of opinion or migration processes of the subpopulations between the L subspaces (e.g. different opinions or different regions). Hence, the total number N_α of each subpopulation remains constant

$$\sum_{i=1}^{L} n_{\alpha i}(t) = N_\alpha, \tag{7.141}$$

and the total population number is given by

$$\sum_{\alpha=1}^{A} N_\alpha = N. \tag{7.142}$$

If we take into account birth- and death processes, the dynamics of the total population sizes of the subpopulations has to be added since neither N_α nor N are conserved by birth- and death processes. In addition, it could be possible that individuals change their subpopulation, e.g., change the political party in case of election processes, or change the status of marriage. Those even more complex systems can be treated using the formalisms and methods described in Chaps. 3 and 4.

7.4.1 Inter-group and Intra-group Interactions

The individual transition rates per unit of time $p_{ij}^{\alpha}(\vec{n})$ for a member of subpopulation α for a transition from region j to region i are assumed to depend on the distribution of all subpopulations $\vec{n} = \{n_{\alpha i}\}$, for $\alpha = 1, 2, \ldots, A$ and $i = 1, 2, \ldots, L$

$$p_{ij}^{\alpha}(\vec{n}) = \nu_{ij}^{\alpha} \exp\Big(u_i^{\alpha}(\vec{n}) - u_j^{\alpha}(\vec{n})\Big), \tag{7.143}$$

where the symmetric flexibility (mobility) matrix for subpopulation α is given by

$$\nu_{ij}^{\alpha} = \nu_{ji}^{\alpha}. \tag{7.144}$$

The measure of attractiveness $u_i^{\alpha}(\vec{n})$ of the alternative i for a member of subpopulation α can be again represented, e.g., by a truncated Taylor expansion

$$u_i^{\alpha}(\vec{n}) = \delta_i^{\alpha} + \sum_{\beta=1}^{A} \widetilde{\kappa}^{\alpha\beta} n_{\beta i} + \ldots. \tag{7.145}$$

Obviously, the following interpretation can be given: the δ_i^{α} describe the preference of subpopulation α for a certain alternative i. The intra-group interactions are described by the diagonal elements of the interaction matrix $\widetilde{\kappa}^{\alpha\alpha}$, where $\widetilde{\kappa}^{\alpha\alpha} > 0$ favours clustering trends or synergy effects. The inter-group interactions are represented by the off diagonal elements of the interaction matrix $\widetilde{\kappa}^{\alpha\beta}, \alpha \neq \beta$. In case of $\widetilde{\kappa}^{\alpha\beta} > 0$ subpopulation α prefers being close to subpopulation β, whereas in case of $\widetilde{\kappa}^{\alpha\beta} < 0$, subpopulation α wants to separate from subpopulation β.

The total transition rate, the number of transitions from j to i per unit of time with respect to subpopulation α is given by

$$w_{ij}^{\alpha}(\vec{n}) = n_{\alpha j} p_{ij}^{\alpha}(\vec{n}) = n_{\alpha j} \nu_{ij}^{\alpha} \exp\Big(u_i^{\alpha}(\vec{n}) - u_j^{\alpha}(\vec{n})\Big), \tag{7.146}$$

where $\alpha = 1, 2, \ldots, A$ and $i, j = 1, 2, \ldots, L$.

After the definition of the total transition rates (7.146) the Master equation for interacting subpopulations can be formulated.

7.4.2 The Master Equation for Interacting Subpopulations

We introduce the configurational probability $P(\vec{n}, t)$ to find a certain population configuration $\vec{n}$ at time t. The configurational probability $P(\vec{n}, t)$ satisfies the probability normalization condition

$$\sum_{\vec{n}} P(\vec{n},t) = 1, \tag{7.147}$$

where the sum extends over all possible configurations $\vec{n}$. According Chap. 4, the Master equations for interacting subpopulation, without birth- and death processes reads

$$\frac{dP(\vec{n},t)}{dt} = \sum_{\alpha=1}^{A} \sum_{\substack{i,j=1 \\ i \neq j}}^{L} \left[w_{ji}{}^{\alpha}\left(\vec{n}^{(\alpha,ji)}\right) P\left(\vec{n}^{(\alpha,ji)},t\right) - w_{ij}^{\alpha}(\vec{n}) P\left(\vec{n},t\right) \right]. \tag{7.148}$$

In the formulation of the Master equation, $\vec{n}^{(\alpha,ji)}$ arises from the configuration $\vec{n}$ by the substitution $n_{\alpha i} \rightarrow (n_{\alpha i}+1)$, $n_{\alpha j} \rightarrow (n_{\alpha j}-1)$, whereas all other $n_{\beta k}$ remain unchanged. If birth and death processes have to be considered as well, (7.148) can be easily extended.

7.4.3 *The Quasi-Closed Equations for Interacting Subpopulations*

Following Chap. 3, the quasi-closed equations of motion for the mean values

$$\langle n_{\alpha j} \rangle = \sum_{\vec{n}} n_{\alpha j} P(\vec{n},t) \tag{7.149}$$

can be derived:

$$\begin{aligned}\frac{d\langle n_{\alpha j}\rangle}{dt} &= \sum\nolimits_{i=1}^{L} \langle n_{\alpha i}\rangle p_{ji}^{\alpha}(\langle \vec{n}\rangle) - \sum\nolimits_{i=1}^{L} \langle n_{\alpha j}\rangle p_{ij}^{\alpha}(\langle \vec{n}\rangle) \\ &= \sum\nolimits_{i=1}^{L} \nu_{ji}^{\alpha} \langle n_{\alpha i}\rangle \exp\left(u_j^{\alpha}(\langle \vec{n} \rightarrow \rangle) - u_i^{\alpha}(\langle \vec{n} \rightarrow \rangle)\right) - \sum\nolimits_{i=1}^{L} \nu_{ij}^{\alpha} \langle n_{\alpha j}\rangle \exp\left(u_i^{\alpha}(\langle \vec{n} \rightarrow \rangle)\right. \\ &\quad \left. - u_j^{\alpha}(\langle \vec{n} \rightarrow \rangle)\right),\end{aligned} \tag{7.150}$$

where $\alpha = 1, 2, \ldots, A; j = 1, 2, \ldots, L$. Equation (7.150) represents a set of $A \cdot L$ coupled nonlinear differential equations. Since we consider only migratory interactions, the population numbers N_{α} of the $\alpha = 1, 2, \ldots, A$ subpopulations remain constant. The number of independent equations is therefore given by $(A \cdot L - A)$. Our further considerations of interacting subpopulations refer to this mean value equations (7.150).

7.4.3.1 The Stationary Solution of the Master Equation

The stationary solution of (7.150) is given by

$$\widehat{n}_{\alpha j} = C^{\alpha}\exp\left(2u_j^{\alpha}(\widehat{n})\right), \tag{7.151}$$

with the normalization constant C^{α} which has to be determined for each subpopulation separately

$$C^{\alpha} = \frac{N_{\alpha}}{\sum_{k=1}^{L} \exp\left(2u_k^{\alpha}(\widehat{n})\right)}. \tag{7.152}$$

The stationary population pattern $\widehat{n}$ can be obtained by solving the coupled system of transcendental equations (7.152). However, this is a nontrivial problem since many stable, unstable and strange attractors may be found, depending on the actual values of the interaction matrix and the preference parameters.

7.4.4 Chaotic Behaviour of Interacting Subpopulations

On our way from simplicity to complexity we are now approaching the field of chaotic population dynamics. Of course, the occurrence of chaos, depends on the whole structure of the system and the precise form of the interaction structure. Therefore, we restrict ourselves to a very specific example, which however, demonstrates how even more general stochastic systems can be treated in principle. This example was first discussed in detail by Haag (1989).

It is well known, and I have demonstrated it in different chapters of this book, that the underlying process always belongs to the field of decision processes. This means we can interpret for example the members of the subpopulation as different groups of decision makers, and L as the number of alternatives they have to decide on, or the subpopulations identify different populations and L means different regions. The development of market shares of different strongly competing products like VHS, Betamax, Video 2000 and others on the video market may demonstrate such complex consumer behaviour (Arthur 1983, 1991).

In the following, we prefer to discuss the migration behaviour of different populations between different regions.

The case $A = L = 2$ was analysed in Weidlich and Haag (1983) in detail. Beside many stable and unstable critical points, there exists a limit cycle solution for a specific choice of the trend parameters, namely in the case when the inter-group interaction is positive and strong and one subpopulation likes the other, but the other does not like the first one.

Chaotic behaviour, however, needs at least one dimension more. Therefore, we consider next the case of three subpopulations and three regions $A=L=3$, characterized by six independent variables $\langle n_{\alpha i}\rangle$.

We further simplify the structure of the system considerably via the assumption $\delta_j^\alpha = 0$, there is no preference for any region by the subpopulations, and $\nu_{ij}^\alpha = \nu$ the mobility is the same for all subpopulations.

Then we obtain the simplified equations of motion

$$\begin{aligned} \frac{dx_{\alpha j}}{d\tau} &= \sum_{i=1}^{3} x_{\alpha i} p_{ji}^{\alpha}(\vec{x}) - \sum_{i=1}^{3} x_{\alpha j} p_{ij}^{\alpha}(\vec{x}) \\ &= \sum_{i=1}^{3} x_{\alpha i} \exp\left(\sum_{\beta=1}^{3} \kappa^{\alpha\beta}\left(x_{\beta j} - x_{\beta i}\right)\right) - \sum_{i=1}^{3} x_{\alpha j} \exp\left(\sum_{\beta=1}^{3} \kappa^{\beta\alpha}\left(x_{\beta i} - x_{\beta j}\right)\right) \\ &= F_{\alpha j}(\vec{x}) \end{aligned} \tag{7.153}$$

for $\alpha = 1,2,3$ and $j = 1,2,3$, where we have introduced scaled population numbers

$$x_{\alpha j} = \frac{\langle n_{\alpha j}\rangle}{N_\alpha}, \qquad \text{with} \qquad 0 \le x_{\alpha j} \le 1 \tag{7.154}$$

the dimensionless scaled time unit

$$\tau = \nu \cdot t \tag{7.155}$$

and scaled interaction parameters

$$\kappa^{\alpha\beta} = N_\beta \widetilde{\kappa}^{\alpha\beta}. \tag{7.156}$$

Now we will use numerical simulation methods to investigate the dynamic behaviour of our system consisting of three different populations. The members of the three populations have to decide to live in one of the three possible regions. Whether the three populations separate each in another area, live together or a much more complicated spatial-temporal development will occur depends on the strength of the inter-group and intra-group interaction. We assume the following interaction matrix

$$\left(\kappa^{\alpha\beta}\right) = \begin{pmatrix} 1.7 & 1.5 & -1.5 \\ -1.5 & 1.7 & 1.5 \\ \kappa^{13} & -1.5 & 1.7 \end{pmatrix}, \tag{7.157}$$

where the inter-group interaction parameter κ^{13} will be changed from 1.5 to -1.5 in order to demonstrate different possible solution pattern. For all populations, the inter-group interaction dominates. This means that the three populations prefer to live together with members of the same population. Different signs in the intra-group interaction imply that one of the populations does not like to live together with members of the other population. The interaction matrix shows for example that population 1 does not like population 3, but it depends on the sign of κ^{13} whether population 3 likes or dislikes population 1.

The results of the numerical simulations of (7.153) with (7.157) are presented in a series of figures belonging to different values of the interaction parameter κ^{13}. The group of figures denoted by Figs. 7.21, 7.24, and 7.27 shows the projection of the trajectories $x_{\alpha i}$ for α and $i = 1, 2, 3$ into the $\{x_{11}, x_{21}\}$-plane. In the group of Figs. 7.22, 7.25, and 7.28 the logarithm of the Fourier transform $\widehat{x}_{11}(f)$ of the time-evolution of the variable $x_{11}(t)$ is depicted, where the Fourier transform is given by

$$\widehat{x}_{11}(f) = \frac{1}{2\pi} \int_{-\infty}^{\infty} x_{11}(t) e^{-2\pi i f t} dt. \tag{7.158}$$

The group of Figs. 7.23, 7.26, and 7.29 shows the time-dependence of the variable $x_{11}(t)$.

Since the dynamics of all variables is strongly interlinked, each other variable $x_{\alpha j}$ could be used instead of x_{11} for the analysis of the dynamics.

In Figs. 7.21, 7.22 and 7.23, the results of the simulations are presented for a strong positive inter-group interaction of population 3 with population 1. This means that population 3 wants to follow population 1. Obviously, Fig. 7.21 shows the projection of a simple limit cycle in the $\{x_{11}, x_{21}\}$-plane. Correspondingly, the Fourier spectrum, Fig. 7.22, represents a few discrete frequencies to this periodic dynamics. The evolution of the $x_{11}(t)$ reflects this statement.

Figures 7.24, 7.25 and 7.26 depict the behaviour of the trajectory x_{11} for a small negative inter-group interaction $\kappa^{13} = -0.55$. Still there exists a limit cycle, which,

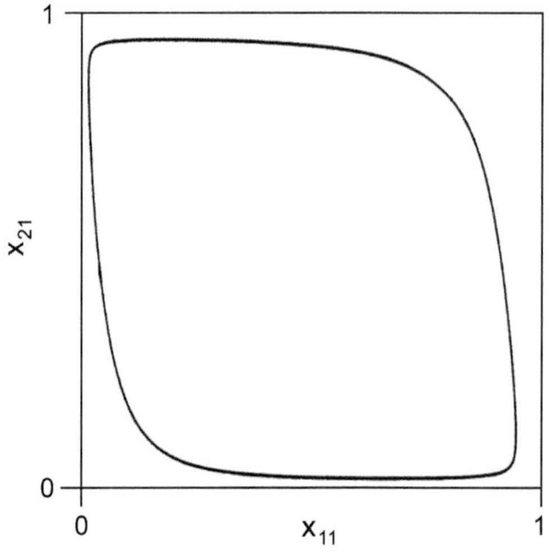

Fig. 7.21 Projection of the "stationary" trajectory for $\kappa^{13} = 1.5$ (Source: Haag 1989)

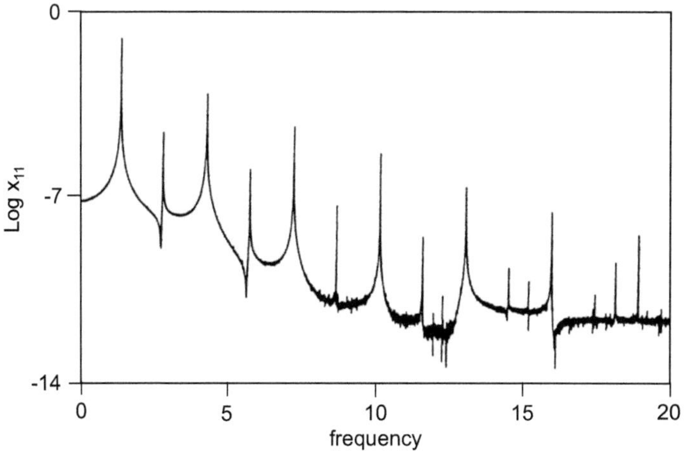

Fig. 7.22 Fourier spectrum of the trajectory for $\kappa^{13} = 1.5$ (Source: Haag 1989)

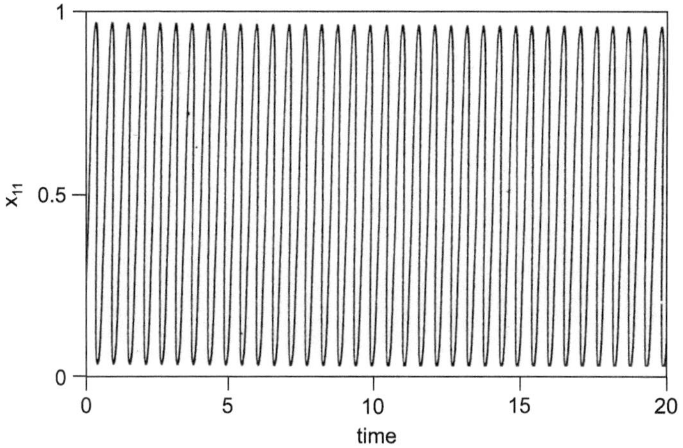

Fig. 7.23 Evolution with time of $x_{11}(t)$ for $\kappa^{13} = 1.5$ (Source: Haag 1989)

however, is much more complex. There appears period doubling in limit cycle, Fig. 7.24 which can also be identified in the Fourier spectrum, Fig. 7.25. The evolution of $x_{11}(t)$, Fig. 7.26, exhibits different cycles.

Finally a chaotic dynamics is obtained for $\kappa^{13} - -1.5$ as shown in the Figs. 7.27, 7.28, and 7.29. The periodic limit cycle has evolved via a series of period doubling processes into a chaotic system, characterized by a broad band continuous Fourier spectrum. The route to chaos pursued from Figs. 7.21, 7.22, 7.23, 7.24, 7.25, 7.26, 7.27, 7.28, and 7.29 is called the Ruelle-Takens picture (Ruelle and Takens 1971).

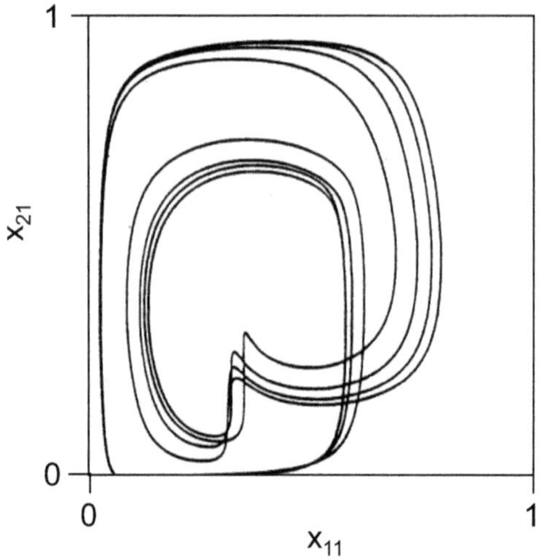

Fig. 7.24 Projection of the "stationary" trajectory for $\kappa^{13} = -0.55$ (Source: Haag 1989)

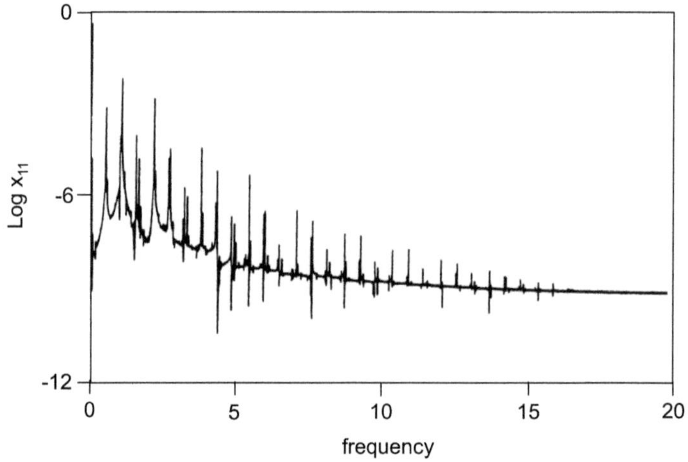

Fig. 7.25 Fourier spectrum of the trajectory for $\kappa^{13} = -0.55$ (Source: Haag 1989)

Lyapunov Exponents, Strange Attractors and Fractal Dimensions
In the following let us introduce some mathematical concepts for a further characterization of chaotic trajectories.

If for certain domains of the state space ($\alpha = 1, 2, 3; i = 1, 2, 3$) the flux in state space is contractive, this means that a small volume element around the attractor

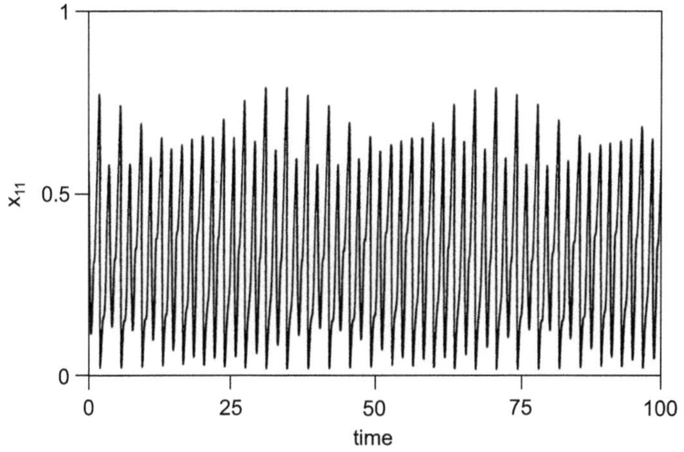

Fig. 7.26 Evolution with time of $x_{11}(t)$ for $\kappa^{13} = -0.55$ (Source: Haag 1989)

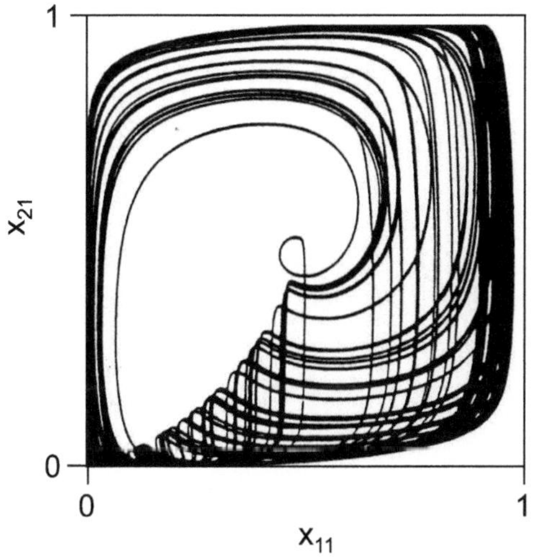

Fig. 7.27 Projection of the "stationary" trajectory for $\kappa^{13} = -1.5$ (Source: Haag 1989)

$$\Delta V = \prod_{\alpha=1}^{3} \prod_{i=1}^{3} \Delta x_{\alpha i} \tag{7.159}$$

contracts in the course of time,

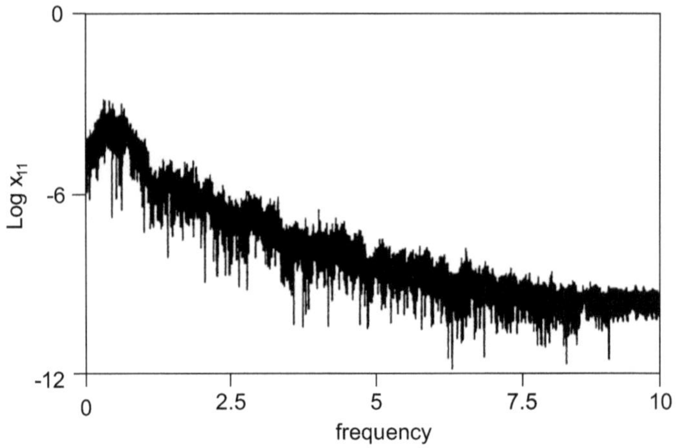

Fig. 7.28 Fourier spectrum of the trajectory for $\kappa^{13} = -1.5$ (Source: Haag 1989)

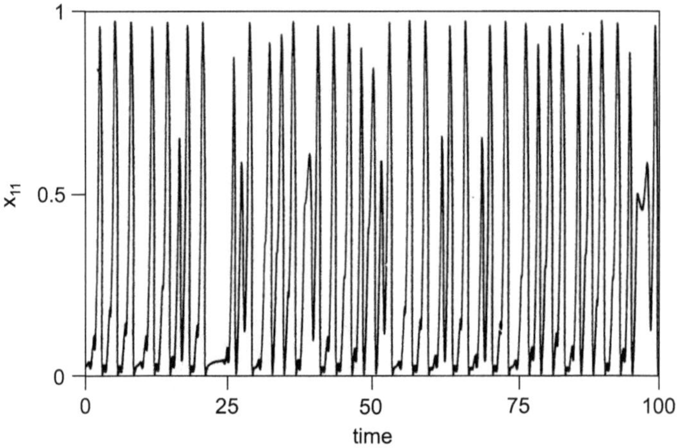

Fig. 7.29 Evolution with time of $x_{11}(t)$ for $\kappa^{13} = -1.5$ (Source: Haag 1989)

$$\frac{1}{\Delta V} \cdot \frac{d\Delta V}{dt} = \left[\prod_{\alpha=1}^{3} \prod_{i=1}^{3} \Delta x_{\alpha i} \right]^{-1} \frac{d}{dt} \left[\prod_{\alpha=1}^{3} \prod_{i=1}^{3} \Delta x_{\alpha i} \right] < 0, \tag{7.160}$$

the system is called dissipative. Therefore, each volume element ΔV out of the dissipative domain is mapped to a zero volume for $t \rightarrow \infty$. The domain of volume zero, to which all trajectories, starting within ΔV are finally attracted for, $t \rightarrow \infty$ is denoted as attractor.

If we now insert the equations of motion (7.153) in (7.160) we obtain the equivalent condition to (7.160)

$$\sum_{\alpha=1}^{3}\sum_{i=1}^{3}\frac{\partial F_{\alpha i}}{\partial x_{\alpha i}} < 0. \tag{7.161}$$

Fixed points and limit cycles are specific examples of such attractors. If, however, the asymptotic dynamics is chaotic, we have the special case of a strange attractor. Moreover, a contraction in all directions in state space is not necessary. It is typical for a strange attractor that there exist simultaneously stretching and folding processes for the volume element. The flux contracts in some directions and stretches in others in this case.

Chaotic trajectories can be further analysed by introducing the concept of Lyapunov exponents (Haken 1983) and fractal dimension (Hentschel and Procaccia 1983). These concepts were introduced for the purpose of characterizing strange attractors in dynamic systems.

In order to visualize this contraction process in detail let us consider an infinitesimal ball in (our 6-dimensional) state space. During the dynamic evolution of our system (7.153), this volume element will be distorted, but, being infinitesimal, it will remain an ellipsoid. Assume, that the principal axes of the ellipsoid are denoted by $\varepsilon_i(t)$, $i=1,2,\ldots,6$. Then it turns out that their evolution with time is given by

$$\varepsilon_i(t) = \varepsilon_0\exp(\lambda_i t), \tag{7.162}$$

where the exponents λ_i are the famous Lyapunov exponents. From (7.162) the definition and the determination concept of the Lyapunov exponents λ_i can be extracted

$$\lambda_i = \lim_{t\to\infty}\lim_{\varepsilon_0\to 0}\left[\frac{1}{t}\ln\left(\frac{\varepsilon_i(t)}{\varepsilon_0}\right)\right]. \tag{7.163}$$

Because the volume of the ellipsoid contracts for dissipative systems, the relation

$$\Delta V = \prod_{i=1}^{6}\varepsilon_i(t) = \prod_{i=1}^{6}\varepsilon_0 e^{\lambda_i t} = \Delta V_0\prod_{i=1}^{6}e^{\lambda_i t} = \Delta V_0\exp\sum_{i=1}^{6}\lambda_i t \tag{7.164}$$

must hold

$$\sum_{i=1}^{6}\lambda_i < 0. \tag{7.165}$$

By means of the Lyapunov exponents λ_i, the attractor can be further characterized:

Table 7.2 Lyapunov spectrum of system (7.153) for the interaction matrix (7.157)

κ_{13}	λ_1	λ_2	λ_3	λ_4	λ_5	λ_6	D_{KY}
1.50	0.00	−3.56	−4.64	−6.15	−6.16	−8.89	1.00
0.00	0.00	−0.34	−0.34	−3.79	−8.30	−8.94	1.00
−0.50	0.00	−0.25	−0.25	−4.12	−7.56	−15.3	1.00
−0.55	0.00	−0.00	−0.79	−4.06	−7.10	−15.3	2.00
−0.60	0.48	0.00	−1.26	−4.11	−7.37	−14.1	2.38
−1.50	0.88	0.00	−7.54	−9.44	−11.9	−27.2	2.15

1. If all Lyapunov exponents are negative, $\lambda_i < 0$, for $i = 1, 2, \ldots, 6$, the attractor is a stable focus
2. If $\lambda_1 = 0, \lambda_i < 0$, for $i = 2, \ldots, 6$, the attractor is a stable limit cycle. The principle axis of the ellipsoid along the limit cycle remains constant, while all other axis are contracted
3. If $\lambda_1 = \lambda_2 = 0, \lambda_i < 0$, for $i = 3, \ldots, 6$, the attractor is a two-dimensional torus in phase space
4. If $\lambda_1 > 0$, a strange attractor is found. Because of (7.160), at least one negative Lyapunov exponent must exist. Hence, simultaneous stretching and contraction processes appear

For the introduced migratory system (7.153) with (7.157), the Lyapunov exponents have been calculated using the algorithm described in Benettin et al. (1976). The Lyapunov exponents for different values of the inter-group interaction parameter κ^{13} are listed in Table 7.2.

The comparison of the results of Table 7.2 with the trajectory analysis presented in Figs. 7.21, 7.22, 7.23, 7.24, 7.25, 7.26, 7.27, 7.28, and 7.29 shows the full agreement between both interpretation methods.

Furthermore, the Kaplan-Yorke dimension (Kaplan and Yorke 1979) of an attractor D_{KY} can be calculated

$$D_{KY} = j + \frac{\sum_{i=1}^{j} \lambda_i}{\left|\lambda_{j+1}\right|}. \tag{7.166}$$

Here, the exponents are ordered in descending order $\lambda_1 \geq \lambda_2 \geq \ldots \geq \lambda_6$, and j is the largest integer, for which

$$\sum_{i=1}^{j} \lambda_i \geq 0. \tag{7.167}$$

Of course, the Kaplan-Yorke dimension D_{KY} is an integer for a limit cycle ($D_{KY} = 1$), or for a torus in phase space ($D_{KY} = 2$), but it can be fractal for a strange attractor.

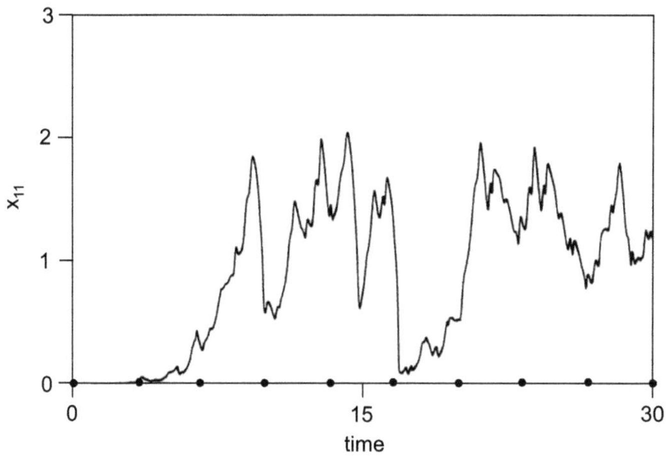

Fig. 7.30 Distance between initially adjacent trajectories in case of a limit cycle (*black circle*) for $\kappa^{13} = 1.5$ and of a strange attractor $\kappa^{13} = -1.5$ (*solid line*) (Source: Haag 1989)

As a consequence of the stretching and folding processes in case of strange attractors, neighbouring trajectories separate exponentially even if they were close to each other at the beginning. This is demonstrated in Fig. 7.30 in case of a limit cycle and for a strange attractor.

Finally, we consider the correlation dimension $C(r)$, which has been introduced by Grassberger and Procaccia (1983). This intuitively appealing dimension is obtained from the correlation between random points on the attractor.

Consider a set $\{\vec{y}_i, i = 1, 2, \ldots, M\}$ of points on the attractor, obtained from a time series $\vec{y}_i = \{x_{ak}(t + i\tau)\}$ with a fixed time increment τ between successive measurements. Due to the exponential divergence of trajectories, the pair of points will be dynamically uncorrelated, but they are spatially correlated, since they are points on the attractor. This spatial correlation can be measured by a sum $C(r)$, defined as

$$C(r) = \lim_{M \to \infty} \frac{1}{2M^2} \left[\sum_{i,j=1}^{M} \Theta\Big(r - |\vec{x}_i - \vec{x}_j|\Big) \right], \tag{7.168}$$

where the number of pairs (i,j) with Euclidian distance $|x_i - x_j| < r$ are summed up. If the attractor densely fills a subspace of the phase space of dimension D_C it is clear that $C(r)$ should grow with r as follows:

$$C(r) \propto r^{D_C} \tag{7.169}$$

since the random points j on the attractor contained in a ball of radius r around the reference point i fill that ball on a subspace of dimension D_C. Calculating $C(r)$ for a sufficiently large number M and plotting $\ln C(r)$ versus $\ln r$ yields the correlation

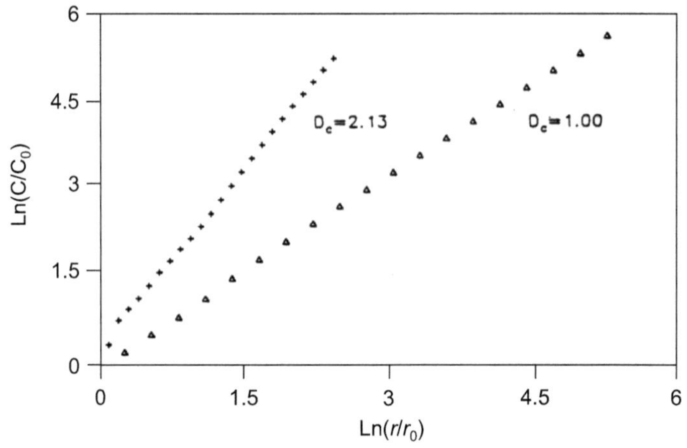

Fig. 7.31 Determination of the correlation dimension D_C in case of a limit cycle for $\kappa^{13} = 1.5$ and of a strange attractor $\kappa^{13} = -1.5$ (Source: Haag 1989)

dimension D_C as slope of the curve, according Fig. 7.31. In case of the strange attractor ($\kappa^{13} = -1.5$) we obtain $D_C \approx 2.13$ indicating that the correlation dimension of the strange attractor is fractal and the attractor lies almost on a two-dimensional surface in the phase space.

Grassberger and Procaccia (1983) have proved that between the different definitions D_{KY} and D_C, the relation holds

$$D_{KY} \geq D_C. \tag{7.170}$$

This relation is confirmed in our case by the numerical results.

Conclusion

The purely theoretical consideration of a system of interacting subsystems shows that depending on the strength and sign of the inter-group and intra-group interaction, a variety of different attractors exist. The dynamics may be characterized by smooth trajectories, limit cycles but also chaotic behaviour. The concepts of phase transitions, strange attractors and deterministic chaos fully apply.

Under such circumstances migratory systems or decision systems composed of interacting subsystems may behave in a highly irregular and unpredictable way. However, the simulations of the last chapter have shown that some conditions with respect to the system parameters are necessary for the occurrence of chaotic trajectories.

1. In order to obtain a tendency of the members of each subpopulation to cluster together with other members of the same subpopulation, a rather strong intra-group interaction is required.

2. The sign of the inter-group interactions between at least two different subpopulations must be different. In this case, there exists a tendency to agglomerate in different parts of the state space.
3. The interaction matrix must be asymmetric. Otherwise, only stable fixpoints or limit cycles exist.

Of course, even in case of only three subpopulations and three alternatives, the number of stable or unstable critical points, limit cycles and strange attractors may be huge and underlines the complexity of interacting subsystems. However, this example has also demonstrated that such systems exhibit chaotic solutions only under very restrictive conditions. In most real-world migratory systems, we expect regular behaviour. In case of decision systems, e.g. selection of different parties, conditions favoring cyclic or even chaotic behaviour cannot be completely excluded.

References

Ahmad S (1993) On the nonautonomous Volterra-Lotka equations. Proc Am Math Soc 117:199–204

Arthur B (1983) Competing technologies and lock-in by historical events. IIASA Paper WP-83-90, Laxenburg

Arthur B (1991) The economy and complexity. In: Stein DL (ed) Lectures in the sciences of complexity. Addison-Wesley, Redwood City

De Angelis DL (1976) Application of stochastic models to a wildlife population. Math Biosci 31(3–4):227–236

Benettin G, Galgani L, Strelcyn J-M (1976) Kolmogorov entropy and numerical experiments. Phys Rev A 14:2338

Campbell D (1991) Introduction to nonlinear phenomena. In: Stein DL (ed) Lectures in the sciences of complexity. Addison-Wesley, Redwood City

Dendrinos DS (1985) Urban evolution, studies in the mathematical ecology of cities. Oxford University Press, Oxford

Dendrinos DS, Haag G (1984) Towards a stochastic theory of location: empirical evidence. Geogr Anal 16:287–300

Goel NS, Richter-Dyn N (1974) Stochastic models in biology. Academic, New York

Grassberger P, Procaccia I (1983) Meassuring the strangeness of strange attractors. Physica D 9:189

Haag G (1989) Dynamic decision theory: applications to urban and regional topics. Kluwer, Dordrecht

Haag G, Dendrinos DS (1983) Toward a stochastic theory of location: a nonlinear migration process. Geogr Anal 15:269–286

Haken H (1978) Synergetics. An introduction. Springer, Berlin

Haken H (1983) Advanced synergetics. Springer, Berlin

Haken H (2004) Synergetics: introduction and advanced topics. Springer, Berlin

Hanson HC (1976) The giant Canada goose. University Press, Carbondale

Hentschel HGE, Procaccia I (1983) The infinite number of generalized dimensions of fractals and strange attractors. Phys D 8:435–444

Hurwitz A (1895) Ueber die Bedingungen, unter welchen eine Gleichung nur Wurzeln mit negativen reellen Theilen besitzt. Math Ann 46:273–284

Kaplan JL, Yorke JA (1979) Chaotic behavior of multidimensional difference equations. In: Peitgen H-O, Walter H-O (eds) Functional differential equations and approximations of fixed points. Lecture notes in mathematics, vol 730. Springer, Berlin

Lorenz EN (1979) On the prevalence of aperiodicity in simple systems. In: Grmela M, Marsden JE (eds) Global analysis. Springer, New York

Lotka AJ (1920) Analytical note on certain rhythmic relations in organic systems. Proc Natl Acad Sci USA 6:410

Ludwig D (1974) Stochastic population theories, Lecture notes in biomath, vol 3. Springer, Berlin

Malthus TR (1798) An essay on the principle of population. Oxford World's Classics, reprint

Matsuo K, Lindenberg K, Shuler KE (1978) Stochastic theory of nonlinear rate processes with multiple stationary states. Relaxation time from a metastable state. J Stat Phys 19:65–75

May RM (1973) Stability and complexity in model ecosystems. Princeton University Press, Princeton

Mosekilde E, Rasmussen S, Joergensen H, Jaller F, Jensen C (1985) Chaotic behaviour in a simple model of urban migration. Preprint. Technical University of Denmark

Papageorgiou YY (1980) On sudden urban growth. Environ Plan A 12:1035–1050

Pearl R, Reed L (1920) On the rate of growth of the population of the United States. Proc Natl Acad Sci 6:275

Ruelle D, Takens F (1971) On the nature of turbulence. Commun Math Phys 20:167–192

van Kampen NG (1977) Stochastic processes in physics and chemistry. J Stat Phys 17:71

van Kampen NG (1978) An introduction to stochastic processes for physicists. In: Garido L, Seglar P, Shepherd PJ (eds) Stochastic processes in nonequilibrium systems. Proceedings, Sitges 1978. Springer, New York

Verhulst PF (1838) Notice sur la loi que la population suit dans son accroissement. Corresp Math Phys 10:113–121

Volterra V (1927) Variations and fluctuations of the numbers of individuals in coexisting animal populations. Mem. R. Comitato talassogr. Ital., mem. 131, reprint in Opere Math. 5, Rome

Weidlich W, Haag G (1983) Concepts and models of a quantitative sociology: the dynamics of interacting populations, Springer series of synergetics, vol 14. Springer, New York

Wheaton WC (1974) A comparative static analysis of urban spatial structure. J Econ Theory 9:223–237

Wilson EO, Bossert WH (1973) Einführung in die Populationsbiologie. Springer, Berlin

Chapter 8
The Master Equation in Migration Theory

Contents

8.1 The Gravity Model

The modelling and understanding of the dynamic development of spatial pattern based on human activities has been a research topic since the beginning of regional sciences.

G. Haag, *Modelling with the Master Equation*, DOI 10.1007/978-3-319-60300-1_8

A first complete description of a "system of cities", which goes beyond a rather superficial analogy to physics was given by Reynaud (1841). He is regarded in the meantime as the real founder of the theory of central places, with all basic principles, including the geometric formulation of market areas as hexagons. His work has been discovered and discussed by Robic (1982) again.

Another pioneer of the use of gravity models in social sciences was Carey (1858), who concluded: "Man Tends, of necessity, to gravitate towards his fellow man".

The (presumably) first application of mathematical migration models took place after the analysis of migration flows by Ravenstein (1885, 1889) and the publication of his famous article "The laws of migration". The analysis of the development of cities during the nineteenth and the beginning of the twentieth century by Weber (1899) and Reilly (1931) was conducted on the basis of the Ravenstein article.

More than 130 years ago, Ravenstein (1885) described the migration flows between English cities in analogy to Newton's law from physics, which states that the attraction force F_{ij} between two masses m_i and m_j in a certain distance d_{ij} is given by

$$F_{ij} = \gamma \frac{m_i m_j}{d_{ij}^2}, \tag{8.1}$$

where γ is the constant of gravitation. The simplest form is a direct transfer of the Newtonian law

$$w_{ij} = k \frac{n_i n_j}{d_{ij}^2} \tag{8.2}$$

where w_{ij} represents the migration flow (number of migratory events in a certain interval of time) form j to i, and n_j, n_i are the population numbers of the two regions or cities, and $d_{ij} = d_{ji}$ represents the distance between the regions.

Hence, in this direct analogy, the migratory flow between a region j and region i is exactly as large as the inverse flow.

The parameter k is often determined by the requirement that the sum of the estimated migration flows w_{ij} must be equal to the sum of the observed flows w_{ij}^e (e = empirical)

$$k = \frac{\sum_{i,j=1}^{L} w_{ij}^e}{\sum_{i,j=1}^{L} \frac{n_i n_j}{d_{ij}^2}} \tag{8.3}$$

The aggregation over a subset of migration flows, e.g., between municipalities should correspond to the migration flows for example between provinces on a higher aggregate level. As shown by Grimmeau (1994), simple gravitational models cannot meet this scale requirement since the big bundle of social, economic and geographical characteristics of the regions are not displayed by this simple model approach.

Often the "gravity model" is also used in other types of spatial interaction processes, such as for estimating the number of rail users or aircraft user between two centres, the number of phone calls between two regions, the number of potential customers to a shopping centre, to mention a few possible applications. The fact that strong correlations with the term $n_i n_j / d_{ij}^2$ exist, even in the above mentioned very different spatial interaction topics was for Stewart (1947, 1948) an indication of the fundamental importance of the gravity approach to the social sciences.

What kind of distance deterrence function seems to be appropriate?
Therefore, in recent decades, this very simple "interaction model" was often modified to map better specific features of the study area (Poulain 1981). There is no convincing reason to hold at a distance effect $\propto d_{ij}^2$.

An interpretation of the spatial dependence as spatial "resistance", impedance function or spatial deterrence function, which may take various analytical forms was discussed with reference to the negative exponential decay and negative power decay in Fotheringham and O'Kelly (1989), Reggiani et al. (2011), and in Wilson (1981a, b, c). We come back to this discussion in the Sects. 8.4.1 and 8.9.3.

Summaries of different gravitational approaches with applications can be found also in Fotheringham and O'Kelly (1989), Haynes and Fotheringham (1984), De La Barra (1990), and Krugman (1993).

A first very simple modification arises from the observation that the distance dependence of migration flows does not corresponds to Newton's law and

$$w_{ij} = k \frac{n_i nj}{d_{ij}^{\alpha}}, \tag{8.4}$$

led mostly to a better adaptation of the migration flows to the empirically observed data, where the coefficient α represents a measure of the distance sensitivity of the population. More generally, the migration flows can then be expressed as follows

$$w_{ij} = k\, n_i n_j f\left(d_{ij}\right) \tag{8.5}$$

with the decay or distance function $f(d_{ij})$, which have to be specified from case to case.

What kind of distance definition seems to be appropriate?
Another point of discussion in literature is the question what is meant by the distance measure d_{ij}. The pure geographical distance between a region i and region j does not always prove significant. Better statistical tests are obtained by considering the effort required to move from one region to another. This can account for transport costs or travel time. Also "social" distances measured e.g., as differences

in social status, income or cultural behaviour, but also in the form of language and ethnic barriers can be crucial in specific applications (Huriot and Thisse 1984; Nijkamp and Reggiani 1992; Straubhaar 1988).

The definition of d_{ij} is as already mentioned rather difficult. Empirical work using different definitions of d_{ij} demonstrates that often the distance between the centre of mass of the population of the regions is more appropriate as distances based on the centre of gravity, especially when consumer behaviour is considered. Instead of distance, measured in length-units (e.g., km), the length of the routes between the centres or even more appropriate, the travel-time between the regions, is proposed and applied by professionals.

Figure 8.1 shows the number of migratory flows depending on the number of regional subdivisions for Germany in 2016. In addition, the average population size is depicted. In case of L regions we have $L(L-1)$ interregional flows and the average population size of a region is N/L when N is the total population size of Germany (82 Mio.). Obviously, on the one hand, with increasing number of regions, the definition of a significant distance d_{ij} becomes more precise. On the other hand, the average population size of a region decreases and the number of flows increases dramatically.

If the distances d_{ij} as well as the distance deterrence function $f(d_{ij})$ are not well defined, they introduce a lot of uncertainty into the model, which leads to a considerable loss of explanatory power. We will see in Sect. 8.3 how we can avoid the distance definition problem and the definition problem of an appropriate distance deterrence function, simultaneously. This improves significantly both, the empirical estimation of the model parameters and explanatory factors of the migration flows and the quantification and understanding of the interaction between regions.

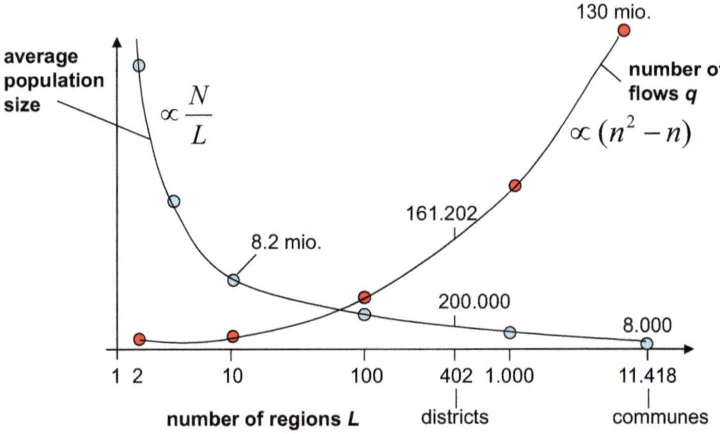

Fig. 8.1 Number of migratory flows, average population size and number of regions. Case Germany

Furthermore, another point in question was the sole use of population as a measure of the "gravitational mass". The sizes n_i, n_j instead should be replaced by effective attraction measures of the regions i and j and thus contribute to the explanation of migration flows.

A simple specification of the gravity approach to describe the interaction between two regions may be obtained by means of the so-called Cobb and Douglas (1928) approach

$$w_{ij} = k\, O_j^{\alpha_1} D_i^{\alpha_2} f\left(d_{ij}\right), \tag{8.6}$$

where O_j and D_i represent measures (stock variables) of the places of origin j and destination i. The exponents (elasticity's) α_1, α_2 can be determined using optimization methods.

As an example, let us consider a typical migration model developed and used by Birg et al. (1983) to model inter-regional migration flows between 79 regions in West Germany

$$w_{ij} = k \frac{(WL_i)^{\alpha_1} (WN_i)^{\alpha_2} (L_i)^{\alpha_3} (ALQ_i)^{\alpha_4}}{\left(WL_j\right)^{\beta_1} \left(WN_j\right)^{\beta_2} \left(L_j\right)^{\beta_3} \left(ALQ_j\right)^{\beta_4}} \frac{n_i^{\alpha_5} n_j^{\alpha_6}}{d_{ij}^{\gamma}} u_{ij}, \tag{8.7}$$

where w_{ij} describes the migration flow $j \rightarrow i$, and n_j, n_i the population numbers of origin and destination region, and $i, j = 1, 2, \ldots, L$. In this example, the set of explanatory indicators belongs to the labour and the housing market: unemployment rate (ALQ_i), income (L_i), quantitative housing supply (WN_i), and qualitative housing supply (WL_i). The set of exponents $\alpha_l, l = 1, \ldots, 6$, and $\beta_k, k = 1, \ldots, 4$, and the exponent γ of the distance d_{ij} between the regions are weighting factors of importance of the different explanatory indicators. k is scaling factor and u_{ij} describes statistical uncertainties.

The model becomes explicit by determination of all exponents and unknown parameters.[1] In this example, applying a log-linear optimization procedure yields

$$F\left(k, \vec{\alpha}, \vec{\beta}, \gamma\right) = \sum_{\substack{i,j \\ i \neq j}}^{L} \left(\ln w_{ij}^{emp} - \ln w_{ij}\right)^2 = \min, \tag{8.8}$$

where w_{ij}^{emp} describes the empirical flows between the L regions. Because of the special functional form of the migration model, the log-linear estimation procedure (8.8) yields a linear algebraic system of equations for the exponents.

For many years, this kind of Gravity model was used for the quantitative description of migration flows. However, some constructive criticism should be mentioned:

[1] Details of the estimation: Number of observations: 6162, degrees of freedom: 6150, $R^2 = 0.43$

Symmetry
Let us consider two completely identical regions, with respect to all parameters. Then we can take as granted that both migration flows w_{ij} and w_{ji} must have the same value $w_{ij} = w_{ji}$. This requires a symmetry in the exponents, e.g., in the case of the above example, that $\alpha_k = \beta_k, k = 1, \ldots, 4$.

Distance definition and distance deterrence function
The definition of an appropriate measure of distance d_{ij} and the selection of an appropriate distance deterrence function are essential for the quality of the migration model.

Zero flow problem
Moreover, the model cannot be used to describe the populating of an empty region, e.g., if in region 5, there is no population $n_5 = 0$ so far, then all flows from this region to any other region and vice versa are zero $w_{5j} = w_{j5} = 0$, for all j.

Incomplete set of explanatory variables
All explanatory variables used in (8.7) and (8.9) must be positive, $z_k > 0$, to guarantee positive and finite migration flows. In practical applications, it may happen that an important explanatory variable is not available for all regions. In this case, one has three possibilities: (1) to exclude this region out of the spatial system, (2) to exclude this variable as a possible explanatory variable in all regions, (3) to perform an estimation for this missing variable.

Test of hypothesis
The selection of an appropriate set of significant explanatory variables requires a complete estimation (optimization) of the model for each potential set of new variables. All necessary statistical tests must therefore be calculated for each test set of variables.

Zero or small empirical migratory flows
The log-linear estimation procedure (8.8) cannot be used in the case of zero empirical migration flows because of the logarithm. This problem can be circumvented in a first approximation by adding one population act to all flows: $w_{ij}^{emp} = w_{ij}^{emp} + 1$. However, if the population numbers in the different regions are very unequal, the estimation procedure introduces a systematic error in the calculation of the exponents. Furthermore, in case of many regions a tremendous number of individuals is added to the system, which leads again to a systematic error. In the case of Germany (Fig. 8.1), on the level of 11,418 communes (2013) more than 120 million individuals would be added to the total population size of about 80 million inhabitants.

However, if we consider the above-described issues and restrictions, the Gravity model may be an appropriate mathematical model for many applications in social sciences. An adapted generalization of this type of Gravity models reads

$$w_{ij} = k \frac{\prod_{m=1}^{G} \left(z_i^m\right)^{\alpha_m}}{\prod_{m=1}^{G} \left(z_j^m\right)^{\alpha_m}} \frac{\left(n_i n_j\right)^{\beta}}{d_{ij}^{\gamma}} = k \prod_{m=1}^{G} \left(\frac{z_i^m}{z_j^m}\right)^{\alpha_m} \frac{\left(n_i n_j\right)^{\beta}}{d_{ij}^{\gamma}} \geq 0, \tag{8.9}$$

where we have introduced a set of (normalized) explanatory variables z_i^m, $m = 1, 2, \ldots, G$, $i = 1, \ldots, L$ related to different significant regional socio-economic influences.

8.2 Gravity Type Spatial Interaction Models

In general terms, a gravity type spatial interaction model as a modification of a gravity approach, can be represented by

$$w_{ij} = k\, h\left(O_j\right) g(D_i) f\left(d_{ij}\right), \tag{8.10}$$

where the "effective attractiveness" functions of origin $h(O_j)$ and destination $g(D_i)$ are composed of characteristics (explanatory variables) of the origin and destination region. This general form (8.10) includes the simple gravity model (8.9) as a special case.

However, the generalized gravity model (8.10) is inconsistent with respect to the aggregation condition of the migration flows, e.g., when it is required that the sum of the single migration flows adds-up to the total migration current (Somermeijer 1966)

$$\sum_{i=1}^{L} w_{ij} = O_j \tag{8.11}$$

and

$$\sum_{j=1}^{L} w_{ij} = D_i. \tag{8.12}$$

By a suitable choice of the pre-factors A_j, B_i, the generalized gravity model (8.13) can be modified appropriately, so that the aggregation requirement will be fulfilled:

$$w_{ij} = A_j B_i g(D_i)\ h\left(O_j\right) f_{ij}\left(d_{ij}\right), \tag{8.13}$$

using

$$A_j = O_j \left[h(O_j) \sum_{i=1}^{L} B_i g(D_i) f_{ij} \right]^{-1} \tag{8.14}$$

and

$$B_i = D_i \left[g(D_i) \sum_{j=1}^{L} A_j h(O_j) f_{ij} \right]^{-1}. \tag{8.15}$$

The hereby resulting spatial interaction model (8.13) is much more complicated than the simple version (8.4) or (8.9). Specifically, the estimation of the system parameters is much more comprehensive for this approach. The scaling factors A_j, B_i can be estimated using recursion procedures (Batty 1970; Batty and Mackie 1972; Wilson 1970) and, taking into account certain conditions of convergence (MacGill 1979; Sen 1982).

8.3 Entropy Maximization and Spatial Interaction Modelling

In thermodynamics, "entropy" is a measure of the order state of matter, namely the number of micro-states related to a given macro-state of the system. In a closed system, the entropy is constantly increasing and reaches a maximum of "disorder" for its equilibrium state. Therefore, the equilibrium state represents the most probable configuration of the system.

In closed physical systems there is a tendency towards increasing disorder of the micro-states of the system. The relation to the probability of finding a macro-state through certain micro-states makes it possible to transfer the entropy concept to different areas of social science (Wilson 1970). Thus the entropy concept has in principle a probabilistic background, related to the statistical distribution of events in an uncertain situation.

Since the distribution of the migration flows within a spatial system can also assume a large number of different configurations, the transmission of the framework concept of the entropy for the analysis of spatial interactions also makes sense.

In the meantime, numerous fields of application have been examined, such as commodity flow and migration flow distribution, shopping trip distribution or traffic flow assignment, to mention a few examples (Couclelis 2017).

The basic idea is always that the spatial distribution of the quantities of interest (e.g., the migratory flows) can be selected as the statistically most likely distribution by means of the entropy principle, taking into account given restrictions.

According to Wilson (1970) and Nijkamp and Reggiani (1992, 1998) two fields of application can be distinguished: firstly, the use of entropy as a descriptive concept in the sense that the most likely state of the system corresponds to a state of maximum entropy and this state will be realized. Secondly, the entropy can be used as an indicator for measuring the spatial organization degree of the system.

With regard to the statistical foundation and analysis of spatial interaction models, we now follow the fundamental theoretical work of Wilson (1970).

Let us consider a spatial system consisting of L regions. The flows w_{ij} between the regions $i, j = 1, 2, \ldots, L$, which can be migratory flows for instant, are initially indeterminate. A central task to be solved consists in finding ways to determine the flows w_{ij}. The problem we are confronted with is to make unbiased estimates leading to w_{ij} which are in agreement with all the possible knowledge available about the system. In this case we would have to consider the constraints for the origin flows in the form

$$\sum_{i=1}^{L} w_{ij} = O_j \tag{8.16}$$

and for the destination flows

$$\sum_{j=1}^{L} w_{ij} = D_i \tag{8.17}$$

Of course, the following consistency conditions must also apply

$$\sum_{i=1}^{L} \sum_{j=1}^{L} w_{ij} = \sum_{j=1}^{L} O_j = \sum_{i=1}^{L} D_i = w \tag{8.18}$$

According to the entropy principle, the most probable state is represented by the spatial distribution to which the greatest number of micro-states can be assigned. The number of possibilities for assigning flows between an origin region j and a destination region i is calculated by the number of combinatorial possibilities

$$\omega(w_{ij}) = \frac{w!}{\prod_{i=1}^{L} \prod_{j=1}^{L} w_{ij}!} \tag{8.19}$$

The maximum number $\omega(w_{ij})$ represents the number of possible realizations (micro-states) of individual flows to a given origin-destination matrix (macro-state). Since the maximum number $\omega(w_{ij})$ dominates the number of micro-states of any alternative arrangements the flow distribution associated with this maximum is the most likely one.

Instead of the maximum of the function, one can also determine the maximum of the logarithmic function, that is, the maximum of the function

$$\ln \omega(w_{ij}) = \ln w! - \sum_{i=1}^{L} \sum_{j=1}^{L} \ln w_{ij}! \tag{8.20}$$

or using Stirling's approximation

$$\ln \omega(w_{ij}) = \ln w! - \sum_{i=1}^{L} \sum_{j=1}^{L} (w_{ij} \ln w_{ij} - w_{ij}) \tag{8.21}$$

The expression

$$S = -k_B \sum_{i,j=1}^{L} w_{ij} \ln w_{ij} \tag{8.22}$$

represents the so called information entropy S of the system with Bolzmann's constant k_B. Without loss of generality, we will set $k_B = 1$.

Since $\ln w!$ is a constant it need not be considered in the maximization procedure. Therefore, the ultimate problem to be solved reads

$$\max \ln \omega(w_{ij}) = - \sum_{i=1}^{L} \sum_{j=1}^{L} (w_{ij} \ln w_{ij} - w_{ij}) \tag{8.23}$$

subject to the constraints (8.16), (8.17) and the assumption

$$\sum_{i=1}^{L} \sum_{j=1}^{L} c_{ij} w_{ij} = C. \tag{8.24}$$

Equation (8.24) underlines the assumption that the spatial system concerned has an upper limit C on its total transportation budget. By taking this constraint into account, however, the statistically most probable spatial distribution of the flows is based on general cost considerations, that is, taking into account transport costs c_{ij} between regions i and j.

The solution to the optimization task given above, taking into account the various constraints, is achieved by assigning a corresponding Lagrangian function L

$$\begin{aligned} L = \ln (\omega(w_{ij})) &+ \sum_{j=1}^{L} \lambda_j \left(O_j - \sum_{i=1}^{L} w_{ij} \right) \\ &+ \sum_{i=1}^{L} \mu_i \left(D_i - \sum_{j=1}^{L} w_{ij} \right) + \beta \left(C - \sum_{j=1}^{L} \sum_{i=1}^{L} c_{ij} w_{ij} \right), \end{aligned} \tag{8.25}$$

where λ_j, μ_i and β are the corresponding Lagrange parameters, respectively. The necessary conditions for a maximum are thus

$$\frac{\partial L}{\partial w_{ij}} = -\ln w_{ij} - \lambda_j - \mu_i - \beta c_{ij} = 0 \tag{8.26}$$

Thus we obtain for the flows, using (8.26)

$$w_{ij} = \exp\left(-\lambda_j - \mu_i - \beta c_{ij}\right) \tag{8.27}$$

and making use of (8.16), (8.17) and the abbreviations (8.28), (8.29)

$$A_i = \left[\sum_{j=1}^{L} B_j D_j \exp\left(-\beta c_{ij}\right)\right]^{-1} \tag{8.28}$$

$$B_j = \left[\sum_{i=1}^{L} A_i O_i \exp\left(-\beta c_{ij}\right)\right]^{-1} \tag{8.29}$$

the spatial interaction model (8.30)

$$w_{ij} = A_i B_j O_i D_j \exp\left(-\beta c_{ij}\right). \tag{8.30}$$

The distribution of the flows w_{ij} thus calculated represents the most likely distribution for the corresponding origin–destination relations. In this way, the "optimal" distribution of the flows within the spatial system can be estimated knowing the set of origin zones O_i, destination zones D_j and the transport cost distribution c_{ij}. The parameters A_i, B_j and β have to be determined recursively (Nijkamp and Reggiani 1992).

The distribution of the flows w_{ij} corresponds to a spatial interaction model of a gravity type with exponential transportation cost dependence. The negative exponential arises naturally in the theory. The concept of entropy thus opens up a new approach to spatial interaction modeling (Wilson 1970).

Furthermore, it is to be noted that the additivity of the flows is ensured by the presented concept and the dependence on transport costs is also reflected as expected. The strong analogy with physics is shown in the subordinate condition, which here assumes the meaning of a given "total energy" (8.24). In contrast to physics, however, this assumption is not based on a "conservation law" and thus represents an approximation.

However, some criticisms of the entropy concept are also appropriate. The striking elegance of the method is limited by the necessary proximity to thermodynamics. Thus, the existence of the entropy can only be shown for equilibrium systems or systems that are close to equilibrium, that is, as long as linear regression laws apply. The treatment of systems that are out of balance, and this is often the case with socioeconomic systems, is, however, rather questionable. This is one reason for a detailed study of the Master equation approach as modelling framework in the next subsection. The Master equation is suitable for the adequate treatment of non-equilibrium states.

8.4 The Master Equation Migration Model (Weidlich-Haag-Model)

Migration processes in all their theoretical and empirical aspects are of overwhelming importance to all spheres of society. Since migration and especially the demographic composition of the stream of migrants and their social affiliations are, among others factors of population growth, fundamental economic and social indicators of our society are directly influenced. Furthermore, even huge migration flows are based on single decisions and therefore on a big set of influencing factors. This means that flows may change their amplitude and direction sometimes in an unexpected manner (Fischer et al. 1990; Haag 1989).

To understand the reasons, agglomeration intensive research work has been initiated during the last few decades (Papageorgiou 1980; Wheaton 1974; Mills 1972). However, it is not the purpose of this book to deal with and investigate the underlying social and economic motivations to migrate into a sometimes already overpopulated area. Instead the modelling purpose as a base of complementary research of the problem refers to the question how migration processes generate the agglomeration trends and how agglomeration trends can be identified and adequately described by the mathematical framework of the Master equation (Bertuglia et al. 1994; Haag and Weidlich 1983, 1984a, b).

Whether or not a person migrates depends on the comparison between the current and future expected costs and benefits associated with such a move. Individuals are assumed to base their migration decision upon observed conditions and expectations. The underlying motivations to migrate from $A \rightarrow B$ are fortunately often well-defined and rationally understandable in the broad sense and thus available for inquiry (Courgeau 1985). On the other hand, all these motivations must always result in a clear decision to maintain or to change location in a given interval of time (Weidlich and Haag 1983). In principle, the number of relocations of a group of individuals (subpopulation, e.g., age group) per time interval can be counted.

In Germany these data are published yearly for different age groups and other subpopulations in form of migration tables on district level (402 districts in 2016). On the level of communes (11.418 communes in 2016) only in-flow and out-flow data are available. Understanding the modelling of the dynamics of these interregional migration processes is the general focus of this chapter. It follows, that those variables that determine a region's out-migration are the same that determine another's region's in-migration. Individuals are assumed to base their migration decision upon observed conditions and expectations.

Lee (1966), Isard (1954), Birg et al. (1983), and Huriot and Thisse (1984) assume that migration is influenced by gravity variables at origin, at destination and the links between them. Regional characteristics that attract migrants are regarded as "pull factors", while those that drive people out are regarded as "push factors". Since the act of migrating imposes costs, distance is seen as impeding migration (Courgeau 1995). Distance is widely accepted as having an impact on migration, as it increases the direct costs of moving, the opportunity and information costs. In this sense, distance-weighted variables represent spatial spillovers (Sardadvar and Rocha-Akis 2016). The closer a favourable regional characteristic is located, the higher its effect on actual migration is.

We formulate migration events as dynamic choice processes of individuals at the micro level, where the number of alternatives is given by the number of regions. Then the realized population distribution is the result of a sequence of decision processes of all individuals of the population. Empirical research has shown that the individual motivations and resulting decisions of the migrations are highly complex (Haag and Gruetzmann 2001). That is why it is reasonable to formulate the theory in probabilistic terms. In other words, for a certain member of the population, there is a specific probability per time unit of moving from one area to another. Now we proceed in analogy to the previous chapters.

(a) Definition of transition rates. These transition rates are formulated in terms of utility functions, which may depend on a set of explanatory variables. Changes in time of those explanatory variables lead to changes in the attractiveness of regions
(b) Formulation of the Master equation on the basis of the defined specific transition rates
(c) Calculation of the stationary solution of the Master equation
(d) Quasi-deterministic equations of motion for mean values
(e) Estimation of trend-parameters
(f) Determination of explanatory variables

8.4.1 *Definition of Transition Rates*

We consider in detail the migration of one homogeneous subpopulation consisting of N members (agents) between L regions. Later, we will subdivide the total population N into subpopulations α, $\alpha = 1, 2, \ldots, A$, where n_i^α is the number of members of subpopulation α living in region i, $i = 1, 2, \ldots, L$. However, in the next chapter we will only consider one subpopulation. This means that the macro state of the migratory system is characterized by the population configuration

$$\vec{n} = (n_1, n_2, \ldots, n_L), \tag{8.31}$$

where n_i is the number of people living in region $i = 1, 2, \ldots, L$ and the total population is given by

$$\sum_{i=1}^{L} n_i = N. \tag{8.32}$$

Our aim again is to understand the dynamics of the population distribution $\vec{n}(t)$.

The Mobility Matrix

The transition rate $p_{ij}(\vec{n})$, in other words, the probability per time unit that an individual living in a certain region j will migrate to region i will be assumed in analogy to Eq. (6.16) and is composed of two factors:

$$w_{ij} = n_j p_{ij}(\vec{n}) = n_j \nu_{ij} \cdot \exp\big(u_i(n_i + 1) - u_j(n_j)\big), \tag{8.33}$$

namely the symmetric mobility matrix, responsible for the strength of migratory interaction

$$\nu_{ij} = \nu_{ji} \geq 0 \tag{8.34}$$

and a push/pull term called regional attractiveness, sometimes also called regional utility $u_i(n_i)$. The migratory flow w_{ij} between $j \rightarrow i$ is assumed to be proportional to the population number n_j of the origin region j multiplied with the probability per time unit $p_{ij}(\vec{n})$ that an individual living in region j will migrate to region i.

It is assumed, that the same combination of explanatory variables determines the attractiveness of the origin and destination region. In Sect. 8.7 we will demonstrate how the attractiveness parameter $u_i(n_i)$ and the mobility matrix will be estimated on the basis of empirical data. It means that the underlying explanatory variables can be determined and tested in a second step. This is a big advantage of the Weidlich and Haag migration model.

The symmetric mobility matrix ν_{ij} replaces the distance deterrence function $f(d_{ij})$ of the Gravity model.

Furthermore, it is possible to define a global mobility ν_0, characterizing the mean mobility of the population

$$\nu_0 = \frac{1}{L(L-1)} \sum_{\substack{i,j=1 \\ i \neq j}}^{L} \nu_{ij}. \tag{8.35}$$

However, a multitude of explanatory socio-economic variables also merge into the mobility matrix ν_{ij}, e.g., structural data related to the transport infrastructure between the regions.

The modelling effort (8.33) aims to structure and condense all the information contained in the empirical data base to a set of well-interpretable indicators on the macro level. The mobility matrix ν_{ij} so far should contain all effects which will either facilitate or impede a transition between two states $i \leftrightarrow j$, independent of any gain of attractiveness or utility. Thus, in particular, the effect of distance d_{ij} (or transportation costs) between two regions i, j in its most general meaning will be reflected by the mobility matrix.

The two Eqs. (8.34) and (8.35) suggest the splitting of ν_{ij} in two factors

$$\nu_{ij} = \nu_0 \cdot f_{ij} \tag{8.36}$$

with a generally time-dependent deterrence factor $f_{ij} = f_{ji}$ describing the strength of migratory interaction and therefore implicitly also the effect of space d_{ij}. It follows from (8.36) and (8.35) that the deterrence factor must fulfil the normalization condition

$$\langle f \rangle = \frac{1}{L(L-1)} \sum_{\substack{i,j \\ i \neq j}}^{L} f_{ij} = 1. \tag{8.37}$$

This means that the spatial mean value of the dimensionless deterrence factor f_{ij} is equal to one. By means of the f_{ij} the strength of different migratory interactions can be compared.

The Distance Deterrence Function or Generalized Cost Function

Generalized costs of transportation $c_{ij} = c_{ji}$ can be introduced in its most general meaning using the definition:

$$\exp\left(-c_{ij}\right) = \frac{1}{L(L-1)} f_{ij} \tag{8.38}$$

It is worth mentioning that only if it turns out that the generalized costs c_{ij} are highly correlated with the geographic distance d_{ij}, in other words, if

$$c_{ij} \approx \beta\, d_{ij} \tag{8.39}$$

is approximately fulfilled, the mobility matrix assumes the well-known exponential form

$$\nu_{ij} \propto \exp\left(\quad \beta\, d_{ij}\right). \tag{8.40}$$

A variety of factors can contribute to what is usually called generalized costs (Wilson 1981a, b, c; Fotheringham and O'Kelly 1989). For example, travel time t_{ij}, transportation costs taken as proportional to distance d_{ij}, and barrier effects (excess time, parking search, ...) e_{ij}, to mention a few possible influencing factors. Such terms are usually combined linearly:

$$c_{ij} = \alpha\, t_{ij} + \beta\, d_{ij} + e_{ij} + \cdots. \tag{8.41}$$

A comparison of empirical data with assumptions about the deterrence effect of distance (see also Sect. 8.9.3) has shown that distance lose its importance in explaining inter-settlement flows for large distances. It is obvious that for shorter distances often the work place, the housing location, the organisation of family related activities can still be organised by an increase of transport activities, i.e., by commuting. If the migration distance exceeds, i.e., $d_{ij} > 100$ km the whole life of the migrants has to be reorganised and whether or not the next destination will be in 200 km or 500 km is no longer of strong importance. Therefore, the following deterrence function seems to be appropriate in case of interregional migration (Haag et al. 1991):

$$\nu_{ij} = a(t)\exp\left(-\frac{\beta\, d_{ij}}{1 + \gamma d_{ij}}\right) \tag{8.42}$$

where the parameter β depicts the strong impact of distance on migratory behaviour, and the parameter γ is associated with a kind of saturation behaviour, $a(t)$ is a scaling parameter.

The Dynamic Utility Function

We characterize the attractiveness of a region i for a member of the population n_i by a function depending on regional properties and characteristics called regional utility $u_i(\vec{n})$. Empirical studies show (Sanders 1992; Pumain 1991) that regions with large population size attract more people than regions with smaller population. This population size effect $s_i(\vec{n})$ or also called agglomeration effect must be distinguished from the size-independent effects δ_i, which also influence the utility functions. Therefore the total regional utilities are decomposed into two factors:

$$u_i(\vec{n}) = s_i(\vec{n}) + \delta_i, \tag{8.43}$$

where the agglomeration term $s_i(\vec{n})$ describes endogenous regional size effects and δ_i comprises all size-independent effects influencing the decision behaviour of the migrants.

Agglomeration Term $s_i(n_i)$

The agglomeration term describes endogenous regional size effects. Theoretical and empirical work demonstrates (Courgeau 1995; Haag and Weidlich 1984a, b) that a positive agglomeration trend will be amplified. Anticipating some results of the analysis on concrete regional systems, it turns out that the functional dependence of $s_i(\vec{n})$ of the population numbers is essential for the agglomeration behaviour. Two reasonable assumptions will be used and tested in the following:

Assumption A

The agglomeration term $s_i(\vec{n})$ is expanded in a truncated Taylor series up to the second order

$$s_i(\vec{n}) = \kappa n_i - \sigma n_i^2, \tag{8.44}$$

where $\kappa > 0$ describes agglomeration effects (the bigger the city's growth, the more attractive it becomes for people to migrate in) and $\sigma > 0$ denotes negative externalities based on saturation effects, e.g., increasing congestion.

Assumption B

A decreasing marginal settlement attractiveness is assumed. In the modelling language, it is realised as a logarithmic dependence (Pumain and Haag 1991; Haag 1993)

$$s_i(\vec{n}) = \kappa \ln n_i. \tag{8.45}$$

In Fig. 8.2, the functional dependence of the agglomeration term $s_i(\vec{n})$ for the French urban system, consisting of 78 settlements (1993) is shown. In this case (Assumption A), the City of Paris has already passed its optimal population size, whereas all other cities still gain from population growth. In the case of Assumption (B), the City of Paris exhibits still an increase of attractiveness due to agglomeration but with decreasing slope.

The perceived city size influences the decisions of migrants. In the long run, simulations have shown (Haag 1993; Haag and Max 1995) that the functional dependence of the agglomeration term $s_i(\vec{n})$ on population size shape the size-distribution of settlements.

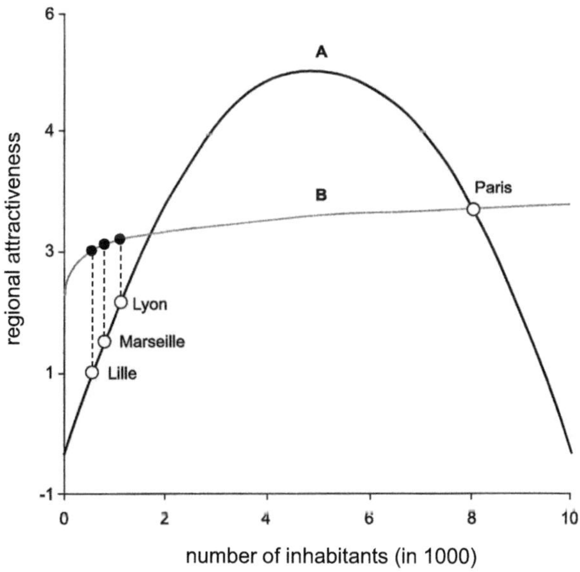

Fig. 8.2 Dependence of the agglomeration term $s_i(\vec{n})$, on population size. Assumptions (A) and (B) for the French urban system, 78 settlements (Source: Haag and Pumain 1991)

Preferences δ_i
Since size effects are concentrated in the term $s_i(\vec{n})$, all effects which are highly correlated with population size (extensive variables) have to be modelled via the term $s_i(\vec{n})$. Hence, the preferences δ_i have to be modelled by intensive indicators, e.g., z-transformed explanatory variables xz_γ, $\gamma = 1, 2, \ldots, G$

$$\langle xz \rangle = 1 \qquad \text{and} \qquad \sigma^2 = \langle xz^2 \rangle - \langle xz \rangle^2 = 1 \tag{8.46}$$

leading to

$$\delta_i = \sum_{\gamma=1}^{G} w_\gamma xz_\gamma, \tag{8.47}$$

where w_γ are the corresponding statistical weights. We will see later (Sect. 8.8) that we are able to estimate the preferences directly using empirical migration data. Then usual regression methods and software programmes can be used to select the appropriate set of significant explanatory variables.

Since in (8.33) only the differences of the utilities of the origin and destination region appear, all utilities are only defined except for an arbitrary common additive constant. This constant can be adjusted in such a way that the utilities always fulfil the condition

$$\langle u \rangle = \frac{1}{L} \sum_{i=1}^{L} u_i(\vec{n}) = 0 \qquad \text{or} \qquad \langle u \rangle = 100. \tag{8.48}$$

A measure for the inhomogeneity of the spatial system with respect to their migratory attractiveness is the regional variance of the attractiveness

$$\sigma_u^2 = \langle u^2 \rangle - \langle u \rangle^2 = \frac{1}{L} \sum_{i=1}^{L} u_i(\vec{n})^2. \tag{8.49}$$

This is a very interesting indicator for spatial disparities.

8.4.2 *The Migratory Master Equation*

On the stochastic level, fully consistent with the probabilistic description of decision processes (Chap. 6), we formulate the Master equation for the probability $P(\vec{n}, t)$ to find a certain population configuration $\vec{n}(t)$ realized at time t

$$\begin{aligned}\frac{dP(\vec{n},t)}{dt} &= \sum_{i,j=1}^{L} w_{ji}\left(\vec{n}^{(ij)}\right)P\left(\vec{n}^{(ij)},t\right) - \sum_{i,j=1}^{L} w_{ij}(\vec{n})P(n,t)\\ &+\sum_{i=1}^{L} w_{i+}\left(\vec{n}^{(i+)}\right)P\left(\vec{n}^{(i+)},t\right) - \sum_{i=1}^{L} w_{i+}(\vec{n})P(\vec{n},t)\\ &+\sum_{i=1}^{L} w_{i-}\left(\vec{n}^{(i-)}\right)P\left(\vec{n}^{(i-)},t\right) - \sum_{i=1}^{L} w_{i-}(\vec{n})P(\vec{n},t)\end{aligned} \tag{8.50}$$

where both, migratory transitions $w_{ij}(\vec{n})$ as well as birth processes $w_{i+}(\vec{n})$ and death processes $w_{i-}(\vec{n})$ are taken into account.

For the individual birth- and death processes in region i, we induce respective transitions

$$\vec{n} = \left(n_1,\ldots,n_i,\ldots,n_j,\ldots,n_L\right) \rightarrow \vec{n}^{(i+)} = \left(n_1,\ldots,n_i+1,\ldots,n_j,\ldots,n_L\right) \tag{8.51}$$

and

$$\vec{n} = \left(n_1,\ldots,n_i,\ldots,n_j,\ldots,n_L\right) \rightarrow \vec{n}^{(i-)} = \left(n_1,\ldots,n_i-1,\ldots,n_j,\ldots,n_L\right) \tag{8.52}$$

where the corresponding configurational transition rates (per unit of time) are denoted as $w_{i+}(\vec{n})$ and $w_{i-}(\vec{n})$, respectively. It is intuitively clear (see also Chap. 7) that these rates can be linked to the individual birth rate $\beta_i(t)$ and individual death rate $\mu_i(t)$ via

$$w_{i+}(\vec{n}) = \beta_i(t)n_i \qquad \text{and} \qquad w_{i-}(\vec{n}) = \mu_i(t)n_i \tag{8.53}$$

in case of a simple linear dependence on its size.

8.4.3 Stationary Solution of the Migratory Master Equation

The stationary solution of the Master equation neglecting birth and death events has been already discussed in detail in Chap. 6. Hence, some of the results of Chap. 6 will be summarized subsequently for the purpose of completeness.

Detailed balance permits an easy construction of the exact stationary solution $P_{st}(\vec{n})$

$$P_{st}(\vec{n}) = \frac{Z^{-1}\delta\left(\sum_{i=1}^{L} n_i - N\right)}{n_1!n_2!\cdots n_L!}\exp\left[2\sum_{i=1}^{L} F_i(n_i)\right], \tag{8.54}$$

where

$$F_i(n_i) = \sum_{m=1}^{n_i} u_i(m) \qquad \text{and} \qquad F_i(0) = 0 \tag{8.55}$$

and

$$\delta\left(\sum_{i=1}^{L} n_i - N\right) = \begin{cases} 1 & \text{if } \sum_{i=1}^{L} n_i = N \\ 0 & \text{otherwise} \end{cases} \tag{8.56}$$

and where the factor Z^{-1} is determined by the normalization condition for the configurational probabilities. It is not a Poisson distribution since the migrants do not behave independent of each other. It is also not a Gaussian distribution as we would expect in case of a Fokker-Planck approximation.

An approximate solution for the most probable population configuration $\widehat{n} = (\widehat{n}_1, \widehat{n}_2, \ldots, \widehat{n}_L)$can be easily obtained (see Sect. 6.5.1) yielding

$$\hat{n}_j = \frac{N\exp\left(2u_j(\hat{n})\right)}{\sum_{i=1}^{L}\exp\left(2u_i(\hat{n})\right)} \tag{8.57}$$

for $j = 1, 2, \ldots, L$ and where $\widehat{n}_j$ is the most probable regional population number.

Depending on the functional form of the utility functions $u_j(\vec{n})$ only one stationary solution exists and the corresponding stationary population distribution $P_{st}(\vec{n})$ is unimodal—or more than one solution exists and the multimodal distribution function exhibits peaks corresponding to the different possible equilibrium states.

Distance from Its Equilibrium State

In Sect. 3.3.3, it was proven that there exists always one stationary solution of the Master equation and that any time-dependent solution $P(\vec{n}, t)$ approaches for $t \to \infty$ the unique stationary solution if all trend parameters remain constant.

Comparing the actually realized population pattern $\vec{n}^e(t)$ with the equilibrium pattern $\widehat{n}(t)$ related to the actually measured trend parameters $xz_\gamma^e(t)$, we are able to define a distance from equilibrium under the current situation. As a quantitative measure for this distance from equilibrium, we introduce the migratory stress (Weidlich and Haag 1988)

$$s(\vec{n}^e(t), \hat{n}(t)) = \frac{1}{2}(1 - r(\vec{n}^e(t), \hat{n}(t))) \quad \text{with} \quad 0 \leq s(\vec{n}^e(t), \hat{n}(t)) \leq 1, \tag{8.58}$$

where $r(\vec{n}^e(t), \hat{n}(t))$ is the correlation coefficient. The migratory pattern clearly demonstrates that more than 25 years after German unification, the whole country is still far from its equilibrium state (see Sect. 8.9.3).

8.5 Quasi-Deterministic Equations of Motion

As in the proceeding chapters, approximate mean value equations for population pattern $\langle \vec{n} \rangle_t$ can be derived on the basis of the Master equation (8.50). We skip the derivation of the mean value equations here since the procedure is straightforward and has already been explained several times in this book. We finally obtain for the approximate mean value equations

$$\begin{aligned} \frac{d\langle n_j \rangle}{dt} &= \sum\nolimits_{i=1}^{L} \langle n_i \rangle p_{ji}(\langle \vec{n} \rangle) - \sum\nolimits_{i=1}^{L} \langle n_j \rangle p_{ij}(\langle \vec{n} \rangle) + \langle w_{j+} \rangle - \langle w_{j-} \rangle \\ &= \sum\nolimits_{i=1}^{L} \nu_{ji} \langle n_i \rangle \exp\left(u_j(\langle \vec{n} \rangle) - u_i(\langle \vec{n} \rangle) \right) \\ &- \sum\nolimits_{i=1}^{L} \nu_{ij} \langle n_j \rangle \exp\left(u_i(\langle \vec{n} \rangle) - u_j(\langle \vec{n} \rangle) \right) + \rho_j(t) \langle n_j \rangle \end{aligned} \tag{8.59}$$

for $j = 1, 2, \ldots, L$, where we have introduced the rate of natural increase

$$\rho_j(t) = \left(\beta_j - \mu_j\right). \tag{8.60}$$

Evidently, (8.59) is a set of L coupled nonlinear first order differential equations for the mean value of the regional population size. Of course, if we introduce explicitly the functional dependence of the regional utility $u_j(\langle \vec{n} \rangle)$ on population size and further explanatory variables, e.g., (8.44) or (8.45), the evolution with time of the population pattern can be solved by numerical simulation using scenario technique. We will come back to this task later.

However, if empirical considerations justify that the rate of natural increase does not depend on its region

$$\rho_j(t) = \left(\beta_j - \mu_j\right) \approx \rho(t), \tag{8.61}$$

it is possible to separate the birth/death processes from the migratory redistribution events by introducing relative population shares

$$x_j(t) = \frac{\langle n_j(t) \rangle}{N(t)} \quad \text{with} \quad 0 \leq x_j(t) \leq 1 \quad \text{and} \quad \sum_{j=1}^{L} x_j(t) = 1. \tag{8.62}$$

The development of the total population $N(t)$ can be directly calculated from (8.59) yielding

$$\frac{dN(t)}{dt} = \rho(t)N(t) \qquad \text{or} \qquad N(t) = N(0)\exp\int_0^t \rho(t')dt'. \tag{8.63}$$

On the other hand, the equation of motion for the relative population shares $x_j(t)$ reads

$$\begin{aligned} \frac{dx_j(t)}{dt} &= \sum_{i=1}^{L} \nu_{ji} x_i(t) \exp\left(u_j(\vec{x}(t)) - u_i(\vec{x}(t))\right) \\ &\quad - \sum_{i=1}^{L} \nu_{ji} x_j(t) \exp\left(u_i(\vec{x}(t)) - u_j(\vec{x}(t))\right) \end{aligned} \tag{8.64}$$

for $j = 1, 2, \ldots, L$ with its stationary solution

$$x_{jst} = \frac{\exp\left(2u_j(\vec{x}_{st})\right)}{\sum_{i=1}^{L} \exp(2u_i(\vec{x}_{st}))}, \tag{8.65}$$

which will be approached in the course of time $t \to \infty$.

$$\lim_{t \to \infty} x_j(t) = x_{j_{st}}. \tag{8.66}$$

8.5.1 Dynamic Phase Transitions and the Settlement Instability

In the next two parts of this chapter, we will discuss fundamental dynamic aspects in the development of human populations, namely

(a) The development and formation of cities and metropoles via a dynamic phase transition
(b) The occurrence and temporal stability of the rank-size distribution of a system of settlements as a self-organization process.

As already mentioned, the development and formation of cities and metropoles can only be understood by taking into account the strong agglomeration effects becoming apparent in its gravitational attraction of metropoles. The cause for this huge increase of metropolitan populations is obviously not the surplus of birth events and the general increase of the total population of the world, but the highly visible universal trends of population, independent of its nationality and education, to agglomerate. This means that the growth of metropoles is primarily due to migration flows into cities. The rich metropole New York gains population as well as a much poorer metropole like Daressalam.

Since our migratory system (8.59) is highly nonlinear, more than one stable equilibrium state may exist, and it depends on the starting condition (initial condition) of the population system, which of those possible equilibrium states will be assumed, everything else will be kept constant. In Weidlich and Haag (1987), a dynamic phase transition model for spatial agglomeration processes was investigated in detail. We will restrict ourselves to the discussion of some fundamental results.

The global nature of the dynamics including the number and stability of equilibrium states may change if certain "control" parameters pass critical values. In migratory systems, the strength of the agglomeration effect acts as such a control parameter. Such a change of the global dynamic structure of the system is denoted as "phase transition" in analogy to similar phenomena in natural sciences. Several authors have investigated nonlinear migration effects and spatial instabilities during the last 30 years (Papageorgiou and Smith 1983; Puu 1983; Hotelling 1978; Andersson and Ferraro 1982; Leonardi and Casti 1986; Haag and Weidlich 1984a, b; Weidlich and Haag 1987; Rosser 2011; Krugman 1994).

However, the mathematical structure of the utility function determines the details of the dynamics of the spatial system, its stationary solution and their stability. A comparison of simulations of different scenarios with empirical data, therefore, may help to decide which mathematical form of the utility seems to be the most appropriate. In other words, we aim to find out how individuals react on changing population sizes (Batty 1970; Batty and Longley 1994).

In so far the next two analysis considerations of the utilities

$$\text{Assumption A:}\quad u_j(\vec{n}) = \kappa\, n_j \tag{8.67}$$

$$\text{Assumption B:}\quad u_j(\vec{n}) = \kappa \ln n_j \tag{8.68}$$

provide fundamental insights in the final formulation of a consistent and appropriate migration model, based on the Master equation framework.

8.5.2 The Development and Formation of Cities and Metropoles as a Dynamic Phase Transition

Yet, it is questionable whether Assumption (A) or (B) seems to be most appropriate for long-term modelling of urban dynamics and the formation of cities and metropoles.

The functional dependence of the regional utilities has to be specified for the simulation, analytical investigation and stability analysis. From the standpoint of comparison with empirical population data a careful regression analysis of the utilities $u_j(\vec{n})$ with a set of significant explanatory variables (8.47) has to be implemented. As our purpose is to go as far as possible in the mathematical analysis, we adopt an idealized model, which nevertheless is designed to contain the main agglomeration effects.

Our choice of the idealized utilities implies

(a) The model describes L equivalent regions, whose migratory dynamics is only governed by the agglomeration trend and not by predetermined regional differentiations. In other words, agglomeration is modelled as a pure self-organizing process of regionally interacting populations breaking the symmetry of primordially equivalent regions.
(b) The saturation of the agglomeration trend in Assumption (A) is neglected, so that the agglomeration effects are enhanced. Of course, this assumption is essential for the structure of the stationary solution and will be discussed and weakened later.
(c) The dependence of the regional utility $u_j(\vec{n})$ on additional explanatory variables is neglected. Since our aim is to investigate some fundamental aspects related to agglomeration effects only, this assumption is necessary and well justifiable.
(d) Distance effects are neglected. Thus, transaction costs and externalities play no rule. However, the stationary solution of (8.65) is completely independent of the mobility matrix, in other words of distance effects. Distance effects determine how the system approaches its stationary state as well as the time-scale, but not the stationary state (see Chap. 4) as such. This means we assume $\nu_{ij} = \nu$.
(e) Birth- and death effects are neglected, too. As long as birth- and death effects only linearly depend on the population size and not on its region, the spatial redistribution effects are described sufficiently by (8.64)

8.5.3 Why Do Populations Agglomerate in Cities?

A central problem is the stability of the homogeneous distribution of the population, in case of L equally populated regions. For which values of the agglomeration parameter κ will this state remain stable or, alternatively, become unstable and what will be the preferred path of the system if the homogeneous solution becomes unstable?

For the investigation of these questions, let us start with the basic Eq. (8.64) and test the stability of the stationary solution for the two chosen functional forms of the regional attractiveness (8.67) and (8.68).

Assumption A: Stability Analysis for Linear Agglomeration Effects
Only a linear dependence of the regional attractiveness on population size will be investigated. This means we simplify the agglomeration term of the utility

$$u_j(\vec{n}) = \kappa\, n_j \qquad \text{for} \quad j = 1, 2, \ldots, L, \tag{8.69}$$

where the agglomeration parameter $\kappa > 0$ favours big settlements. Inserting (8.69) in (8.64) leads to the equations of motion

$$\frac{dx_j(\tau)}{d\tau} = \sum_{i=1}^{L} x_i(\tau)\exp\left(\kappa N\left(x_j - x_i\right)\right) - \sum_{i=1}^{L} x_j(\tau)\exp\left(\kappa N\left(x_i - x_j\right)\right) \tag{8.70}$$

with the scaled time $\tau = \nu t$, where $x_j(\tau)$ represents the scaled regional population distribution with

$$\sum_{j=1}^{L} x_j(\tau) = 1. \tag{8.71}$$

A homogeneous population distribution represents the stationary state x_{jst} of this migratory system, where

$$x_{jst} = \bar{x} = \frac{1}{L}. \tag{8.72}$$

In order to decide about the stability of the homogeneous population distribution we perform a linear stability analysis

$$x_j(\tau) = \bar{x} + \delta x_j(\tau), \tag{8.73}$$

where $\delta x_j(\tau)$ represents small perturbations of the homogeneous stationary solution. In the next step, the equation of motion (8.70) is expanded into a Taylor series in $\delta x_j(\tau)$, up to first order

$$\begin{aligned}\frac{d}{d\tau}\delta x_j &= \sum_{i=1}^{L}(\bar{x}+\delta x_i)\exp\left(\kappa N\left(\delta x_j-\delta x_i\right)\right)-\sum_{i=1}^{L}(\bar{x}+\delta x_j)\exp\left(\kappa N\left(\delta x_i-\delta x_j\right)\right)\\ &= \sum_{i=1}^{L}\left((\bar{x}+\delta x_i)\left(1+\kappa N\left(\delta x_j-\delta x_i\right)+\ldots\right)-(\bar{x}+\delta x_j)\left(1+\kappa N\left(\delta x_i-\delta x_j\right)+\ldots\right)\right)\\ &= L(2\kappa N\bar{x}-1)\delta x_j-2\kappa N\bar{x}\sum_{i=1}^{L}\delta x_i+\sum_{i=1}^{L}\delta x_i+\cdots\end{aligned} \tag{8.74}$$

or with

$$\sum_{i=1}^{L}\delta x_i(\tau)=0, \tag{8.75}$$

we finally obtain the linear first-order equation for the eigenvalue λ.

$$\frac{d}{d\tau}\delta x_j = L\left(2\kappa\frac{N}{L}-1\right)\delta x_j=\lambda\delta x_j \tag{8.76}$$

with the eigenvalue λ

$$\lambda = L(2\kappa\bar{n}-1)=2\kappa N-L. \tag{8.77}$$

Depending on the value of the agglomeration parameter κ, the eigenvalue λ can be positive, zero or negative. The critical value of κ_c

$$\kappa_c=\frac{1}{2\bar{x}}=\frac{1}{2}\frac{L}{N} \tag{8.78}$$

separates stable and unstable modes.

For sufficiently small values $\delta x_j(\tau)$, the perturbations will be diminished for $\lambda < 0$ or enhanced for $\lambda > 0$ exponentially, according to (8.79)

$$\delta x_j(\tau)=\delta x_j(0)\exp\left(\lambda\delta x_j(\tau)\right). \tag{8.79}$$

If the agglomeration parameter exceeds its critical value $\kappa > \kappa_c$, the homogeneous stationary distribution becomes unstable and one or several regions may grow. The eigenvalue λ is degenerated since λ does not depend on the index j of the region. As a consequence, each small deviation from the homogeneous stationary state $x_{jst}=\bar{x}$ is either damped or enhanced in the same exponential manner. In order to decide which initial mode of deviation $\delta x_j(\tau)$ is preferred in the sense of fastest growth a nonlinear mode selection analysis is required as shown in Weidlich and Haag (1987) for this example. We will skip this analytical step and just combine some of the results. The fastest mode of growth corresponds to one region becoming more densely populated while all other $L-1$ regions begin to deplete. The remaining stationary state is therefore characterized by one big region and $L-1$ almost depleted regions (see Fig. 8.3).

The stability analysis and the simulations clearly show that the implementation of endogenous agglomeration effects in the regional utilities may cause that a homogeneous population distribution becomes unstable and a phase transition towards a more complicated regional population distribution sets in. However, the development of a system of unequally populated settlements requires that the agglomeration parameter exceeds its critical value $\kappa > \kappa_c$. If we also take into account the saturation parameter σ, not only one dominating city will evolve, but depending on σ, two or more equally populated big cities will develop and all other settlements will become small in the course of time. To demonstrate this, a nonlinear stability analysis is required.

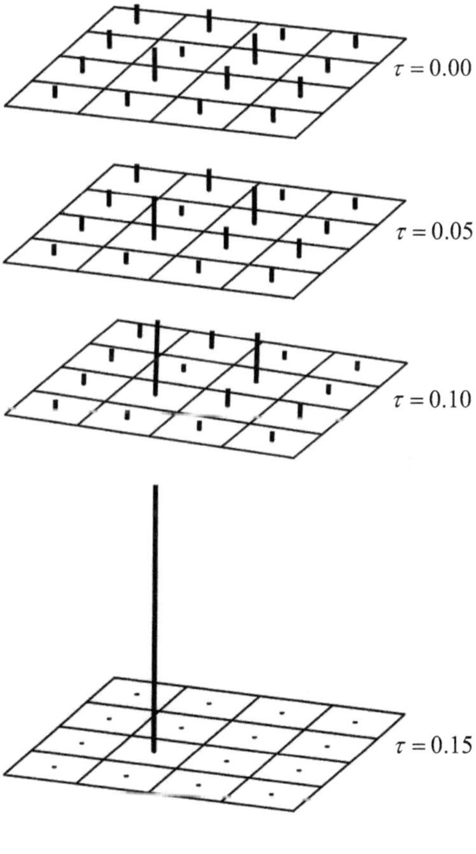

Fig. 8.3 Evolution with time of an initially homogeneous distribution into one metropolitan area and 15 almost depleted regions ($\widehat{\kappa} = 1.2\kappa_c$) (Source: Weidlich and Haag 1987)

Assumption B: Stability Analysis for Marginal Decreasing Agglomeration Effects
What happens if decreasing marginal settlement attractiveness is assumed? This means, if we perform a linear stability analysis assuming an agglomeration term of the utility

$$u_j(\vec{n}) = \kappa \ln n_j \qquad \text{for} \qquad j = 1, 2, \ldots, L, \tag{8.80}$$

where the agglomeration term $\kappa > 0$ favours big settlements. By insertion of (8.80) in (8.64) we obtain the equations of motion

$$\begin{aligned} \frac{d}{d\tau}\delta x_j &= \sum_{i=1}^{L}(\bar{x}+\delta x_i)\exp\left(\kappa \ln\left(\delta x_j - \delta x_i\right)\right) - \sum_{i=1}^{L}(\bar{x}+\delta x_j)\exp\left(\kappa \ln\left(\delta x_i - \delta x_j\right)\right) \\ &= \sum_{i=1}^{L}\left((\bar{x}+\delta x_i)\left(1+\kappa\left(\delta x_j - \delta x_i\right)+\ldots\right) - \left(\bar{x}+\delta x_j\right)\left(1+\kappa\left(\delta x_i - \delta x_j\right)+\ldots\right)\right) \\ &= L(2\kappa-1)\delta x_j - 2\kappa\sum_{i=1}^{L}\delta x_i + \sum_{i=1}^{L}\delta x_i + \cdots \end{aligned} \tag{8.81}$$

or with

$$\sum_{i=1}^{L}\delta x_i(\tau) = 0 \tag{8.82}$$

and finally the linear first-order equation for the eigenvalue λ

$$\frac{d}{d\tau}\delta x_j = L(2\kappa-1)\delta x_j = \lambda\delta x_j \tag{8.83}$$

with the eigenvalue λ

$$\lambda = L(2\kappa - 1). \tag{8.84}$$

Depending on the value of the agglomeration parameter κ, the eigenvalue λ can be positive, zero or negative. The critical value is $\kappa_c = 1/2$. For $\kappa < \kappa_c$ the homogeneous population distribution remains stable. If $\kappa > \kappa_c$, the symmetry is broken and an unequal population development sets in.

In this case, κ_c is independent of the number of grids L and the total population size N. The parameter κ reflects the preference of the population to agglomerate, whereas, of course, a multitude of factors may influence the agglomeration behaviour of the individuals.

However, empirical research of scientists all over the world demonstrates that the size-distribution of settlements exhibit a similar structure, which can be well approximated by a Pareto or Zipf distribution. Therefore, it is reasonable to expect and to require that the mathematical structure of the approximate equations of motion for the population distribution in the long-run should approximate a rank-size distribution. This will be investigated next.

8.6 Self-Organization Processes, Phase Transitions and the Rank-Size Distribution of Settlements

The Pareto distribution is fascinating scientists since many years, not only because of its seemingly very simple mathematical structure but also because scaling laws are some particular simple cases of more general relations. It should be emphasized that scaling never appears by accident (Barenblatt 2003; White 1978) but reveals an important property of the system under consideration, namely self-similarity. This means that the structure of the system reproduces itself on different scales. In physics, many systems which exhibit scaling behaviour over many orders can be found. In social sciences, e.g., for the city size distribution, a potential law can be found with acceptable accuracy only for a few rank-size decades.

The understanding of the concentration of human populations in settlements, villages, towns, cities and metropolitan areas is of general interest to geographers and regional scientists all over the world (Fig. 8.4). In particular, the stability of the rank-size distribution in most countries of the world demonstrate the importance of this issue. The urbanisation process is mainly considered as a transformation of rural into urban population. The causes and consequences, the search for possible

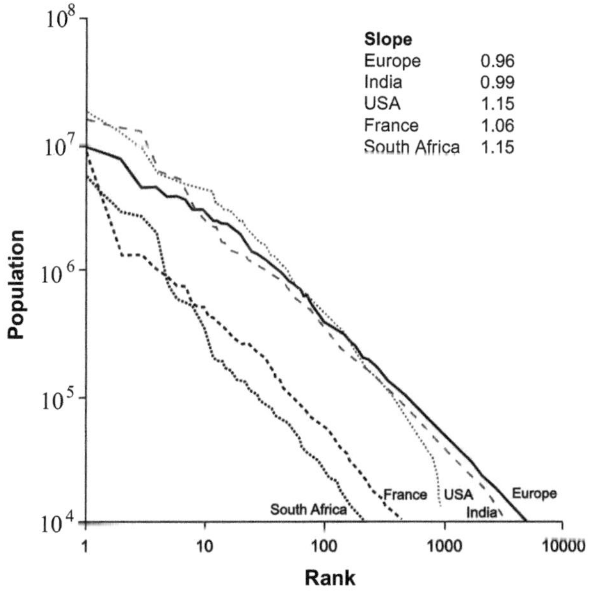

Fig. 8.4 Hierarchical differentiation in city-sizes (Sources: Europe: Moriconi-Ebrard F., 1994, GEOPOLIS/India: Census of India 2001/USA: United States Census 2000/France: INSEE, Recensement de la Population 1999/South Africa: Statistics South Africa, Census 2001, Base CVM)

explanations of how empirically observed size distributions are generated have focused the empirical and theoretical interest of many researchers since Auerbach (1913), Gibrart (1931), and Zipf (1949). Zipf assumes that the rank-size rule is the result of two counteracting forces, a "force of concentration" in order to avoid transportation costs during production and a "force of diversification" locating production and consequently consumption in the vicinity of the natural resources under exploitation, which are rather dispersed. Berry and Garrison (1958a, b), Curry (1964) and others suggested that the distribution would correspond to a state of maximum entropy within the urban system. However, theoretical distributions are rather different from the observed ones (Haran and Vining 1973). The explanation of the distribution is thus shifted to the constraints which have to be defined to force the theory in conformity with the process (Pumain 1982, 1990). Many other researchers have related the rank-size distribution to deterministic or stochastic growth processes within a set of cities (Gibrart 1931; Steindl 1968; Roehner and Wiese 1982). The application of nonlinear dynamic models (Allen and Sanglier 1979; Wilson 1981a, b, c; Pumain et al. 1989; Guerin-Pace 1990) has led to better knowledge about the underlying interaction principles. Portnov (2011) focused his research activities on the question why the largest cities of countries do not always follow Zipf's law, generally obeyed by smaller settlements. According to his findings, high development levels are likely to reduce this gap, while the first city being the national capital is likely to widen the gap between the largest city and the next-ranked city.

Some insights have been gained about the urbanisation process by testing those ideas on sets of empirical data, but there are still no general agreements regarding their interpretation (Pumain 1993, 2006). Descriptions of the size distribution by those mathematical models which attempt to replicate the data suffer from the fact that most of those models are easily adjusted to the observations. This problem is known as "over-identification of models" (Quandt 1964; Robson 1973). Since the rank-size distribution reflects just one dimension of a highly dimensional dynamic process of urban growth, it is not surprising that all the different approaches come out with size distributions more or less similar to the observed ones (Pumain 1982, 2006).

This has the following consequences: the quest for an appropriate dynamic modelling of the urbanization process and its projection to the rank-size distribution can only be decided by a comparison of the complex underlying empirical and modelled theoretical interaction processes and not via looking at the size-distribution. This means, we do not need a further theory for the explanation of the size-distribution but rather a good framework for the modelling of dynamic urbanization process of highly interacting settlements. See also Pumain (1993), who calls for integrating the geographic specificity of a set of cities into mathematical models.

8.6.1 *The Rank-Size Distribution of Settlements as a Dynamic Multifractal Phenomenon*

The above-mentioned obstacles can be largely circumvented by incorporating a stochastic framework, passing from the micro level of individual migration acts to a macroscopic consideration of a system of cities via the Master equation (Pumain and Haag 1991; Haag 1993; Haag and Max 1995). As a result, the evolution of the most probable state of the urban pattern is obtained. However, urban pattern formation is subject to external perturbations from an active environment and internal variations and fluctuations at the level of the parameters. The detailed information about the evolution of each interlinked settlement of the urban system of cities is strongly reduced by considering the rank-size distribution only. The corresponding reduction of information leads to the above-mentioned over-identification problem.

Therefore, we start with an integrated dynamic model for a system of L settlements, where the parameters and functions involved are clearly defined and can be related to empirical observations. The evolution of this system of settlements should then lead to the observed hierarchical organization process characterized by the rank-size distribution.

Of course, the self-organization process, the skewness and complexity of the distribution are related to the exogeneous and endogeneous parameters involved and fully determined by their behaviour. Therefore, the hierarchically skewed distribution of the whole network of settlements is related to the migration of people, birth- and death events as well as flows of material and information, to mention a few but relevant parameters. Interactions among the different settlements depend, in general, on the spatial distribution pattern of the settlements. In the long run, self-accelerating processes due to agglomeration or synergy effects as well as saturation effects must be considered to approach a typical rank-size distribution.

A settlement system cannot be considered as an isolated system but, instead, in its relation to other settlements and its hinterland. Specific properties of the urban system, which are both geographical and historical, are responsible for the evolution of settlements reflected in their complicated dynamics (Zipser et al. 2011). The settlement/settlement and the settlement/hinterland interaction as well as the emigration and immigration rates can be considered and modelled in the Master equation framework.

Let the urban system be composed of $i = 1, 2, \ldots, L$ settlements and the hinterland h. In complete analogy to Chap. 3, we describe the population configuration

$$\vec{n} = (n_1, n_2, \ldots, n_L) \tag{8.85}$$

of the urban system by the Master equation (8.50), where n_i is the population size of settlement i. We assume a hierarchical ordering of the settlements such that n_1 is the population of the largest place, n_i is the population of the i-th ranked place, and $n_L$corresponds to the smallest urban place.

What happens if we take into account negative externalities $\sigma > 0$, based on increasing congestion—in other words, if we assume the functional form (A) $u_i(\vec{n}) = \kappa n_i - \sigma n_i^2$ with an optimum size of attraction for the regional utility or if we assume the functional form (B) $u_i(\vec{n}) = \kappa \ln n_i$, describing decreasing marginal agglomeration effects?

From the Master equation, we derive the approximate mean value equations

$$\frac{d\langle n_j\rangle}{dt} = \sum_{i=1}^{L} \nu_{ji}\langle n_i\rangle \exp\Big(u_j(\langle\vec{n}\rangle) - u_i(\langle\vec{n}\rangle)\Big) - \sum_{i=1}^{L} \nu_{ij}\langle n_j\rangle \exp\Big(u_i(\langle\vec{n}\rangle) - u_j(\langle\vec{n}\rangle)\Big)$$
$$+ \langle W_{jh}\rangle - \langle W_{hj}\rangle + \langle w_{j+}\rangle - \langle w_{j-}\rangle \tag{8.86}$$

for $j = 1, 2, \ldots, L$, where the change in population size of settlement j is due to inter-settlement migration as well as to interactions with the hinterland $\langle W_{jh}\rangle, \langle W_{hj}\rangle$, the environment and birth/death events $\langle w_{j+}\rangle, \langle w_{j-}\rangle$. A considerable class of urban systems can approximately described by

$$\langle W_{jh}\rangle - \langle W_{hj}\rangle + \langle w_{j+}\rangle - \langle w_{j-}\rangle \approx \rho(t)\langle n_j\rangle, \tag{8.87}$$

where $\rho(t)$ does not depend on the settlement under consideration. However, this assumption can easily be relaxed if necessary. Assumption (8.87) is contrary to the model of Gibrart (1931), where he assumed that the growth rates of the settlements $\rho_j(t)$ are spatially and temporally independent.

Using (8.87) it is easy to separate the birth/death and the settlement/hinterland interactions from the migratory processes as already described in Chap. 4. This yields

$$\frac{dx_j(t)}{dt} = \sum_{i=1}^{L} \nu_{ji} x_i(t) \exp\big(u_j(\vec{x}(t)) - u_i(\vec{x}(t))\big)$$
$$- \sum_{i=1}^{L} \nu_{ji} x_j(t) \exp\big(u_i(\vec{x}(t)) - u_j(\vec{x}(t))\big) \tag{8.88}$$

for $j = 1, 2, \ldots, L$, where we have again introduced the relative population shares

$$x_j(t) = \frac{\langle n_j(t)\rangle}{N(t)} \quad \text{with} \quad 0 \le x_j(t) \le 1 \quad \text{and} \quad \sum_{j=1}^{L} x_j(t) = 1 \tag{8.89}$$

and the mobility matrix

$$\nu_{ij} = a(t)\exp(-c_{ij}), \tag{8.90}$$

where costs $c_{ij} = c_{ji}$ are introduced in its most general meaning.

The total urban population changes over time depending on the evolution of $N(t)$. Despite the dramatic growth of the urban population and the number and size of the settlements, the shape of the rank-size distribution remains remarkable stable.

The spatial pattern of settlements suggests that the dynamics of a system of settlements at the most aggregate level could be formalized by a time-dependent Pareto distribution

$$n_k(t) = n_1(t)k^{-q(t)}, \tag{8.91}$$

where $n_k(t)$ is the mean population of the k-th ranked settlement, $n_1(t)$ the population of the largest place, and $q(t)$ the Pareto coefficient. This power law has shown to be an acceptable approximation to empirical observations in different countries (Fig. 8.4). Therefore, it is reasonable to assume, that variations of the Pareto coefficient $q(t)$ will depend on the parameters involved in the complex urban dynamics. In other words, the "scale symmetry" involved in the dynamics of the nested system of settlements will be a function of the inherent parameters (Guerin-Pace 1990, 1993; Pumain 2006). Self-similarity provides the basis for the understanding of the hierarchical organized structure of settlements resulting in a corresponding "power law". In this sense settlements may be treated as fractal objects (Frankhauser and Guerin-Pace 1993; Frankhauser 1998).

The Pareto distribution (see Sect. 2.3.5) is a continuous probability distribution. Zipf's law, also sometimes called the zeta distribution, may be seen as a discrete counterpart of the Pareto distribution.

Comparing (8.91) with the hyperbolic distribution of word frequencies (8.92) investigated by Mandelbrot (1991)

$$p_k(t) = F(k + V)^{-1/D} \tag{8.92}$$

shows that for $V \to 0$ the Pareto coefficient $q(t)$ is the inverse of the fractal dimension D

$$q(t) = 1/D(t). \tag{8.93}$$

The dynamic formation of such a scaling law of settlement systems is qualitatively pointed out by Pumain (1993).

It cannot be assumed that the temporal evolution of the settlement systems starts from the very beginning as a Pareto distribution. On the contrary, we expect that settlements are sparsely distributed in historical dimensions and in the course of hundreds of years certain settlements grow faster than others and the formation of a rank-size distribution was evolving. This requires from a mathematical point of view the introduction of more degrees of freedom. Instead of the pure rank-size Eq. (8.91) we introduce (8.94) in (8.86)

$$n_k(t) = n_1(t)k^{-q_k(t)} \tag{8.94}$$

and transform the equations for the population numbers of the settlements into equations for varying Pareto coefficients $q_k(t)$ (Haag 1993; Haag and Max 1995). Then the Pareto distribution would be a stable attractor of the spatial system of settlements if

$$\lim_{t\to\infty} q_k(t) = \bar{q} \tag{8.95}$$

holds, at least approximately. This seems to be a rather restrictive condition, since the dynamics depends on the functional form of the utility function, in other words on how agglomeration effects as well as other economic effects will be modelled mathematically.

This chapter aims to investigate under what conditions the Pareto distribution is most likely to be found as a stationary solution of our migratory system of settlements (8.86). The dynamics of the settlement/hinterland system connects the slow dynamics of the rank-size distribution of settlements with the much more dynamic migratory behaviour of individuals.

The equations for the settlement/hinterland become fully explicit by insertion of the analytical form of the settlement utilities $u_k(\vec{n})$. In Figs. 8.5 and 8.6 the

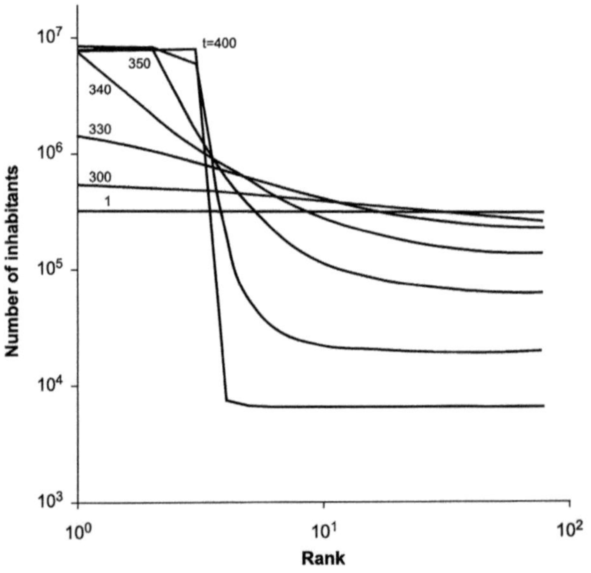

Fig. 8.5 Simulation of the French settlement system using Assumption (A): $\kappa = 0.596$; $\sigma = 0.188$; $\nu = 0.001$; $N = 25.6 \times 10^6$; $L = 78$ (Source: Haag 1993)

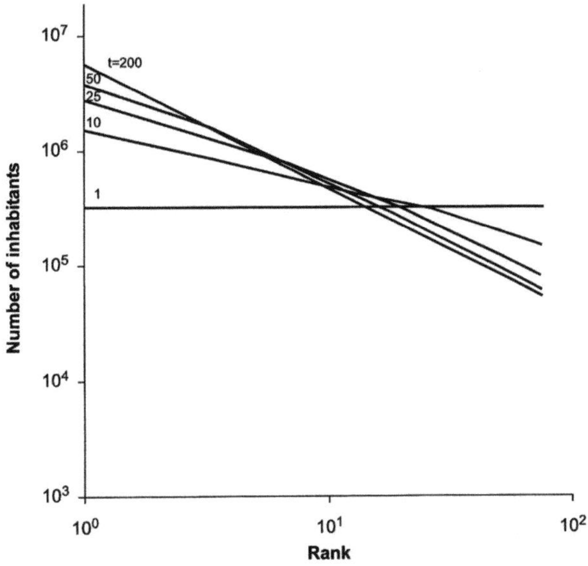

Fig. 8.6 Simulation of the French settlement system using Assumption (B): $\kappa = 0.500\,;\, \nu = 0.001\,;$ $N = 25.6 \times 10^6\,;\, L = 78\,;$ $q = 1.02$ (Source: Haag 1993)

simulations of two hypothetical settlement systems (Assumptions A and B) are shown, respectively. In both cases the preference parameters are fixed to $\delta_j = 0$. This means that effects related to varying and changing preferences are not explicitly taken into account. In other words, the dynamics investigated and treated as responsible for the development of the rank-size distribution fully depends on the self-organization process of settlements caused by decisions of individuals to migrate.

In both simulations, (A) and (B), we start with an almost homogeneous distribution

$$n_k(0) = N/L + \varepsilon_k = \bar{n} + \varepsilon_k \tag{8.96}$$

of settlements, where $\varepsilon_1 > \varepsilon_2 > \ldots > \varepsilon_L$, and $|\varepsilon_k| << 1$ for all $k = 1, 2, \ldots, L$, and $\sum_{k=1}^{L} \varepsilon_k = 0$.

Furthermore, the system of settlements is treated as a closed system. Hence, the ranking of settlements is the result only of the dynamic self-organization processes involved. The total population size $N(t)$ remains constant.

In Fig. 8.5, Assumption (A) is used for the simulation. The urban system remains close to the initially chosen homogeneous distribution for a rather prolonged time period ($\approx$300 time units). When the hierarchy starts to develop, the evolution dramatically accelerates and within the next 40 time units as an intermediate state, a Pareto-like distribution appears, which finally ends up in a stationary distribution characterized by a few dominating cities and a set of almost equally populated small settlements. The reason for this strong bifurcation behaviour of settlements with respect to population is based on the functional structure of the regional utilities. The assumed second-order polynomial favours an optimal city size.

In Fig. 8.6, the simulation with a decreasing marginal attractiveness $u_k(\vec{n})$ on settlement size is shown (Assumption B). The hierarchical ordering process starts immediately. Within a much shorter time interval of approximately 100–200 units, a Pareto distribution is approached. The dynamics leads independently from the initial conditions to a unique stationary solution. The Pareto distribution is the result of a dynamic self-organization process with the Pareto distribution as a stable attractor.

Stationary Settlement Distribution

The stationary settlement size distribution n_k^{st} can easily be obtained as solution of the stationary Master equation via

$$n_k^{st} = C \exp(2u_k(\vec{n}_{st})), \tag{8.97}$$

where the constant C is given by the normalization condition. It is worth mentioning that the stationary state (8.97) is independent of the specific structure of the deterrence function ν_{ij}. However, the temporal path of the urban system towards its stationary state depends on the regional interdependencies.

From (8.97), it immediately follows

$$\frac{n_1^{st}}{n_k^{st}} = \exp[2(u_1 - u_k)] \tag{8.98}$$

or with (8.94) and taking the logarithm and after some minor manipulations

$$q_k^{st} = \frac{2}{\ln k}(u_1 - u_k) \qquad \text{for} \qquad k = 2, 3, \ldots, L. \tag{8.99}$$

However, the mean rank-size coefficient $\bar{q}$ is of empirical interest:

$$\bar{q} = \frac{1}{L}\sum_{k=1}^{L} q_k = \frac{2}{L}\sum_{k=1}^{L}\frac{1}{\ln k}(u_1 - u_k), \tag{8.100}$$

which can be seen as a first approximation of the Pareto coefficient. Hence, the Pareto coefficient is directly linked to the differences of regional attractiveness of the settlements.

Let us now go one step further to introduce and test the impact of both Assumptions (A) and (B), respectively.

Assumption A

Inserting (8.67) in (8.100) leads to

$$\begin{aligned}\bar{q} &= \frac{2}{L}\sum_{k=1}^{L}\frac{1}{\ln k}(u_1 - u_k) = \frac{2}{L}\sum_{k=1}^{L}\frac{1}{\ln k}(\kappa n_1 - \kappa n_k) \\ &= \frac{2\kappa}{L}\sum_{k=1}^{L}\frac{1}{\ln k}(n_1 - n_1 k^{-q_k}) = \frac{2\kappa n_1}{L}\sum_{k=1}^{L}\frac{(1-k^{-q_k})}{\ln k}\end{aligned} \tag{8.101}$$

an inconsistent relationship between the mean Pareto coefficients $\bar{q}$ and the other parameters related to the attractiveness of the settlements. This is an important critique to the usage of Assumption (A) for long-time simulation of interregional behaviour of settlement systems.

Assumption B

Introducing Assumption (B) (8.68) into (8.100) leads to

$$\begin{aligned}\bar{q} &= \frac{2}{L}\sum_{k=1}^{L}\frac{1}{\ln k}(u_1 - u_k) = \frac{2}{L}\sum_{k=1}^{L}\frac{1}{\ln k}(\kappa \ln n_1 - \kappa \ln n_k) \\ &= \frac{2}{L}\sum_{k=1}^{L}\frac{1}{\ln k}(\kappa \ln n_1 - \kappa \ln n_1 k^{-q_k}) = \frac{2}{L}\sum_{k=1}^{L}\frac{1}{\ln k}(\kappa \ln n_1 - \kappa \ln n_1 + \kappa q_k \ln k) \\ &= \frac{2\kappa}{L}\sum_{k=1}^{L}\frac{1}{\ln k} q_k \ln k = 2\kappa\,\bar{q}\end{aligned} \tag{8.102}$$

This Eq. (8.102) is only consistent if we require that the agglomeration parameter assumes its critical value $\kappa = \kappa_c = 1/2$. This means that the system of settlements fulfils the condition of self-organized criticality (SOC).

Self-organized criticality (SOC) is a concept, introduced by Bak et al. (1987) in physics. SOC is considered one of the important mechanisms by which complexity arises in nature (Bak and Paczuski 1995). SOC is typically observed in slowly driven non-equilibrium systems with extended degrees of freedom and a high level of nonlinearity.

SOC is a property of a class of dynamical systems that have a critical point as an attractor, in our case of a system of settlements $\kappa = \kappa_c = 1/2$. The macroscopic behaviour of our settlement system thus displays the spatial and/or temporal scale-invariance characteristic of the critical point of a phase transition, but without the need to tune control parameters to precise values.

The Pareto distribution is the result of a dynamic self-organization process in our modelling approach. Starting with a spatial instability of the homogeneous population distribution different spatial modes compete with each other and drive the spatial system towards its critical point as an attractor, namely the Pareto distribution.

8.7 Model Implications for the Weidlich-Haag Migratory Model

What are the consequences with respect to the modelling of interregional migration?

The described stability analysis of the Weidlich-Haag migration model and the performed simulation experiments of the self-organization process (8.88) favour Assumption (B). Only in this case (8.86) the Pareto distribution can be obtained in a dynamic self-organization process. Of course, these results have some consequences with respect to the structural form of our migration model.

Let us first introduce the regional attractiveness of Assumption (B) (8.68) in the Weidlich-Haag model (8.33). This leads to

$$\begin{aligned} w_{ij} &= n_j \nu_{ij} \exp\left(u_i - u_j\right) \\ &= n_j \nu_{ij} \exp\left(\kappa \ln n_i - \kappa \ln n_j + \delta_i - \delta_j\right) \\ &= n_j \nu_{ij} n_i^{\kappa} n_j^{-\kappa} \exp\left(\delta_i - \delta_j\right) \\ &= \nu_{ij} n_i^{\kappa} n_j^{1-\kappa} \exp\left(\delta_i - \delta_j\right) \end{aligned} \tag{8.103}$$

where w_{ij} describes the migration flow from region or settlement j to i, and $\nu_{ij} = \nu_{ji}$ the so-called mobility matrix, in other words, the spatial interaction term, κ the strength of the agglomeration effects and finally δ_i the preference effects of the population for a specific region or settlement i. Of course, the regional preferences depend on a set of explanatory variables. We will come back to this point later and discuss some consequences of (8.104) now:

$$w_{ij} = \nu_{ij} n_i^{\kappa} n_j^{1-\kappa} \exp\left(\delta_i - \delta_j\right) \quad \text{for} \quad i, j = 1, 2, \ldots, L. \tag{8.104}$$

Let us assume that we want to investigate the migratory flows between two completely identical regions with respect to all parameters. We mark the two regions with i and j. Then we have the same population $n_i = n_j$, the same regional preferences $\delta_i = \delta_j$, and of course as always $\nu_{ij} = \nu_{ji}$, respectively.

We can take as granted under this conditions that also the migratory flows between both regions must have the same value, in other words $w_{ij} = w_{ji}$. Furthermore, the parameter κ has to be fixed to the value $\kappa = 1/2$, the SOC-value. Our migratory model fulfils the required symmetry condition and the rank-size requirement only if our model fulfils simultaneously the condition of self-organized criticality (SOC).

If we insert $\kappa = 1/2$ into (8.104), we obtain the final form of our migratory model, namely:

$$w_{ij} = \nu_{ij} \sqrt{n_i n_j} \exp\left(\delta_i - \delta_j\right) = \nu_0 f_{ij} \sqrt{n_i n_j} \exp\left(\delta_i - \delta_j\right), \tag{8.105}$$

where $i, j = 1, 2, \ldots, L$.

Obviously (8.105) shows that the back and forth migratory flows between two regions i,j will have the same size $w_{ij} = w_{ji}$ if the regional preferences of both regions agree, $\delta_i = \delta_j$, independent of the population size of both regions.

For the analysis of migration flows and the use of migration flows for the prediction of population development of a system of regions, the spatial preference indicators δ_i as part of regional attractiveness $u_i(\vec{n})$ play the decisive role.

Indicators related to the labour market system, supply facilities, transport infrastructure, leisure quality and environmental quality and housing market are frequently used in literature to describe the different attractiveness components.

Via multiple-regression analysis, it can be determined which set of sub-indicators significantly affect the regional attractiveness components of a settlement or region.

However, the attractiveness of a location is generally determined not solely by its equipment with facilities, number of jobs, rent level and land uses, to mention a few indicators, but partly also by accessible facilities, jobs and land use in neighbouring regions.

Therefore, it makes sense to use two types of indicators. One type of indicators will be used if the equipment of a region only in the region itself has an attractiveness influencing effect due to a very limited scope of use, such as the equipment with primary schools.

The second type of indicators is used when the features have an effect that goes beyond the boundaries of a region such as the equipment with railway stations, universities or airports. In this case, the indicators describing the attractiveness components are designed in a way that they take into account both: indicators related to the region as well as indicators referring to neighbouring regions.

Explanatory Variables of Regional Preferences δ_i

Since the migratory flows depend only on differences of regional preferences, $(\delta_j - \delta_i)$, we can scale the preferences without of any loss of generality

$$\sum_{i=1}^{L} \delta_i = 1. \tag{8.106}$$

In all practical examples data pre-processing plays a crucial role. Therefore, the first step concerns the normalization or standardization of data. This is a very important step, since in all applications we have to deal with explanatory variables of different scales and units.

Normalization requires to scale all numeric variables in a certain fixed range usually in the interval [0, 1].

Standardization of data imposes a transformation of the data set of potential explanatory variables xz_i^γ. $i = 1, 2, \ldots, L; \gamma = 1, 2, \ldots, G$, with the aim to have zero mean and unit variance:

$$xz_i^\gamma = \frac{z_i^\gamma - \langle z^\gamma \rangle}{\sigma^\gamma}. \tag{8.107}$$

The drawbacks of both methods are clear. In case of normalization, outliers in the data set may certainly scale the working range of the data to a very small interval.

Standardized variables are, however, not bounded, unlike normalization. Therefore, in all practical examples, it depends on the structure of the data set, whether a simple normalization or standardization seems to be adequate, or if more data advanced data transformations should be applied.

In the next step, we assume a selection, e.g., by multiple regression analysis, of a set of (standardized) explanatory variables xz_γ, $\gamma = 1, 2, \ldots, G$, related to different significant socio-economic influences as indicated above.

This leads to the representation of the preference indicators

$$\delta_i = \sum_{\gamma=1}^{G} a_\gamma xz_i^\gamma = a_1 xz_i^1 + a_2 xz_i^2 + a_3 xz_i^3 + \cdots, \tag{8.108}$$

where $i = 1, 2, \ldots, L; \gamma = 1, 2, \ldots, G$ and the statistical weights $a_\gamma \geq 0$ can be seen as the corresponding elasticity's of the explanatory variables xz_i^γ.

8.7.1 Final Structure of the Weidlich-Haag Model

If we insert (8.108) in the Weidlich-Haag model (8.105), we obtain the gravity-like structure for the flows between region $j \to i$:

$$w_{ij} = \nu_{ij}\sqrt{n_i n_j}\exp(\delta_i - \delta_j) = \nu_{ij}\sqrt{n_i n_j}\exp\left(\sum_{\gamma=1}^{G} a_\gamma xz_i^\gamma - \sum_{\gamma=1}^{G} a_\gamma xz_j^\gamma\right)$$

$$w_{ij} = \nu_0 f_{ij}\sqrt{n_i n_j}\frac{\exp\left(\sum_{\gamma=1}^{G} a_\gamma xz_i^\gamma\right)}{\exp\left(\sum_{\gamma=1}^{G} a_\gamma xz_j^\gamma\right)} \geq 0 \qquad . \tag{8.109}$$

Let us now compare the final structure of the Weidlich–Haag model with the structure of the Gravity model and the points and requirements we defined in Sect. 8.1:

Symmetry
The symmetry requirement is fulfilled. Let us consider two completely identical regions with respect to all parameters. Since $\nu_{ij} = \nu_{ji}$, and $\delta_i = \delta_j$, is always fulfilled the symmetry $w_{ij} = w_{ji}$ can be taken as granted.

Distance definition and distance deterrence function
The interaction matrix or mobility matrix $\nu_{ij} = \nu_{ji}$ plays the role of the distance deterrence function. However, neither an appropriate measure of distance d_{ij} has to be defined nor an appropriate distance deterrence function.

Zero flow problem
If we compare the Weidlich-Haag model (8.109) with the Gravity model (8.9) it seems that the zero flow-problem still remains and at the beginning, empty regions cannot be filled because of the factor $\sqrt{n_i n_j}$. Since the mobility matrix $\nu_{ij} = \nu_{ji}$ itself depends on $1/\sqrt{n_i n_j}$according to (8.148), transitions into up to now empty regions may happen.

Incomplete set of explanatory variables
All parameters of the Weidlich-Haag model, the interaction matrix $\nu_{ij} = \nu_{ji}$ and the preferences δ_i will be estimated on the basis of empirical data. This will be shown in the next part of this chapter. The preferences δ_i can be further matched to empirical data by a multiple regression analysis (8.108). Therefore, it is easily possible to test different sets of explanatory variables, even if some data sets are incomplete.

Test of hypothesis
The selection of an appropriate set of significant explanatory variables requires not a completely new estimation (optimization) of the trend parameters $\nu_{ij} = \nu_{ji}$ and δ_i but only a new performance of an econometric regression analysis (Greene 2003).

Zero or small empirical migratory flows
The estimation of the Weidlich-Haag model is based on a nonlinear estimation procedure and accepts zero flows. The log-linear estimation procedure (see next chapter) can be used as a starting point in the nonlinear estimation to accelerate the convergent process, but this performance is not a necessary condition for the nonlinear estimation. Therefore, systematic errors are not expected.

However, the Weidlich-Haag migration model (8.109) is based on the modelling of the fundamental decision processes of individuals, households or agents. In addition, fundamental symmetry considerations and the requirement that the rank-size distribution will be approximated in the long-run, determine the final modelling structure.

8.7.2 Final Structure of the Approximate Mean Value Equations of the Weidlich-Haag Model

In Sect. 8.5, quasi-deterministic equations of motion for spatial population pattern $\langle n_j \rangle_t$ have been derived. For simplicity, we use $\langle n_j \rangle_t = n_j$ in the following. If we insert the specific form of the Weidlich-Haag model (8.105) in the approximate mean value Eq. (8.86), we obtain a compact system of equations of motion for the regional dynamics

$$
\begin{aligned}
\frac{dn_j}{dt} &= \sum_{i=1}^{L} \left(w_{ij}(\vec{n}) - w_{ji}(\vec{n}) \right) + \langle w_{j+} \rangle - \langle w_{j-} \rangle \\
&= \nu_0 \sum_{i=1}^{L} f_{ji} \sqrt{n_i n_j} \exp\left(\delta_j - \delta_i\right) - \nu_0 \sum_{i=1}^{L} f_{ij} \sqrt{n_i n_j} \exp\left(\delta_i - \delta_j\right) + \rho_j(t) n_j \\
&= 2\nu_0 \sum_{i=1}^{L} f_{ij} \sqrt{n_i n_j} \sinh\left(\delta_j - \delta_i\right) + \rho_j(t) n_j
\end{aligned}
\tag{8.110}
$$

where $i, j = 1, 2, \ldots, L$. In a compact formulation, the approximate mean value equations read:

$$
\frac{dn_j}{dt} = 2\nu_0 \sum_{i=1}^{L} f_{ij} \sqrt{n_i n_j} \sinh\left(\delta_j - \delta_i\right) + \rho_j(t) n_j, \tag{8.111}
$$

where $\rho_j(t)$ describes spatially dependent birth- and death effects.

The development and formation of a spatial system consisting for example of a network of cities and rural regions, taking into account agglomeration effects as well as a bundle of explanatory socio-economic variables summed up in regional preferences $\delta_i(zx)$ can be simulated using (8.111). This nonlinear system of differential equations is compatible with the development of hierarchical settlement structures as observed in the rank-size distribution for $\rho_j(t) \approx \rho(t)$. All practical applications, however, require the adaptation of our model to empirical data. Therefore, the next section is devoted to the estimation of model parameters from empirical data.

8.8 Determination of Utilities and Mobilities from Empirical Data

The trend parameters, i.e., the utilities and mobilities are determined by minimizing the sum of the squared deviations between the empirical flows $w_{ij}^e(t)$ and the flows $w_{ij}(t)$ calculated on the base of our migratory model. In other words, we apply the method of least squares (Haag and Weidlich 1986; Weidlich and Haag 1988). But also the log-linear estimation of trend parameters, that means the minimization of the sum of square deviations between $\ln w_{ij}^e(t)$ and $\ln w_{ij}(t)$ leads to comparable good results in many cases. We shall begin with the log-linear estimation since this procedure provides closed analytical expressions of all trend parameters in terms of empirical quantities.

We assume that the following empirical data (index e) are available:

$$\left(n_i^e(t), w_{ij}^e(t)\right) \quad \text{for} \quad i,j = 1,2,\ldots,L,\ i \neq j \text{ and } t = 1,2,\ldots,T. \tag{8.112}$$

Our model concerning the migration matrix now has to be matched to the empirical migration flows $w_{ij}^e(t)$, by an optimal adaptation of the utilities $u_i(t)$ and the mobility's $\nu_{ij}(t)$.

In principle, we have to deal with an optimization problem. This means, we have to define an objective function or cost function. Many different forms of objective functions and optimization procedures have been developed and applied in the last decades.

In social sciences, a log-linear objective function $F(\nu, \vec{u})$ is often applied, based on the fact that the maximum of a positive definite function and the maximum of the logarithm of that function agree. Therefore, we start with the log-linear estimation procedure.

8.8.1 Log-Linear Estimation of Utilties and Mobilities

The log-linear estimation of the model parameters consists in minimizing the objective function

$$F(\underline{\nu}, \vec{u}) = \sum_{\substack{i,j=1 \\ i \neq j}}^{L} \left(\ln(w_{ij}^e) - \ln\left(n_j^e \nu_{ij} \exp(u_i - u_j)\right)\right)^2 = \min \tag{8.113}$$

by a best possible choice of the model parameters $\nu_{ij} = \nu_{ji}$ and u_i for all possible combinations of $i,j = 1,2,\ldots,L$. Because the probability transition rates $p_{ij}(\vec{n})$ depend only on the difference of the utilities $\vec{u}$, we impose without any restriction the constraint

$$\sum_{i=1}^{L} u_i = 0, \tag{8.114}$$

in order to obtain uniquely determined values of the utilities. This means, the utilities are only determined up to an additive constant, which has no influence on the dynamics of the system.

The necessary condition for minimizing the objective function (8.113) requires that its variation at its minimum vanishes

$$\delta F\left(\underline{\nu}, \vec{u}\right) = \sum_{i=1}^{L} \left[\frac{\partial F}{\partial u_i} + \lambda\right] \delta u_i + \frac{1}{2} \sum_{\substack{i,j=1 \\ i \neq j}}^{L} \left[\frac{\partial F}{\partial \nu_{ij}} + \frac{\partial F}{\partial \nu_{ji}}\right] \delta \nu_{ij} = 0, \tag{8.115}$$

where the Lagrange parameter λ takes into account the constraint (8.114). The variations δu_i and $\delta \nu_{ij}$ are considered independent of each other. Hence, each of the corresponding terms must vanish independently. This leads to the following set of equations:

$$\frac{\partial F}{\partial u_i} + \lambda = 0 \quad \text{for} \quad i = 1, 2, \ldots, L \tag{8.116}$$

$$\frac{\partial F}{\partial \nu_{ij}} + \frac{\partial F}{\partial \nu_{ji}} = 0 \quad \text{for} \quad i, j = 1, 2, \ldots, L \text{ but } i \neq j. \tag{8.117}$$

In the next step, one has to introduce the explicit forms (8.113) of the objective function

$$\begin{aligned} \frac{\partial F}{\partial u_i} &= \sum_{\substack{j=1 \\ j \neq i}}^{L} 2\left[\ln\left(\nu_{ij}\right) + \left(u_i - u_j\right) - \ln\left(\frac{w_{ij}^e}{n_j^e}\right) - \ln\left(\nu_{ji}\right) + \left(u_i - u_j\right) + \ln\left(\frac{w_{ji}^e}{n_i^e}\right)\right] \\ &= \sum_{\substack{j=1 \\ j \neq i}}^{L} 4\left[\left(u_i - u_j\right) - \ln\left(\frac{w_{ij}^e n_i^e}{w_{ji}^e n_j^e}\right)\right] \end{aligned} \tag{8.118}$$

and

$$\frac{\partial F}{\partial \nu_{ij}} = \frac{2}{\nu_{ij}} \left[\ln\left(\nu_{ij}\right) + \left(u_j - u_i\right) - \ln\left(\frac{w_{ji}^e}{n_i^e}\right)\right], \tag{8.119}$$

where we have used the symmetry $\nu_{ij} = \nu_{ji}$ in (8.119). Using (8.118) and (8.119), the conditions for an optimal set of model parameters (8.115) obtain the form

$$\sum_{\substack{j=1\\j\neq i}}^{L} 4\left[(u_i - u_j) - \frac{1}{2}\ln\left(\frac{w_{ij}^e n_i^e}{w_{ji}^e n_j^e}\right)\right] + \lambda = 0 \tag{8.120}$$

and

$$\frac{2}{\nu_{ij}}\left[\ln\left(\nu_{ij}^2\right) - \ln\left(\frac{w_{ij}^e\, w_{ji}^e}{n_j^e\, n_i^e}\right)\right] = 0. \tag{8.121}$$

It is evident that both Eqs. (8.120), (8.121) are uncoupled. This means that the determination of the optimal set of utilities and the determination of the interaction matrix are independent of each other.

Summation of (8.120) over all $i = 1, 2, \ldots, L$ shows that the Lagrange parameter λ vanishes

$$\lambda = \frac{1}{L}\sum_{\substack{j=1\\j\neq i}}^{L}\left[\ln\left(\frac{w_{ij}^e n_i^e}{w_{ji}^e n_j^e}\right)\right] = 0. \tag{8.122}$$

Inserting (8.122) in (8.120) finally yields for the utilities

$$u_i = \frac{1}{2L}\sum_{\substack{j=1\\j\neq i}}^{L}\ln\left(\frac{w_{ij}^e\, n_i^e}{w_{ji}^e\, n_j^e}\right) = \frac{1}{2L}\sum_{\substack{j=1\\j\neq i}}^{L}\ln\left(\frac{p_{ij}^e}{p_{ji}^e}\right) \qquad \text{for } i = 1, 2, \ldots, L \tag{8.123}$$

and the mobility matrix

$$\nu_{ij} = \nu_{ji} = \sqrt{\frac{w_{ij}^e\, w_{ji}^e}{n_j^e\, n_i^e}} = \sqrt{p_{ij}^e \cdot p_{ji}^e} \qquad \text{for } i,j = 1, 2, \ldots, L \text{ and } i \neq j \tag{8.124}$$

where we have introduced the empirically determined transition rates

$$p_{ij}^e = \frac{w_{ij}^e}{n_j^e}. \tag{8.125}$$

The attractiveness values or utilities (8.123) and the mobility matrix or interaction terms (8.124) are expressed in terms of the empirically determined transition rates (8.125).

At a first glance, it is obvious that averaged differences in the transition rates $\left(p^e_{ij}/p^e_{ji}\right)$ determine the utilities. If more transitions between $j \to i$ have been made as backwards in average, the level of the corresponding utility is above its average value $u_i > 0$.

If on the other hand the transition rates are comparable, $p^e_{ij} \approx p^e_{ji}$, then this may be an indicator for a strong interaction or exchange between those states i, j.

The average interaction (or global mobility) can be easily computed

$$\nu_0 = \frac{1}{L(L-1)} \sum_{\substack{i,j=1 \\ i \neq j}}^{L} \nu_{ij}. \tag{8.126}$$

Moreover, it is helpful for practical discussions to introduce an interaction term $0 \leq f_{ij} \leq 1$ describing the strength of interaction between state i and state j

$$\nu_{ij} = \nu_0 f_{ij}. \tag{8.127}$$

Since by definition (8.127), the average value of interaction $\langle f_{ij} \rangle = 1$ equals one

$$\left\langle f_{ij} \right\rangle = \frac{1}{L(L-1)} \sum_{\substack{i,j=1 \\ i \neq j}}^{L} f_{ij} = 1, \tag{8.128}$$

the strength of interaction can be compared with its average value. This means, for example, that a value of $f_{ij} = 5$ corresponds to a 5 times higher interaction or exchange of population between state i and state j as expected in the average.

The utilities fulfil

$$\langle u \rangle = \frac{1}{L} \sum_{i=1}^{L} u_i = 0, \tag{8.129}$$

and the variance of the utility is given by

$$\sigma_u^2 = \left\langle u^2 \right\rangle - \langle u \rangle^2 = \frac{1}{L} \sum_{i=1}^{L} u_i^2. \tag{8.130}$$

Let us now discuss a few questions related to the Weidlich-Haag model: What is the benefit of introducing utilities u_i or preferences δ_i instead of using a classical regression term such as (8.109)?

(a) Let us first deal with data mining aspects, namely an efficient concentration of information obtained in the empirical data base

$$w_{ij}^e, n_j^e \quad \text{for} \quad i,j = 1, \ldots, L \text{ and } i \neq j. \tag{8.131}$$

In this case (8.131), the number of observations q is

$$q = L^2 - L \tag{8.132}$$

since the diagonal of the matrix (8.131) has to be subtracted.

The number of model parameters k which has to be determined is given by $(L-1)$ utilities, since the sum of the utilities is fixed to zero, and $(L^2 - L)/2$ values for the symmetric interaction matrix (without diagonal elements). In total

$$k = (L-1) + (L^2 - L)/2 = \frac{1}{2}(L^2 + L - 2). \tag{8.133}$$

The number of degrees of freedom, defined as $f = q - k$ thus amounts to

$$f = \frac{1}{2}(L^2 - 3L + 2) \tag{8.134}$$

since it is assumed that the elements of the transition matrix and the utilities have to be determined completely independent of each other, as the log-linear estimation indicates. For illustrational purposes, let us consider interregional migration within the 402 districts (year 2016) of Germany as a specific example. In this concrete case, we have $q = 161{,}202$ observed migration events per year and per subpopulation, compared with 401 utilities (one utility value is fixed by the normalization condition) and 80,601 values of interaction. This is still a rather big number of parameters, but a reduction factor of about two in the number of data. The outcome of those statistical tests considers both, the number of available data and the number of parameters which have to be determined in relation to the quality of the results.

(b) A further benefit concerns the unique determination and easy interpretation of all model parameters u_i, ν_{ij} using (8.123) and (8.124).

(c) Another advantage concerns the interpretation of the model parameters u_i, ν_{ij} compared with the meaning and interpretation of the pure transition rates w_{ij} between states i, j. Since the transition rates w_{ij} obtain a lot of unfiltered information and uncertain data, e.g., between smaller and far distant regions, it is common practice to calculate the total in-flow D_i and total out-flow O_i of a specific region

$$D_i^e = \sum_{j=1}^{L} w_{ij}^e \quad \text{and} \quad O_i^e = \sum_{j=1}^{L} w_{ji}^e \quad \text{for } i = 1, 2, \ldots, L \tag{8.135}$$

and to apply directly regression techniques in order to determine appropriate explanatory variables. This is a typical engineering procedure, without principle thoughts.

The fundamental mathematical form of the transition rates (8.33) supports and facilitates the interpretation of huge data sets related to the transition rates.

8.8.2 Non-Linear Estimation of Utilities and Mobilities

The log-linear estimation of the utilities and the interaction matrix have provided an analytical and unique expression. However, even better results can be expected when we apply the optimization procedure directly to the residuals

$$\varepsilon_{ij} = w_{ij}^{e} - w_{ij} \qquad \text{for } i \neq j \tag{8.136}$$

and determine the model parameters by minimizing an appropriate objective function. We suppose the following least square procedure

$$F\left(\underline{\nu}, \vec{u}\right) = \sum_{\substack{i,j=1\\ i\neq j}}^{L} \left(w_{ij}^{e} - n_j^{e}\nu_{ij}\exp\left(u_i - u_j\right)\right)^2 = \min, \tag{8.137}$$

which again leads to the variation principle

$$\delta F\left(\underline{\nu}, \vec{u}\right) = \sum_{i=1}^{L}\left[\frac{\partial F}{\partial u_i} + \lambda\right]\delta u_i + \frac{1}{2}\sum_{\substack{i,j=1\\ i\neq j}}^{L}\left[\frac{\partial F}{\partial \nu_{ij}} + \frac{\partial F}{\partial \nu_{ji}}\right]\delta\nu_{ij} = 0 \tag{8.138}$$

under the constraint (8.114).

In complete analogy to the log-linear procedure, we obtain the Eqs. (8.139), (8.140)

$$\frac{\partial F}{\partial u_i} + \lambda = 0 \qquad \text{for } i = 1, 2, \ldots, L \tag{8.139}$$

and

$$\frac{\partial F}{\partial \nu_{ij}} + \frac{\partial F}{\partial \nu_{ji}} = 0 \qquad \text{for } i, j = 1, 2, \ldots, L \text{ and } i \neq j \tag{8.140}$$

for the determination of the model parameters $\nu_{ij} = \nu_{ji}$ and u_i. However, the next step is much more sophisticated. For the explicit forms (8.139) and (8.140) of the objective function (8.137), we need to calculate

$$\frac{\partial F}{\partial u_i} = \sum_{\substack{j=1 \\ j\neq i}}^{L} 2\left(w_{ji}^e n_i^e \nu_{ji} \exp\left(u_j - u_i\right) - n_i^{e2} \nu_{ji}^2 \exp 2\left(u_j - u_i\right)\right)$$
$$-\sum_{\substack{j=1 \\ j\neq i}}^{L} 2\left(w_{ij}^e n_j^e \nu_{ij} \exp\left(u_i - u_j\right) - n_j^{e2} \nu_{ij}^2 \exp 2\left(u_i - u_j\right)\right) \tag{8.141}$$

and

$$\frac{\partial F}{\partial \nu_{ji}} = 2\left(n_i^{e2} \nu_{ji} \exp 2\left(u_j - u_i\right) - n_i^e w_{ji}^e \exp\left(u_j - u_i\right)\right). \tag{8.142}$$

By inserting (8.141) and (8.142) in (8.139), it can be seen that the Lagrange parameter again disappears, which means that $\lambda = 0$ and (8.139) reduces to

$$\frac{\partial F}{\partial u_i} = 0 \qquad \text{for } \ i = 1, 2, \ldots, L \tag{8.143}$$

and

$$\frac{\partial F}{\partial \nu_{ij}} + \frac{\partial F}{\partial \nu_{ji}} = 0 \qquad \text{for } \ i, j = 1, 2, \ldots, L \ \text{ and } \ i \neq j. \tag{8.144}$$

Furthermore, (8.144) can be solved explicitly for $\nu_{ij} = \nu_{ji}$ yielding for the mobility matrix

$$\nu_{ij} = \frac{n_i^e w_{ji}^e \exp\left(u_j - u_i\right) + n_j^e w_{ij}^e \exp\left(u_i - u_j\right)}{n_i^{e2} \exp 2\left(u_j - u_i\right) + n_j^{e2} \exp 2\left(u_i - u_j\right)}. \tag{8.145}$$

We now have to insert (8.145) into the objective function (8.137) and obtain a system of $L - 1$ transcendental equations for the utilities u_i, $i = 1, 2, \ldots, L$

$$F(\vec{u}) = \sum_{\substack{i,j=1 \\ i\neq j}}^{L} \left[w_{ij}^e - n_j^e \frac{n_i^e w_{ji}^e \exp\left(u_j - u_i\right) + n_j^e w_{ij}^e \exp\left(u_i - u_j\right)}{n_i^{e2} \exp 2\left(u_j - u_i\right) + n_j^{e2} \exp 2\left(u_i - u_j\right)} \exp\left(u_i - u_j\right) \right]^2$$
$$= \min. \tag{8.146}$$

Different solution algorithms can be applied to solve (8.146). In order to accelerate the optimization procedure, it is useful to start with utilities obtained via the log-linear estimation procedure (8.123). The optimization procedure (8.146) does not depend on a specific functional form of the utilities u_i.

However, it has been shown that the transition rates assume the specific form (8.105) for the Weidlich-Haag migration model. In this special case, inserting

$$u_i = \frac{1}{2} \ln n_i + \delta_i \tag{8.147}$$

in (8.145) yields for the mobility matrix

$$\nu_{ij} = \frac{1}{\sqrt{n_i^e n_j^e}} \frac{w_{ji}^e \exp(\delta_j - \delta_i) + w_{ij}^e \exp(\delta_i - \delta_j)}{\exp 2(\delta_j - \delta_i) + \exp 2(\delta_i - \delta_j)}, \tag{8.148}$$

which can be inserted into the objective function (8.137). This finally leads to a transcendental system for the $L-1$ spatial preference parameters δ_i in case of the Weidlich-Haag migration model

$$F\left[\vec{\delta}\right] = \sum_{\substack{i,j=1 \\ i \neq j}}^{L} \left[w_{ij}^e - \frac{w_{ji}^e \exp(\delta_j - \delta_i) + w_{ij}^e \exp(\delta_i - \delta_j)}{\exp 2(\delta_j - \delta_i) + \exp 2(\delta_i - \delta_j)} \exp(\delta_i - \delta_j) \right]^2$$
$$= \min. \tag{8.149}$$

Now the difference to the log-linear estimation procedure (8.113) becomes obvious. The equations for the determination of the utilities and the interaction matrix are no longer uncoupled. Furthermore, they are highly nonlinear equations and there exists only numerical methods of solution. It can be assumed that they will have unique solutions, based on the fact that the solutions of the nonlinear estimation procedure and the log-linear procedure differ only slightly in most empirical tests. But an explicit proof of this hypothesis does not currently exist.

However, the statistical tests for the nonlinear estimation are of course always better than for the log-linear estimation procedure (see the following Sect. 8.9).

At a first glance, it is surprising why Eq. (8.149) for the estimation of the regional preferences δ_i does not depend on population numbers. Only empirical migration flows are required. How can we understand this? In Sect. 8.4, it was shown that the regional attractiveness values $u_i(\vec{n})$ show a size effect. In other words, the population tends to agglomerate. The separation of the size effects leads to the so-called preferences δ_i. If we assume that the size effect is rather well excluded, it is not surprising that the estimation condition (8.149) for the regional preferences does not depend on size values n_i. However, we cannot assume that all size-relevant components are completely excluded via the $\log n_i$- term. Other possible explanatory variables could also exhibit and contribute to an agglomeration trend via correlations with regional size. The regional preferences have the dimension [1]. Therefore, dimensionless indicators should be used as explanatory variables in the subsequent multi-regression analysis.

What is the advantage of using the nonlinear estimation procedure?

(a) The most important point, however, is that the interaction matrix ν_{ij} is a function of the data and the unknown utilities u_j. Therefore, the $\nu_{ij} = \nu_{ij}\left(\vec{n}_k^e, w_{kl}^e, u_k\right)$ can directly be inserted into the objective function (8.137). This leads to a system of $(L-1)$ coupled, highly nonlinear transcendental equations for the utilities u_i, $i = 1, \ldots, L$ (8.146) or preferences (8.149) only. This simplifies and stabilizes the numerical optimization procedures considerable.
(b) The outcome of the log-linear parameter estimation (8.123) can be used as a starting point for the nonlinear optimization (8.146) or (8.149), leading to an acceleration of the numerical search procedure.
(c) Let us consider again the data mining aspect, namely the efficient concentration of information obtained in the empirical data base

$$w_{ij}^e, n_j^e \quad \text{for} \quad i,j = 1, \ldots, L \quad \text{and} \quad i \neq j. \tag{8.150}$$

In this case (8.150), the number of observations q is again

$$q = L^2 - L \tag{8.151}$$

since the diagonal of the matrix (8.150) has to be subtracted.
The number of model parameters k which has to be determined is now given by only

$$k = (L - 1) \tag{8.152}$$

utilities or preferences since the interaction terms (interaction matrix)

$$\nu_{ij} = \nu_{ij}\left(\underline{w}, \vec{n}, \vec{u}\right) \qquad \text{for} \quad i \neq j \tag{8.153}$$

are completely determined by the data set w_{ij}^e, n_j^e and the utilities u_i or preferences δ_i. The degrees of freedom, defined as $f = q - k$ thus amount to

$$f = q - k = L^2 - 2L + 1, \tag{8.154}$$

which leads to a considerable improvement of the statistical tests in practical applications compared to the log-linear estimation.

Based on (8.145), the average interaction term, called global mobility in migratory applications, can be directly determined via utilities

$$\nu_0 = \frac{1}{L(L-1)} \sum_{\substack{i,j=1 \\ i \neq j}}^{L} \frac{n_i^e w_{ji}^e \exp(u_j - u_i) + n_j^e w_{ij}^e \exp(u_i - u_j)}{n_i^{e2} \exp 2(u_j - u_i) + n_j^{e2} \exp 2(u_i - u_j)} \tag{8.155}$$

or preferences

$$\nu_0 = \frac{1}{L(L-1)} \sum_{\substack{i,j=1 \\ i \neq j}}^{L} \frac{1}{\sqrt{n_i^e n_j^e}} \frac{w_{ji}^e \exp(\delta_j - \delta_i) + w_{ij}^e \exp(\delta_i - \delta_j)}{\exp 2(\delta_j - \delta_i) + \exp 2(\delta_i - \delta_j)}, \tag{8.156}$$

leading to the determination of the interaction matrix

$$f_{ij} = \frac{v_{ij}}{\nu_0} \quad \text{for} \quad i \neq j \tag{8.157}$$

with its normalization

$$\langle f \rangle = \frac{1}{L(L-1)} \sum_{\substack{i,j=1 \\ i \neq j}}^{L} f_{ij} = 1. \tag{8.158}$$

Centrality Index
The centrality index in network theory describes the number of existing links of a knot k compared with the total number of links (Katz 1953; Aggarwal 2011)

$$C_k^D = \frac{1}{L-1} \sum_{\substack{j=1 \\ j \neq k}}^{L} a_{kj} \quad \text{where} \quad a_{kj} = \begin{cases} 1 & \text{link exist} \\ 0 & \textit{no} \text{ link exists} \end{cases} . \tag{8.159}$$

Strength of Regional Interdependence
The indicator strength of regional interdependence can be seen as a weighted centrality indicator that not only the links are important but also the strength of the different links

$$F_k = \frac{1}{L-1} \sum_{\substack{j=1 \\ j \neq k}}^{L} f_{kj} \tag{8.160}$$

with f_{kj} according (8.157). Because of (8.158), the average value of the strength of regional interdependence is one:

$$\langle F \rangle = \frac{1}{L} \sum_{k=1}^{L} F_k = \frac{1}{L(L-1)} \sum_{\substack{j,k=1 \\ j \neq k}}^{L} f_{kj} = 1. \tag{8.161}$$

Therefore, F_k is an appropriate measure for the centrality of a region/settlement within the spatial system.

Accessibility
The accessible population with respect to a certain region k is defined as a weighted sum of population of different regions j taking into account the weighted interconnectedness f_{ij} of the different regions (Batten et al. 1995; Rabino and Occelli 1999; Schuermann et al. 1997)

$$E_k = \sum_{j=1}^{L} n_j f_{kj}. \tag{8.162}$$

In analogy to (8.162) it is possible to define accessibility variables of other variables.

8.9 Interregional Migration in Germany

The Federal Republic of Germany (FRG) comprises 16 federal states (NUTS-1-level) (Bundesländer), which are largely autonomous regarding their internal organisation. The areas of the states and their population numbers vary considerably. Some of the states such as Bavaria and the Free Hanseatic Cities of Bremen and Hamburg have their long history. Other states such as Niedersachsen and Rheinland-Pfalz were established after the end of the Second World War. In the year 1952, Baden-Württemberg was formed after a plebiscite by the fusion of the former states Baden and Württemberg (Cornelius 1996). The Saarland joint the FRG in 1957, also after a plebiscite. The reunification of Germany was established on 3rd October 1990. United Germany has about 82 million inhabitants with Berlin as capital.

Since 2013, Germany is subdivided in addition into 402 districts (NUTS-3-level) at a municipal level, namely 295 rural districts and 107 urban districts. The districts are further subdivided into 11,418 communes (LAU-2-regions).

The local government of each federal state has well-defined responsibilities and fiscal resources. The local government is elected in a 4-year turn by the inhabitants of their federal state. Intervention of the Federal Government (Bundesregierung) concerning state affairs takes place only in exceptional cases.

The local administration organised in communes, cities, districts and administrative districts (Regierungsbezirke) is responsible for territorial planning and for the regional development policies.

It is not my intension to comprise an almost complete analysis of migration events in Germany in the next chapter. It is rather the idea to show how the Master equation framework, especially the Weidlich-Haag migration model, can be applied and used, in comparison with the usual Gravity model approach. In so far, we use the empirical migration data between the 402 German districts as benchmark test.

We consider Germany as a system of $L = 402$ regional economies with unrestricted migration. However, the system is not closed since immigration and emigration acts as well as birth and death events have influence on the regional population sizes and their demographic profile (regional age pyramid). These impacts in turn affect the migration flows and provoke a reallocation of the population.

8.9.1 The German Data Base

According to German law, everyone who permanently lives in Germany has to be registered in one of the communes of the country. The move of migrants is recorded in both, in the state of origin and the state of destination. The time of the migration act refers to the year of registration. However, a certain uncertainty results by a possible time lag in the registration leading to a small discrepancy in the population registered. Since multiple migration acts can occur during one year, the number of real relocations exceeds the number of registered migrants slightly. The population numbers are corrected by census data about every 10 years.

The migration flows are based on records from municipal population registers and collected in the Statistisches Bundesamt, Wiesbaden. Data on internal migration and external migration by country of destination and origin and by citizenship of immigrants and emigrants are available as well as information on refugees and people seeking refuge. In addition current population data on the numbers of inhabitants of the municipalities, communes and various historical time series are available, among other things, by age and sex. By definition, migration flows correspond to the act of crossing an administrative boarder.

Data Base on District Level

For illustrational purposes, let us consider the interregional migration behaviour within the 402 districts of Germany (NUTS-3-level). The migration flows and the population data are averaged over the years 2011, 2012, 2013 to smooth out statistical uncertainties in case of very small flows.

The data base used on the level of 402 districts in Germany comprise the complete empirical interregional migration matrix w_{ij}^{e}, $i, j = 1, 2, \ldots, L$, where $L = 402$ together with n_i^e, $i = 1, 2, \ldots, L$ population numbers. However, it is important for the understanding of the development of regional disparities in different sectors also to consider the migration behaviour of sub-populations (age groups by sex and total population). In different ages the motivation and the propensity to migrate vary considerable. Below, for different age-groups, some supposable motivations are listed.

Age of migrants	***Correlated with***
<18 years	migration of young families with children, low mobility of this age group, short distance movements prevail, mostly out of city centres towards suburbs or rural areas
$18 < 25$ years	students, trainee, young people leaving her parents´ home, age group with the highest mobility and a considerable portion of long distance migration acts
$25 < 30$ years	young employees with high mobility, migration towards centres of production and services prevail
$30 < 50$ years	people in the middle of their life, households migrating outwards of city centres, migration towards suburbs and rural areas
$50 < 65$ years	older professionals with very low mobility, children have already left the families, no considerable movements
>65 years	retired persons, lowest mobility of the considered age groups, more rural areas are preferred

Following Sect. 8.8, we have we have $q = 161{,}202$ observed migration events per year and per subpopulation. The number of model parameters which has to be determined by the nonlinear parameter estimation procedure is given by $k = 401$ unknown utilities because of the normalization. All 80,601 values of the interaction matrix $\nu_{ij} = \nu_{ij}\left(\underline{w^e}, \vec{n}^e, \vec{u}\right)$ are completely determined by the estimated utilities and the empirical flows and population numbers. The resulting degrees of freedom, defined as $f = q - k$ thus amount to $f = 160{,}801$. This remarkable high value of degrees of freedom underlines the explanatory power of the Weidlich-Haag migration model, especially for complex regional systems when the deterrence effects are inhomogeneously distributed.

8.9.2 Flow Chart for Parameter Estimation

Figure 8.7 provides a flow chart for the estimation procedure of the trend parameters of the Weidlich-Haag model. The utilities estimated via the log-linear estimation procedure (8.123) are used as initial values for the non-linear estimation procedure to accelerate the parameter estimation process. This is not necessary, but helpful since the log-linear procedure has a unique solution and is in general very close to the non-linear estimation procedure, too.

Subsequently, the results of the application of the Weidlich-Haag model to German migration data will be discussed.

8.9.3 Results of the German Case Study

It can be seen, Fig. 8.8, that the use of the interaction matrix ν_{ij} instead of a pure and rather simple deterrence term (8.42) leads to a considerable improvement of the statistical tests. In order to capture the connectivity of regions, a weight-function $f(d_{ij})$ is defined in the usual Gravity approximation. This underlines the critical remark of by LeSage and Pace (2014), concerning the arbitrariness of choosing a spatial weight matrix as a weak point.

However, the almost perfect agreement between empirical migration flows w_{ij}^e and migration flows w_{ij} using the Weidlich-Haag migration model (8.109) has been criticized by some researchers working with Gravity theory. It is sometimes argued

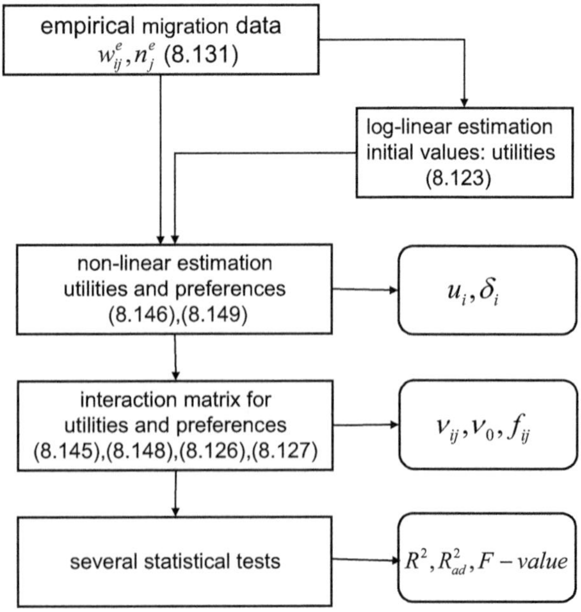

Fig. 8.7 Flow Chart of the estimation procedure of the Weidlich-Haag model

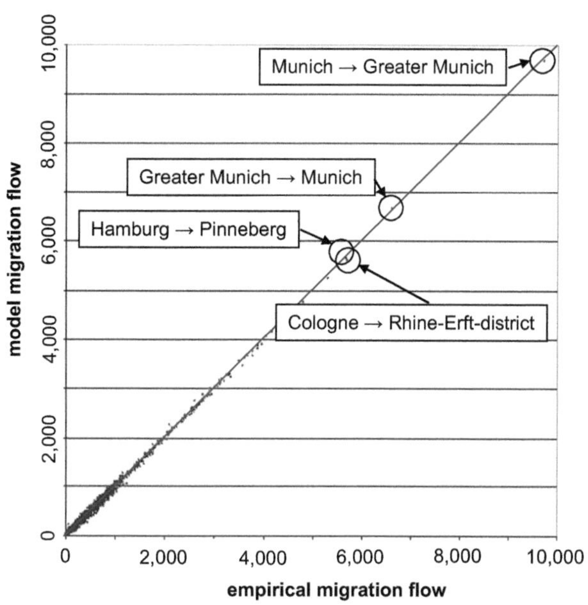

Fig. 8.8 Empirical migration flows versus model migration flows: Weidlich-Haag model, $R^2 = 0.99$, $F = 90{,}239$, total population. Each point represents a comparison of empirical migration flow versus model migration flow for one district (Source: STASA 2017)

that the number of parameters to be estimated allows to fit an elephant. But we should not forget that in case of German districts the huge amount of data of the interaction matrix is calculated only by using the 402 regional attractiveness values of the districts. The interaction matrix does not have to be estimated independently. Only 401 parameters have to be fixed and were estimated and we have an overwhelming big data set of $q = 161{,}202$ migration flows per sub-population and year.

In Fig. 8.9, we introduced the distance deterrence function (8.42) in the migration model and hence in the non-linear estimation of trend-parameters. Instead of an interaction matrix ν_{ij} a distance decreasing function is introduced which suppresses the spatial interactions between the districts in dependence of distance d_{ij}, but regardless of whether these are rural or urban districts or whether the districts are well linked via economic relationships or excellent traffic connections.

The parameter β in (8.42) clearly depicts the strong impact of distance on migratory behaviour. Obviously, up to a distance of about 80–100 km, the distance effect dominates the decision behaviour of the population (Fig. 8.10).

Compared with the pure distance depending interaction term (deterrence function) (8.42) the interaction matrix ν_{ij} exhibits a much more detailed map of the spatial interaction of districts. It is plausible that a stronger spatial interaction may happen especially between urban districts compared to not so distant rural districts. Those interactions may be further supported by a well-developed transportation network and/or functioning firm related interactions or between clusters of firms and corresponding activities.

This explains why instead of a simple structured deterrence function a much more detailed interaction matrix ν_{ij} is needed to describe the complicated structure of interaction of regions. If we do not accept the full complexity of spatial

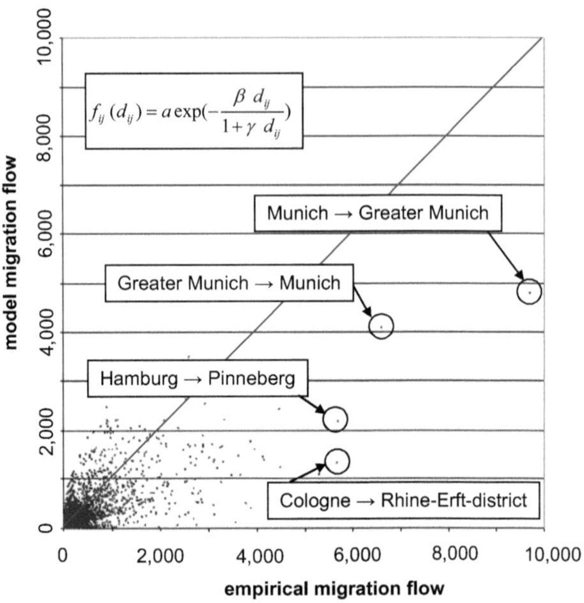

Fig. 8.9 Empirical migration flows versus model migration flows with distance deterrence function, $R^2 = 0.60$, total population, $\beta = 0.132$, $\gamma = 0.0163$ (Source: STASA 2017)

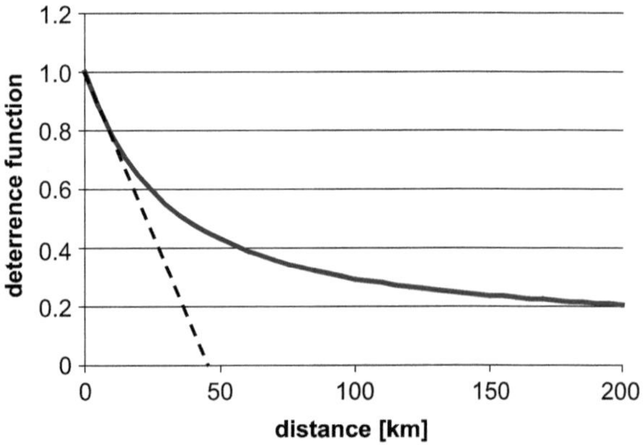

Fig. 8.10 Distance deterrence function (8.42): total population, $\beta = 0.132$, $\gamma = 0.0163$ (year 2013) (Source: STASA 2017)

interactions and hold on at a pure distance effect d_{ij} between the regions i, j, the partly big deviations of the actual interaction term with the distance interaction term leads to significant systematic errors in the estimation of explanatory values of the regional utilities. Therefore, it becomes obvious why the agreement between empirical flow data w_{ij}^e and model flows w_{ij} significantly deviate in case of a Gravity model.

A convincing analysis of the migratory interaction of regions is needed, in combination with a detailed regression analysis of the spatial preferences for any significant and strategic policy advice. This will be shown in the next chapter.

Using (8.35) and (8.145) the strong variations of the global mobility ν_0 characterizing the average mobility to migrate of the population is depicted in Fig. 8.11 for the different age groups. Especially the younger German populations of the age groups of 18–25 years and 25–30 years attract attention by their high mobility compared to the more established group of population in the age between 30 and 50 years. This is the age group of people buying real estate. This means under the same economic conditions $\delta_i = \delta_j$ the average mobility to migrate of members of the older group between 30 and 50 years is less than 40% of the younger people, and the global mobility divides again in half when the group of older people between 50 and 65 years is considered.

After separation of the global mobility according to (8.35) and (8.36), the normalized spatial interaction term $f_{ij} = \nu_{ij}/\nu_0$ with $f_{ij} = f_{ji}$ describing the strength of migratory interaction between the both districts i and j is obtained. The interaction matrix contains implicitly the effect of space d_{ij} as well.

Figure 8.12 shows the distribution of the spatial interaction term $f_{ij} = \nu_{ij}/\nu_0$ related to the City of Stuttgart.

The Region of Stuttgart is situated in the south-west of Germany and covers five state districts called Boeblingen, Esslingen, Goeppingen, Ludwigsburg and Rems-Murr, and the City of Stuttgart. The overall population of the Region of Stuttgart is about 2.7 million inhabitants. Together with its state capital, the City of Stuttgart, it represents the economic and cultural centre of the state of Baden-Wuerttemberg. With an area of 3700 km^2, this region is one of the most densely populated regions of Germany. There are about 600,000 inhabitants concentrated in the City of Stuttgart. This corresponds to 22% of the total population of the Stuttgart Region. Taking into account the adjacent communities (Sindelfingen, Boeblingen, Esslingen, Leonberg, Leinfelden-Echterdingen, Ludwigsburg, Schorndorf) within a

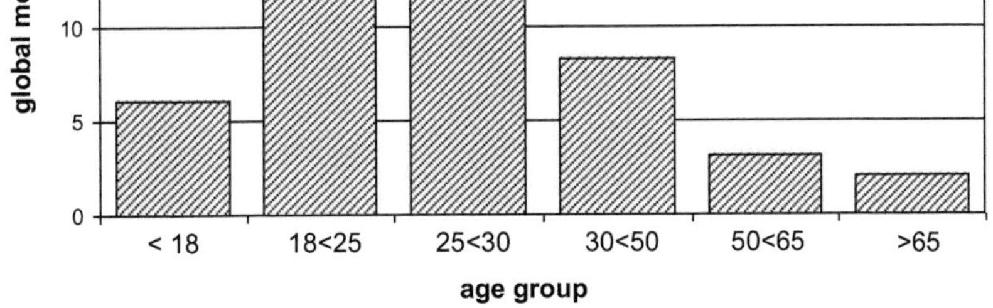

Fig. 8.11 Global Mobility in dependence of the age group (Germany, year 2013) (Source: STASA 2017)

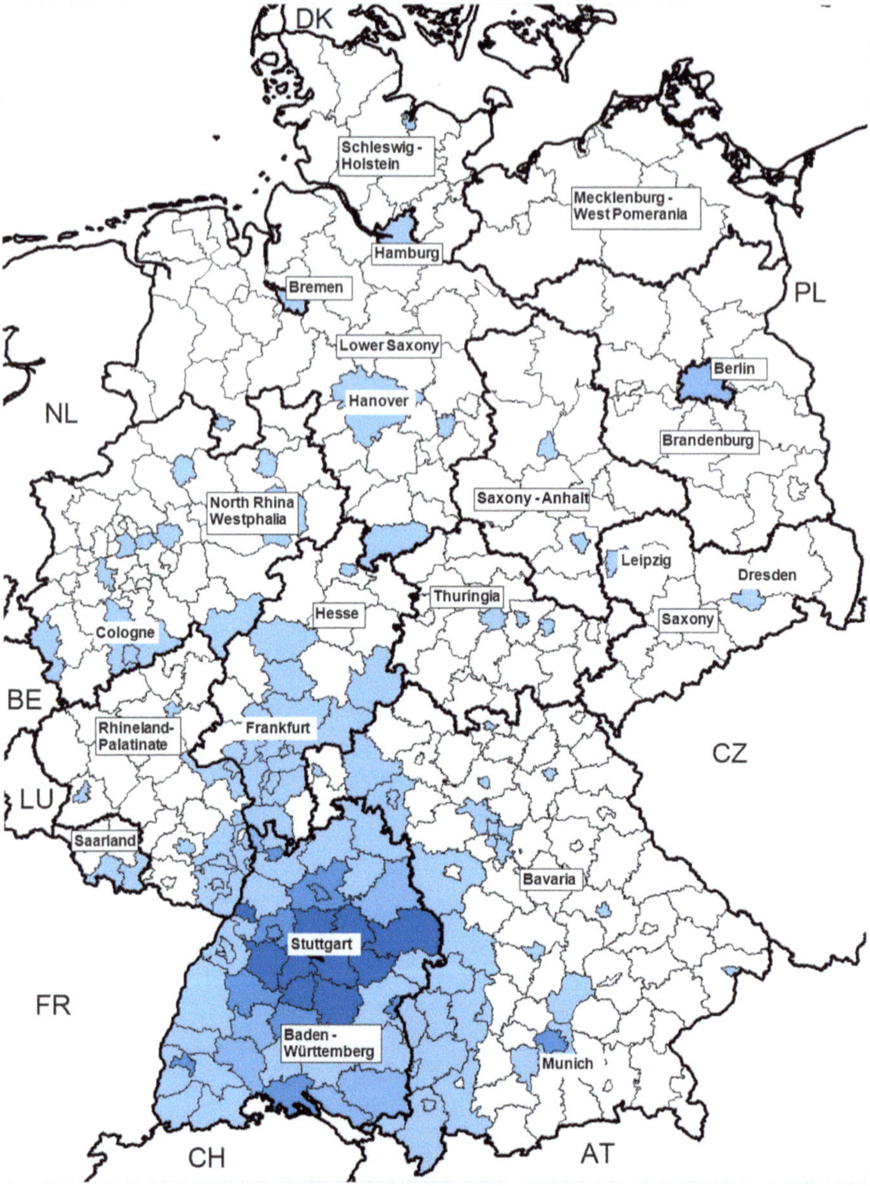

Fig. 8.12 Spatial interaction of the City of Stuttgart with other districts (total population), *dark blue* means strong interaction (Source: STASA 2017)

small circle (15 km) around the centre of the City of Stuttgart, about 38% of the total population of Baden-Wuerttemberg can be found.

Obviously, the City of Stuttgart is strongly linked to all districts of the Federal State of Baden-Wuerttemberg. The strongest links exist to the districts of the

Region of Stuttgart. However, there exist also important links to all capitals of Germany, especially to Munich, Berlin, Hamburg and Frankfurt, which are also marked by persistent economic interactions. Furthermore, links to districts along the boarder-line of Baden-Wuerttemberg are also above-average.

In contrast to the City of Stuttgart exhibits the urban district of Weimar a much more concentrated interaction pattern as depicted in Fig. 8.13.

The City of Weimar is a beautiful tourist village with about 63,000 inhabitants. The city is located in the State of Thuringia between the cities of Erfurt and Jena. The old part of Weimar is added to the list of the UNESCO World Heritage Sites.

Compared with the City of Stuttgart, Weimar possesses less strong links to far distant districts of Germany, but it is well connected with all districts of Thuringia. Its migratory interactions are mainly concentrated to the nearest neighbour districts. Far reaching interactions above-average can be found to the big cities of Berlin, Hamburg, Dresden, Nuremberg, and Munich and some other big cities.

Evidently, Figs. 8.12 and 8.13 demonstrate that a deterrence function, describing the strength of spatial interaction only by distance d_{ij} is really not appropriate for understanding and modelling the full complexity of spatial migratory interactions.

The strength of regional interdependence F_k (8.160) or the centrality index of the different districts is depicted in Fig. 8.14. The more a specific district is connected with the total system of the German districts, the higher the value of the strength of regional interdependence is, or in the sense of network theory, its centrality.

The regional utilities and preferences of the districts are estimated according to (8.146) and (8.149). By definition, the utilities reflect the overall spatial attractiveness of the districts with respect to migration. The preferences show the migratory attractiveness of the districts without agglomeration effects. Of course, depending on the age of the migrants not only their migration mobility is different (see Fig. 8.11) but also their propensity to migrate to different districts, that means the preferences.

The spatial preferences for the German districts are shown in Fig. 8.15. The mean value of the spatial preferences is scaled to 100. Of course, this scaling does not affect the migratory dynamics. Of course, agglomeration effects favour big cities and metropolitan regions. Even without agglomeration trends, the preferences exhibit preferred centres of spatial development, i.e., parts of the metropolitan region Berlin-Brandenburg, the metropolitan regions of Hamburg, Colon, Frankfurt, Stuttgart, Munich and big parts of the Free State of Bavaria. Spatial preferences below average can be found in the Federal States of Saxony-Anhalt, parts of Niedersachsen, Saarland, the Free State of Thuringia, and parts of the Free State of Saxony. Therefore, the map of the preferences shows big spatial disparities.

The age-dependent values of the spatial preferences for the district Cottbus and Munich are depicted in Fig. 8.16. It is obvious that the spatial preferences of the urban districts of Cottbus and Munich differ considerably during the life cycle of the residents.

Both cities seem to be attractive for young people in the process of education and training (age group between 18 and 25 years). Cottbus gains attractiveness for people after about 50 years and young families, especially because of the much

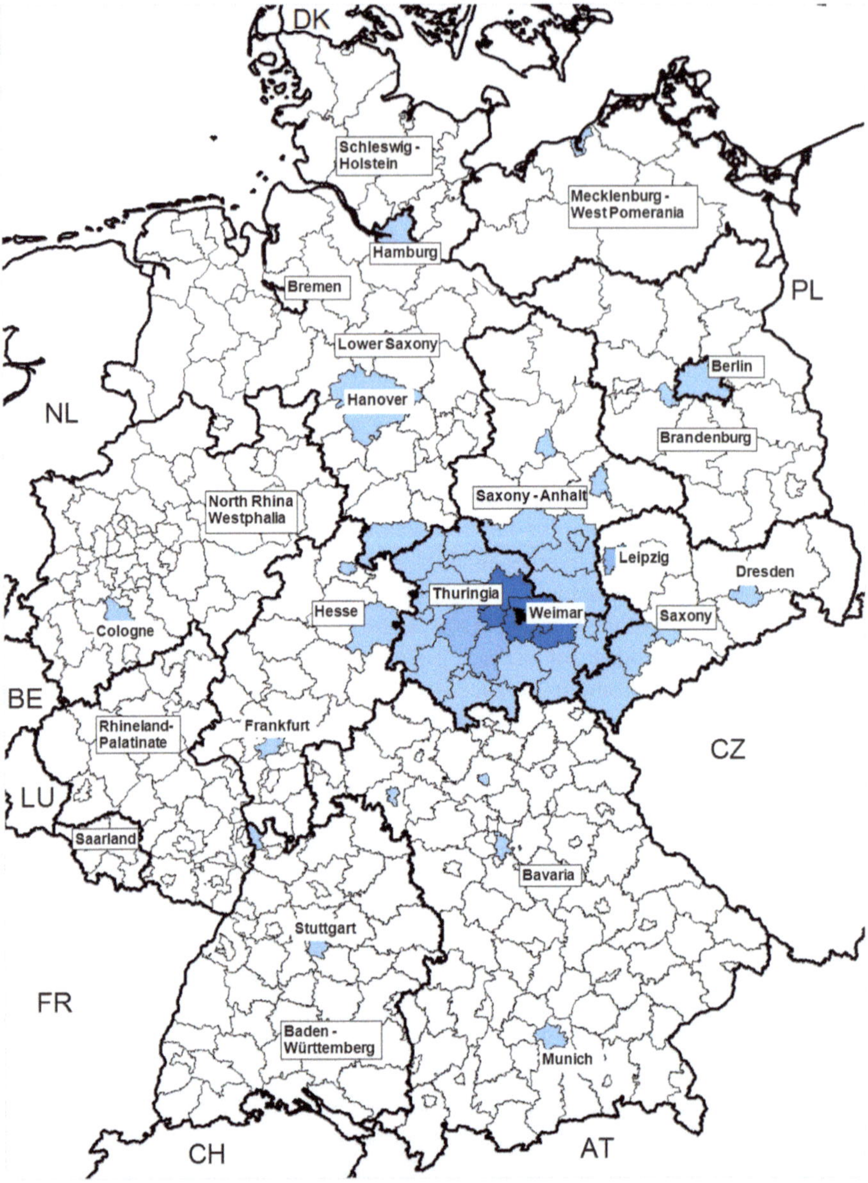

Fig. 8.13 Spatial interaction of the City of Weimar with other districts (total population), *dark blue* means strong interaction (Source: STASA 2017)

cheaper housing opportunities, cheaper costs of living and cheaper land prizes. Munich attracts young people (age-groups 18–25, and 30–50 years) because of the city's lifestyle and the well functioning labour market. The very high prices for land and rents in the city of Munich foster migration towards the suburbs.

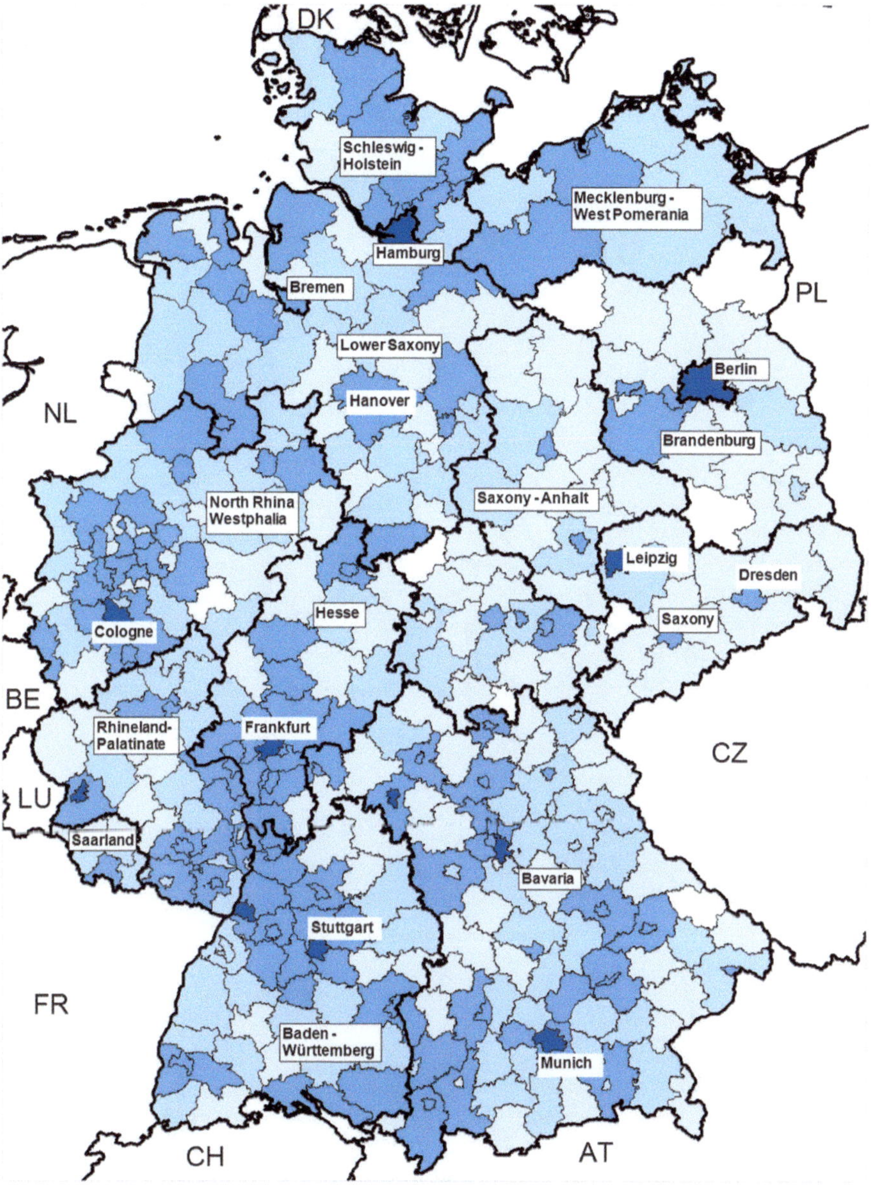

Fig. 8.14 Centrality index or strength of regional interdependence determined for the total population. *Dark blue* indicates a high value of centrality (Source: STASA 2017)

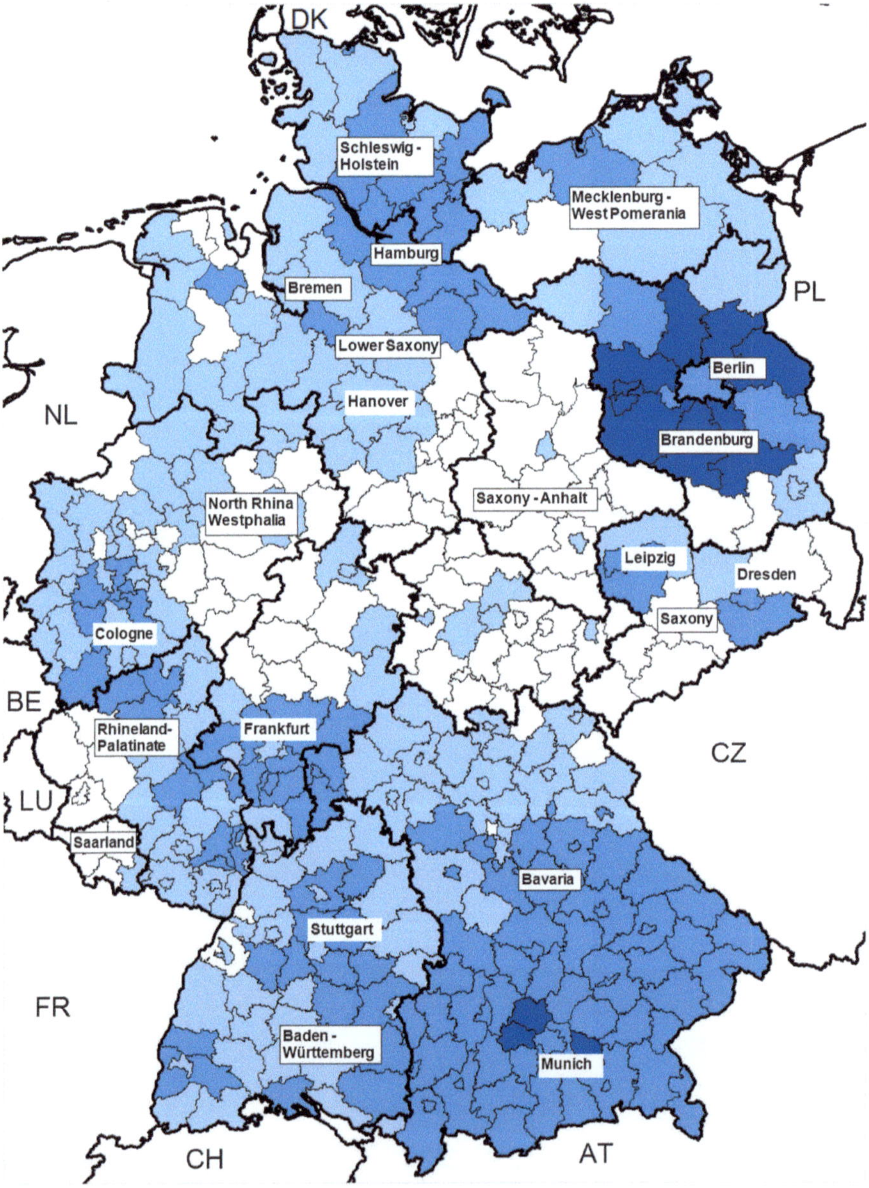

Fig. 8.15 Spatial preferences (total population). *Dark blue* indicates a high value of preference, *white* colour refer to less preferred regions (Source: STASA 2017)

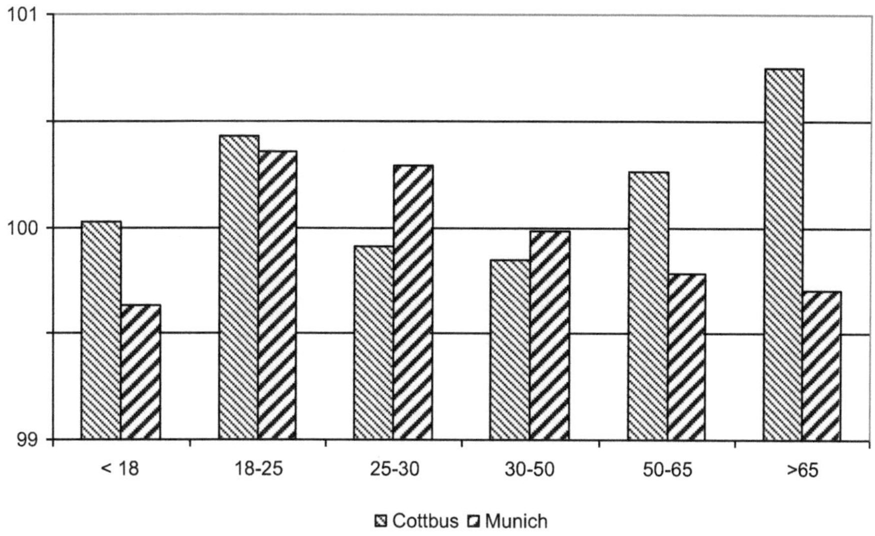

Fig. 8.16 Age-dependence of spatial preferences for the districts Cottbus and Munich (Source: STASA 2017)

8.9.4 Regression Analysis

A multiple regression was performed for the different levels of spatial preferences $\delta_i, i = 1, \ldots, 402$ of the German districts for the year 2015 for the total population. However, if we perform a regression analysis for different age groups, the explanatory variables may change as well as their weights and level of significance. For example, to explain the migration behaviour of the age group of young people (18 to <25 years) the accessibility to education opportunities become important, but seems to be not significant for the total population.

It is of utmost importance to remember that the spatial preferences do not contain any size effect that means dependence of population size. Therefore, the preferences reflect the socio-economic power of a region as perceived by the migrants without any size effect.

In the following Table 8.1 three different regression models (Model 1, Model 2, Model 3) for the spatial preferences are presented.

The following variables have been used and tested in the regression analysis of the spatial preferences. Abbreviations have been assigned to all variables turning out to be significant explanatory variables.

Labour Market

- Rate of unemployment, ALQ
- Income per capita, BLS
- Total labour force
- Accessible work places per 45 min by car

Table 8.1 Regression results of the regional preferences (total population)

Variables	Model 1	Model 2	Model 3
Const.	−0.680***	−0.142***	−0.040
	(−10.044)	(−2.935)	(−0.667)
GfK	2.406E-5***		
	(8.609)		
NDW	0.009***		
	(4.955)		
SB	0.014*	0.026***	0.036***
	(1.898)	(3.587)	(4.644)
ALQ		−0.003	−0.008***
		(−1.383)	(−2.649)
BLS		1.784E-6*	1.603E-6
		(1.500)	(1.132)
CP		0.016***	
		(8.252)	
PD			0.001***
			(3.154)
TCAR			−0.001
			(−0.464)
R^2_{ad}	0.346	0.283	0.179

Note: Values in parenthesis are t-values. ***, **, and * indicate the significance levels at the 1, 5, and 10% using t-statistic. GfK = available income per capita, NDW = new dwellings per 1000 existing dwellings, SB = shop balance: number of business openings and closings per 1000 capita), ALQ = rate of unemployment, BLS = income per capita, CP = construction permit per 1000 existing dwellings, PD = patent density: patents per 1000 capita, TCAR = travel time per car to the next highway

- Available income, GfK
- Apprentice per 1000 employees
- Students per 1000 capita

Housing Market

- Number of new dwellings per 1000 already existing dwellings, NDW
- Construction permission per 1000 already existing dwellings, CP
- Rent per m^2
- Population
- Job Qualification: percentage of employees with high school graduation

Investment and Industrial Sector

- Patent density per 100,000 employees, PD
- Shop balance: number of business openings and closings per 1000 capita, SB
- Overnight stays per capita
- Number of engineers per 1000 employees

- R&D in % on GDP (gross domestic product)
- Capital spending

Others

- Accessibility per car: average travel time to the next highway connection, TCAR
- Accessibility per public transport: average travel time to the next three agglomeration centres
- Criminal acts per 100,000 capita

The regression results of the three models well confirm that variables related to the housing market, the labour market, the innovation power of the region under consideration and accessibility measures are important explanatory variables for the regional preferences in Germany. The great demographic changes in Germany, triggered by migration and falling birth and death rates are causing strong regional changes (Piontek and Wyrwich 2017). Therefore it can be supposed that the regional preferences for the different age groups may also have different explanatory variables or at least different statistical weights in the regression analysis.

According to model 1 the available income, the opening of new dwellings and the balance of the number of business openings and closings per capita explain already about 35% of the regional variation of the preferences ($R^2_{ad} = 0.346$, $F = 68.257$).

Model 2 and 3 confirm that the rate of unemployment, ALQ, and the income per capita, BLS, are important driving forces as well. The patent density, PD (model 3), or the number of construction permitted per 1000 existing dwellings, CP, can be seen as a variable of future expectations and therefore enhance the migratory effects. The gross domestic product (GDP) could be another relevant driving force of regional preferences. Unfortunately, the GDP was not among the list of available variables. However, in economic research the inverse relationship between unemployment and the gross domestic product (GDP) has been analysed. This so called Okun's law indicates that the GDP could be a relevant variable for the explanation of regional differences (Melguizo 2017). As model 3 also shows there is a very small effect of the accessibility measure: travel time by car to the next highway. This also confirms our expectations that a good accessibility of a district has a positive impact on its preference.

Nevertheless, there may be substantial economic and social differences between urban and neighbouring rural districts. The analysis underlines that it is difficult to diminish spatial disparities by means of interregional migration, only, since the explanatory variables are difficult to influence by political authorities.

The complexity of interregional migration can only be described and understood on the level of individuals and firms (micro-level) interacting with each other. The inherent uncertainties in all decision processes are represented by stochastic components. Migration of individuals and the location choice process of firms are ultimately generated by motivation factors, describing the intensity of different "reasons" for changing the place of residence or for selecting a certain location for a firm. It is obvious, that very different explanatory variables or combinations of them may finally result in the same decision to migrate. This fundamental ambiguity bears the potential for the understanding of the diversity of the different driving forces (explanatory variables) which appear in the different regression results.

References

Aggarwal CC (2011) Social network data analysis. Springer, New York

Allen P, Sanglier M (1979) A dynamic model of growth in a central place system. Geogr Anal 11:256–272

Andersson AE, Ferraro G (1982) Accessibility and density distributions in metropolitan areas: theory and empirical studies. Pap Reg Sci Assoc 52:141–158

Auerbach F (1913) Das Gesetz der Bevoelkerungskonzentration. Petermanns Mitteilungen 59:1

Bak P, Paczuski M (1995) Complexity, contingency, and criticality. Proc Natl Acad Sci 92:6689–6696

Bak P, Tang C, Wiesenfeld K (1987) An explanation of the 1/f noise. Phys Rev Lett 59:381

Barenblatt GI (2003) Scaling. Cambridge University Press, Cambridge

Batten DF, Casti J, Thord R (eds) (1995) Networks in action. Springer, Berlin

Batty M (1970) Models and projections of the space economy: a subregional study in Northwest England. Town Plan Rev 41:121–148

Batty M, Longley P (1994) Fractal cities. Academic, London

Batty M, Mackie S (1972) The calibration of gravity, entropy and related models of spatial interaction. Environ Plan 4:205–233

Berry BJK, Garrison WL (1958a) The functional bases of the central place hierarchy. Econ Geogr 34:145–154

Berry BJK, Garrison WL (1958b) Alternate explanations of urban rank-size relationships. Ann Assoc Am Geogr 1:83–91

Bertuglia CS, Clark GP, Wilson AG (1994) Modelling the city, performance, policy and planning. Routledge, London

Birg H, Filip D, Hilge K (1983) Verflechtungsanalys der Bevölkerungsmobilität zwischen den Bundesländern von 1950–1980. IBS-Materialien, 8, Universität Bielefeld

Carey HC (1858) Principles of social aciences 3. Lippincott, Philadelphia

Cobb CW, Douglas PH (1928) A theory of production. Am Econ Rev 18(1):139–165

Cornelius I (1996) Bevoelkerungsbilanz und natuerliche Bevoelkerungsentwicklung. In: Statistisches Landesamt Baden-Wuerttemberg (Hrsg) Baden-Wuerttemberg in Wort und Zahl, Statistische Monatshefte, 44/9, Stuttgart

Couclelis H (2017) Climbing on a milestone for a better view: Goodchild's "Geographical Information Science" paper as vantage point and ground for reflection. Int J Geogr Inf Sci 26:2291–2300

Courgeau D (1985) Migrants and migrations. Population selected papers, INED, Paris

Courgeau D (1995) Migration theories and behavioural models. Int J Popul Geogr 1:19–27

Curry L (1964) The random spatial economy. Ann Assoc Am Geogr 7:138–146

Fischer MM, Nijkamp P, Papageorgiou YY (1990) Spatial choices and processes. North-Holland, Amsterdam

Fotheringham AS, O'Kelly ME (1989) Spatial interaction models: formulations and applications. Kluwer Academic Publishers, Dordrecht

Frankhauser P (1998) The fractal approach. A new tool fort he spatial analysis of urban agglomerations. Population 10:205–240

Frankhauser P, Guerin-Pace F (1993) Measuring the hierarchical organization of an urban system (á paraître)

Gibrart R (1931) Les inégalités économiques. Sirey, Paris

Greene WH (2003) Econometric analysis. Prentice Hall, Upper Saddle River
Grimmeau JP (1994) Le Moodéle gravitatire et le facteur d'échelle. Applications aux migrations intérieures de la Belgique 1989-1991. Espace Popul Soc 1:131–141
Guerin-Pace F (1990) Dynamique lente et dynamique rapide dans le development d'un système de villes. Paper presented at Colloque de l'AIDELF, Rabat
Guerin-Pace F (1993) Deux siécles de croissance urbaine. Anthropos, Paris
Haag G (1989) Dynamic decision theory: Application to urban and regional topics. Kluwer Academic Publishers, Dordrecht
Haag G (1993) The rank-size distribution of settlements as a dynamic multi-fractal phenomenon. Chaos Solitons Fractals 4:519–534
Haag G, Gruetzmann K (2001) Transport and urban development. In: Clark G, Madden M (eds) Regional science in business. Springer, Berlin
Haag G, Max H (1995) Rank-size distribution of settlement systems: a stable attractor in urban growth. J Reg Sci Assoc 74:243–258
Haag G, Weidlich W (1983) An evaluable theory for a class of migration problems. Collaborative paper, CP-83-58, IIASA, Laxenburg
Haag G, Weidlich W (1984a) A stochastic theory of interregional migration. Geogr Anal 16:331–357
Haag G, Weidlich W (1984b) Model construction in dynamic spatial theory. Proceedings of the Task Force Meeting in Dynamic Analysis of Spatial Development. IIASA, Laxenburg
Haag G, Weidlich W (1986) A dynamic migration theory and its evaluation for concrete systems. Reg Sci Urban Econ 16:57–80
Haag G, Munz M, Sanders L, Saint-Julien T, Pumain D (1991) Interurban migration and the dynamics of a system of cities: the stochastic framework with an application to the French urban system. Environ Plan A 24:181–198
Haran EGP, Vining DR (1973) On the implications of a stationary urban population distribution for the size distribution of cities. Geogr Anal 4:194–200
Haynes KE, Fotheringham AS (1984) Gravity and spatial interaction models. Sage, Beverly Hills
Hotelling H (1978) A mathematical theory of migration. Republished in Environ Plan A 10
Huriot JM, Thisse J (1984) Distance in spatial analysis. Institute de Mathematiques Economques. LATEC LA, CNRS 342, Université de Dijon
Isard W (1954) Location theory and trade theory: short-run analysis. Q J Econ 68:305–322
Katz L (1953) A new status index derived from sociometric index. Psychometrica 18:39–43
Krugman P (1993) First nature, second nature and metropolitan location. J Reg Sci 33:129–144
Krugman P (1994) Fluctuations, instability and agglomeration. NBER working paper 4616
La Barra D (1990) Integrated land use and transport modelling. Cambridge University Press, Cambridge
Lee ES (1966) A theory of migration. Demography 3:47–57
Leonardi G, Casti J (1986) Agglomeration tendencies in the distribution of populations. Reg Sci Urban Econ 16:43–56
LeSage JP, Pace RK (2014) The biggest myth in spatial econometrics. Econometrics 2:217–249
MacGill SM (1979) Convergence and related properties for a modified biproportional matrix problem. Environ Plan A 11:499–506
Mandelbrot BB (1991) Die fraktale Geometrie der Natur. Birkhäuser, Berlin
Melguizo C (2017) An analysis of Okun's law for the Spanish provinces. Rev Reg Res 37:59–90
Mills E (1972) Studies in the structure of the urban economy. Johns Hopkins University Press, Baltimore
Nijkamp P, Reggiani A (1992) Interaction, evolution and chaos in space. Springer, Berlin
Nijkamp P, Reggiani A (1998) Entropy, spatial interaction modelling and discrete choice analysis. Eur J Oper Res 36(1):186–196
Papageorgiou YY (1980) On sudden urban growth. Environ Plan A 12:1035–1050
Papageorgiou YY, Smith TR (1983) Agglomeration as local instability of spatial uniform steady states. Econometrica 51:242–252

Piontek M, Wyrwich M (2017) The emergence of entrepreneurial ideas at universities in times of demographic change: evidence from Germany. Rev Reg Res 37:1–37
Portnov BA (2011) Does Zipf's law hold for primate cities? Some evidence from a discriminant analysis of world countries. Jahrbuch für Regionalwissenschaften 31:113–129
Poulain M (1981) Contribution á l'analyse spatiale d'une matrice de migrations internes. Cabay, Bruyland, Brussels
Pumain D (1982) La dynamique des Villes. Economica, Paris
Pumain D (1990) City size dynamics in urban systems. In: Proceedings of the conference on dynamic modelling and human systems, Cambridge
Pumain D (ed) (1991) Spatial analysis and population dynamics. John Libbey Eurotext, Paris
Pumain D (1993) Geography, physics and synergetics. In: Haag G, Mueller U, Troitzsch RG (eds) Economic evolution and demographic change, Lecture notes in economics and mathematical systems, vol 395. Springer, Heidelberg
Pumain D (ed) (2006) Hierarchy in natural and social sciences. Springer, Dordrecht
Pumain D, Haag G (1991) Urban and regional dynamics-towards an integrated approach. Environ Plan A 23:1301–1313
Pumain D, Sanders L, Saint-Julien T (1989) Villes et auto-organisation. Economica, Paris
Puu T (1983) A simplified model of spatio-temporal population dynamics. Umea Economic Studies, 139
Quandt RE (1964) Statistical discrimination among alternate hypothesis and some economic regulations. J Reg Sci 2:1–23
Rabino GA, Occelli S (1999) Understanding spatial structure from network data: theoretical considerations and applications. Cybergeo, 29
Ravenstein E (1885) The laws of migration. J R Stat Soc 48:167–235
Ravenstein E (1889) The laws of migration, second paper. J R Stat Soc 52:241–305
Reggiani A, Bucci P, Russo G (2011) Accessibility and impedance forms: empirical applications to the German commuting networks. Int Reg Sci Rev 34(2):230–252
Reilly WJ (1931) The law of retail migration. Pilsbury, New-York
Reynaud J (1841) Villes encyclopédie nouvelle, tome VIII, Gosselin, 670–687
Robic MC (1982) Cent an savant Christaller: une thérie des lieux centraux. L'Espace Géogr 11:5–12
Robson B (1973) Urban growth: an approach. Methuen, Paris
Roehner B, Wiese KE (1982) A dynamic generalization of Zipf's rank-size rule. Environ Plan 14:1449–1467
Rosser JB (2011) Complex evolutionary dynamics in urban-regional and ecologic-economic systems: from catastrophe to chaos and beyond. Springer, New York
Sanders L (1992) Systéme de Villes et Synergétique. Anthropos, Paris
Sardadvar S, Rocha-Akis S (2016) Interregional migration within the European Union in the aftermath of the eastern emlargements. Rev Reg Res 36:51–59
Schuermann C, Spiekermann K, Wegener M (1997) Accessibility indicators. Universitaet Dortmund, Berichte aus dem Institut fuer Raumplanung (IRPUD) 39, Dortmund
Sen A (1982) Modeling spatial flows. Unpublished manuscript. School of Urban Planning and Policy, University of Illinois at Chicago, IL
Somermeijer WH (1966) Specificatie van economische relaties. Reprint 126, Econometrisch Instituut, Erasmus University Rotterdam
Steindl J (1968) Size distributions in economics. In: Sills DL, Merton RK (eds) International encyclopedia of the social science. Macmillan, New York
Stewart JQ (1947) Suggested principles of social physics. Science 106:179–180
Stewart JQ (1948) Demographic gravitation: evidence and application. Sociometry 11:31–58
Straubhaar T (1988) On the economics of international labor migration, Beiträge zur Wirtschaftspolitik. Paul Haupt, Bern
Weber AF (1899) The growth of cities in the nineteenth century: a study in statistics. Macmillan, New York

Weidlich W, Haag G (1983) Concepts and models of a quantitative sociology: the dynamics of interacting populations, Springer-series synergetics, vol 14. Springer, Berlin
Weidlich W, Haag G (1987) A dynamic phase transition model for spatial agglomeration processes. J Reg Sci 27:529–569
Weidlich W, Haag G (eds) (1988) Interregional migration. Springer, Berlin
Wheaton WC (1974) A comparative static analysis of urban spatial structure. J Econ Theory 9:223–237
White R (1978) Simulation of central place dynamics and the rank size distribution. Geogr Anal 10:201–208
Wilson AG (1970) Entropy in urban and regional modelling. Pion Books, London
Wilson AG (1981a) Geography and the environment. Wiley, New York
Wilson AG (1981b) The evolution of spatiale structure. In: Bennett RJ (ed) European progress in spatial analysis. Pion Books, London
Wilson AG (1981c) Catastrophe theory and bifurcations: applications to urban and regional systems. Croom Helm, London
Zipf GK (1949) Human behaviour and the principle of least effort. Addison-Wesley, Reading, MA
Zipser T, Mlek M, Zipser W (2011) Zipf's law in hierarchically ordered open systems. Jahrbuch für Regionalwissenschaften 31:93–112

Chapter 9
Applications in Urban Dynamics and Transport

Contents

9.1 The STASA Integrated Transport and Land-Use Model

Since the year 2000, more than half of the world's population lives in cities. This means that not only an extensive exchange of population, goods and information in a globalized world can be expected but also that the transport of goods, population and information must be effectively managed. A city which strives to represent an important node in this network must provide beside appropriate economic, social and cultural conditions also political stability, qualified labour, as well as urban and internationally operating interurban and intermodal system of transport (Fotheringham and O'Kelly 1989; Bertuglia et al. 1994; De La Barra 1990; Haag 1989a, b). Moreover, all cities face a common problem: they must possess the capacity to sustain unprecedented numbers of citizens within limited budgets and severe environmental constraints (Isard 1954; Pumain 1982). Therefore, transport systems can only be seen in their regional context (Stasa 1998, 2002; Haag and Gruetzmann 2001; Transecon 2000; Steierwald and Künne 1993; Eurosil 1997).

G. Haag, *Modelling with the Master Equation*, DOI 10.1007/978-3-319-60300-1_9

Hence, the interactions between the transport system consisting of different transport modes and the settlement pattern have to be modelled. Of course, the volume of transport and the mix between public and private provision have an impact on the migration behaviour of households and firms.

Investments into the transport sector and communication networks improve the accessibility and attractiveness of suburban areas (Batten et al. 1995; Haynes and Fotheringham 1984). This may lead to a redistribution of migration and traffic flows and is discussed as one possible reason for urban sprawl (Haag 2002; Scatter 2001; Gayda et al. 2003). The quantitative treatment of these nested processes of the different subsystems (transport-, population-, and communication-subsystems) and their interactions require an integrated modelling approach (Rosser 2011).

The number of real-world applications of integrated transport and land use models has increased steadily during the last decades as witnessed by Hutchinson et al. (1985), Webster et al. (1988), by reviews of Harris (1985), Wegener (1986, 1987, 1994, 1998), Boyce (1988), Aoyama (1989), Batty (1994), Wilson (1997) and many others. Among the many candidates, the Tranus model, the Meplan model and the Stasa model which is described next (Stasa 1998; Eurosil 1997; Haag and Gruetzmann 2001; Weidlich and Haag 1999; Albeverio et al. 2008) have shown their leading competence in the summary of model capabilities.

Changes in the location of population and workplaces can be classified as a slow adjustment process. This means that commuting patterns should be expected to adjust to new conditions on a much faster time scale than household location patterns.

In order to understand the complex processes and interactions between traffic and urban development, both systems will be analysed in a first step by stochastic decision models. For that purpose, the considered area will be divided into regions (cells or locations). The development of these areas is described by "macro-variables". Examples for explanatory variables are above all those which describe the traffic infrastructure of the cells but also the distribution of the population, the distribution of lodgings, factories, shopping centres, leisure centres, to mention a few possible explanatory variables. The modelling concept establishes a connection between the dynamics of these macro-variables and the behaviour of the individuals of the considered population ("micro-level").

The interactions between micro- and macro-level are captured as follows: on the one hand, the dynamics on the macro-level—i.e., the development of the traffic and the regional settlement structure—are determined by the behaviour of individuals on the micro-level. On the other hand, the "attractiveness" differences between the traffic cells which depend on the macro-variables influence the decisions of individuals as well (Haag and Weidlich 1984). Apart from the rational motives of actors, several elements of uncertainty, e.g., irrational behaviour as a result of insufficient information, have to be taken into account. Hence, the description of decision processes is based on a stochastic and dynamic decision model within the Master equation approach.

Empirical data such as traffic flows between the regions, population numbers or migration flows are based on the traffic system as well as on the development of the population distribution. System parameters as "attractiveness" and "distance" (or resistance) parameters are then determined with the help of a non-linear estimation procedure. In a further step the attractiveness values can be connected with the macro-variables or key-variables making use of a multiple regression. At the same time the relevance of the single key variables is determined. The two last steps consist in the numerical simulation of the results and the recommendations arising from these simulations.

Traffic demand is the result of social and individual activities as well as the evolution of land use patterns. Accordingly, individual decision processes of the agents form the basis (individuals, companies, etc.) of all transport phenomena. In this way, the traffic volume is influenced by: the total number of activities, the places and starting time of activities, the means of transport and route selection, the coordination of activities with other agents (e.g., formation of car pools) as well as feedback effects.

In total, therefore, traffic demand is the result of a great number of individual decision processes. These complex processes can take place on a short-term or long-term basis. In order to represent the demand behaviour of the agents in a model one must take into account the following: "*The important point here is that travel demands on any particular route or network are not pre-ordained. They arise out of tens of thousands of individual choices being made every day, with each person in the population deciding where, when and how they wish to travel. Quite reasonably, they make these decisions in their own interest, rather than that of the community as a whole, and on the basis of the best information available at the time, which is often rather poor. The purpose of building a mathematical model of the travel demand has to take into account all these choices in order to be able to predict how that demand might change, as the circumstances determining those choices change*." (Sactra 1994a, b, 1998).

9.1.1 The Micro-level

The micro level of the society is determined by the individual decision behaviour of the different agents of the economic system. So-called "attitudes" are in the foreground of every individual concerning specific "aspects" of the society, such as the present earnings, the kind of work, the present living conditions, the family background, leisure facilities, politics, the traffic situation, to mention a few performance-influencing factors. Let

$$p_{ji}^{\alpha r(l)}\left(\vec{E}, \vec{\kappa}_l\right) \tag{9.1}$$

be the probability per unit of time that an individual $i, i = 1, 2, \ldots, I$ carries out a trip from area i to area j, given a certain population distribution $\vec{E}$ in the considered region. Individuals might be allocated to trip purpose specific subpopulations α.

The different attitudes of each individual are summarised in the "attitude vector" $\vec{\kappa}_l$. In principle, the corresponding individual transition rates $p_{ji}^{\alpha r(l)}\left(\vec{E},\vec{\kappa}_l\right)$ can be determined via panel data.

Dynamic micro models, based on panel data, have the advantage that they can represent the modification of policy relevant variables in the course of time. However, modelling traffic flows requires the preparation of a comprehensive (representative) data base, combined with corresponding costs (Bundesminister für Verkehr 1996).

9.1.2 The Macro-level

The number of the system variables necessary for the description of the society is reduced drastically if one passes to the macro level. Our aim is to derive the trip frequency via a behavioural decision model describing individuals and households on the basis of sub-populations. These groups are characterised by comparable socio-economic features so that the traffic behaviour of individual decision makers within the same group is approximately uniform, yet, quite different between individuals of different groups. Correspondingly, the entire population of the region under investigation is subdivided into groups (Kutter 1972; Meier 1990). The activity pattern and the corresponding trip distribution can therefore be taken as group-related (see Sect. 6.5).

9.1.3 The Modelling

The region under investigation is subdivided into L non-overlapping traffic cells $i, i=1,2,\ldots,L$. The total population might be further disaggregated (persons, budgets and so on) into trip purposes (home—work, home—education, home—shopping, etc.) $\alpha,\alpha=1,2,\ldots,P$. A subdivision according to traffic modes (on foot, bicycle, bus, car, etc.) $r,r=1,2,\ldots,R$ is also introduced since modifications of the modal split have to be considered and evaluated.

The number of trips for the trip purpose α from traffic cell i to traffic cell j at time t (hour of the day) using traffic mode r is denoted by $F_{ji}^{\alpha r}(t)$. Then

$$F_{ji}^{\alpha}(t)=\sum_{r=1}^{R}F_{ji}^{\alpha r}(t) \tag{9.2}$$

is the number of trips between i and j for trip purpose α, and

$$F_{ji}^{r}(t) = \sum_{\alpha=1}^{P} F_{ji}^{\alpha r}(t) \tag{9.3}$$

is the number of trips between i and j using mode r, and

$$F_{ji}(t) = \sum_{r=1}^{R} \sum_{\alpha=1}^{P} F_{ji}^{\alpha r}(t) \tag{9.4}$$

is the total number of trips between i and j. Therefore, accordingly, the total number of trips in the system is given by

$$F(t) = \sum_{i=1}^{L} \sum_{j=1}^{L} F_{ji}(t). \tag{9.5}$$

The traffic volume $O_i^{\alpha}(t)$ (source volume) of the traffic cell i is gained by summation from the traffic flows:

$$O_i^{\alpha}(t) = \sum_{j=1}^{L} F_{ji}^{\alpha}(t). \tag{9.6}$$

The number of trips $D_j^{\alpha}(t)$ for the trip purpose α into the traffic cell j at time t reads correspondingly:

$$D_j^{\alpha}(t) = \sum_{i=1}^{L} F_{ji}^{\alpha}(t). \tag{9.7}$$

The population distribution is denoted by

$$\vec{E} = \{E_1, \ldots, E_i, \ldots, E_L\}, \tag{9.8}$$

where E_i is the number of agents of traffic cell i. E_i will be modified by the decisions of the individual agents to carry out a trip between cell i and anyone from the other cells. Therefore, the population distribution $\vec{E}$ is connected via traffic related activities of the agents with a great number of individual decision processes and finally with the attractiveness of the specific traffic cells.

The micro-level (individuals, households) seems to determine the dynamics of the macro level completely, while there is no feedback of the macro-level to the micro-level. However, this is not the case. Rather, a mutual influence of the two levels occurs: among other things, the actions (activities) of the individual agents of the economic system express themselves in the dynamics of the traffic flows and therefore in the time-dependent population redistribution of the system (Pumain and Haag 1991; Weidlich and Haag 1983). Therefore, micro behaviour and macro dynamics are strongly interlinked. Mathematically, the dependence of the individual trip decisions on the macro state appears in the fact that the attractiveness of the traffic cells among other things depends on the respective macro state, characterised, i.e., by the time-dependent population distribution.

9.1.4 The Stochastic Framework

In addition to rational motives, uncertainties have to be considered in the case of decision processes as well. Therefore, the Master equation framework is used for modelling the integrated transport and urban system. For this reason, the configurational probability $P(\vec{E}, t)$ to find a certain spatial population distribution $\vec{E}$ at time t is introduced. Of course, $P(\vec{E}, t)$ must satisfy the probability normalisation condition at all times:

$$\sum_{\vec{E}} P(\vec{E}, t) = 1, \tag{9.9}$$

where the sum extends over all possible population-configurations $\vec{E}$. If the configurational transition rates $F_t\left(\vec{E} + \vec{k}, \vec{E}\right)$ from any $\vec{E}$ to all neighbouring $\vec{E} + \vec{k}$ are given, the Master equation for $P(\vec{E}, t)$ can be derived according to Chap. 3:

$$\frac{d}{dt} P(\vec{E}, t) = \sum_{\vec{k}} F_t\left(\vec{E}, \vec{E} + \vec{k}\right) P\left(\vec{E} + \vec{k}, t\right) - \sum_{\vec{k}} F_t\left(\vec{E} + \vec{k}, \vec{E}\right) P(\vec{E}, t), \tag{9.10}$$

where the sum on the right-hand side extends over all $\vec{k}$ with non-vanishing configurational transition rates $F_t\left(\vec{E} + \vec{k}, \vec{E}\right)$ and $F_t\left(\vec{E}, \vec{E} + \vec{k}\right)$, respectively. The Master equation has a very direct and intuitively appealing interpretation. The change in time of $\frac{dP\left(\vec{E}, t\right)}{dt}$ is due to two effects of opposite direction: firstly to the probability flux from all neighbouring configurations $\vec{E} + \vec{k}$ into the considered configuration $\vec{E}$, namely $\sum_{\vec{k}} F_t\left(\vec{E}, \vec{E} + \vec{k}\right) P\left(\vec{E} + \vec{k}, t\right)$, and secondly to the probability flux out of the configuration $\vec{E}$ into all neighbouring configurations $\vec{E} + \vec{k}$, namely $\sum_{\vec{k}} F_t\left(\vec{E} + \vec{k}, \vec{E}\right) P(\vec{E}, t)$. Consequently the Master equation represents a balance equation for probability fluxes. The transition rates occurring in the Master equation (transition probabilities per unit of time) are associated directly with the short-term evolution of the conditional probability. Further details are described in Chaps. 3 and 4. The solution of the Master equation, namely the time-dependent solution contains all the information about the dynamics of the population redistribution in detailed manner, and therefore indirectly also about the trip frequencies. The configuration assigned to the time-dependent maximum of the distribution represents the most probable population distribution at time t.

Since the probability distribution $P(\vec{E}, t)$ contains such a tremendous amount of information compared with the empirically available information (data base) concerning transport events and/or urban/regional processes, a less exhaustive description is favoured. Therefore, self-contained equations of motion for the mean population in traffic cell i at time t are derived via:

$$E_i(t) = \sum_{\vec{E}} E_i P(\vec{E}, t), \tag{9.11}$$

where again the summation extends over all possible population-configurations.

To derive equations of motion for the mean values, the Master equation is multiplied by $\vec{E}$ from the left and subsequently summed up via all states $\vec{E}$. However, the equations obtained are not self-contained since the determination of the right side demands the knowledge of the probability distribution $P(\vec{E}, t)$. If it can be assumed that the probability distribution is well behaved and sharply peaked, quasi-closed approximate mean value equations can be derived, as described in Chap. 4.

In this line, first the respective process-specific transition rates have to be defined. This is carried out in the following for the traffic sub-system, the same procedure has to be applied for the urban sub-system as well.

The configurational transition rate $F_t\left(\vec{E} + \vec{k}, \vec{E}\right)$ from population distribution $\vec{E}$ to the neighbouring distribution $\vec{E} + \vec{k}$ is the sum of all the contributions

$$F_t\left(\vec{E} + \vec{k}, \vec{E}\right) = \sum_{r=1}^{R} \sum_{\alpha=1}^{P} \sum_{i,j=1}^{L} F_{ij}^{\alpha r}\left(\vec{E} + \vec{k}, \vec{E}\right), \tag{9.12}$$

where $F_{ij}^{\alpha r}\left(\vec{E} + \vec{k}, \vec{E}\right)$ indicates the number of trips between the traffic cells $j \rightarrow i$ for trip purpose α using traffic mode r. In the total transition rate $F_t\left(\vec{E} + \vec{k}, \vec{E}\right)$ we have neglected contributions referring to simultaneous changes of trip purpose, traffic mode and location. The explicit dependence of the individual terms on $\vec{E}$ indicate that merely those contributions related to a change of the population distribution $\vec{E} \rightarrow \vec{E} + \vec{k}$ are summed up. In this way, a summation via all such terms yields the total transition rate. The index t indicates the possibility of explicit time dependence. After insertion of the transition rates into the Master equation and carrying out of the above mentioned approximations, one receives equations of motion for the short-run population redistribution during the day due to transport processes:

$$E_i(t+1) = E_i(t) + \sum_{\alpha} \sum_{r} \sum_{j} \left(F_{ij}^{\alpha r} - F_{ji}^{\alpha r}\right), \tag{9.13}$$

where $E_i(t)$ is the population stock of traffic cell i and t specifies the daily hour group. Because empirical data are available only for certain hour groups t, a transition to a difference equation in place of a differential equation is adequate.

The term $F_{ij}^{\alpha r}\left(\vec{E}+\vec{k},\vec{E}\right)$ describes purpose specific trips from $j \rightarrow i$ using traffic mode r. Therefore, this transition rate corresponds directly to the empirically countable traffic flows. Now, the question about the driving forces or motives which makes an individual carry out a trip must be raised.

If, however, panel data are available on the trip-decision behaviour of the different agents of the system (micro-level), the configurational transition rates can directly be computed via:

$$F_{ij}^{\alpha r}\left(\vec{E}+\vec{k},\vec{E}\right)=\sum_{l\in\Gamma_i} p_{ij}^{\alpha r(l)}\left(\vec{E},\vec{\kappa}_l,t\right), \tag{9.14}$$

where one has to sum up over all individual contributions of all agents belonging to subpopulation α and travelling from traffic cell j to traffic cell i using mode r. Yet, this procedure is very extensive because of the required immense data base (Courgeau 1985, 1995). Therefore, a less exhaustive procedure which is more amenable to the analysis and simulation work is preferred.

Since $E_i(t)$ agents are in traffic cell i at time t the probability for a trip to another traffic cell will be proportional $E_i(t)$. Let $p_{ji}^{\alpha r}\left(\vec{E},\vec{\kappa},t\right)$ be the transition rate from i to j for trip purpose α. Of course, this transition rate depends among others on the explicit distribution of the agents $\vec{E}$ and cell specific characteristics $\vec{\kappa}$.

The attractiveness values of the traffic cells show a typical time-specific behaviour. For cells dominated by housing units, the attractiveness will be high during the night and low during working hours. The contrary will be found for cells dominated by production units providing many work places: here, it can be expected that the attractiveness of those cells will be high during the working hours of the day and rather low in the late afternoon and at night. However, what determines the temporal daily variation of the cell specific attractiveness values is the mixture of housing specific units, work places, shopping facilities, leisure time facilities and different kinds of services and relating shop opening and closing times:

$$F_{ji}^{\alpha r}(t)=E_i(t)p_{ji}^{\alpha r}\left(\vec{E},\vec{\kappa},t\right), \tag{9.15}$$

where $\vec{k}=\left\{0,\ldots,1_j^{\alpha r},\ldots,0,\ldots,(-1)_i^{\alpha r},\ldots,0,\ldots\right\}$ represents trip probabilities from one traffic cell i into j, and $F_{ji}^{\alpha r}\left(\vec{E}+\vec{k},\vec{E}\right)=0$ for all other $\vec{k}$.

The dynamics of the population redistribution on the macro-level can be derived directly by an averaging procedure from the Master equation:

$$\frac{dE_i(t)}{dt}=\sum_{\alpha,r}\sum_{j=1}^{LMAX}E_j(t)p_{ij}^{\alpha r}\left(\vec{E},\vec{\kappa},t\right)-\sum_{\alpha,r}\sum_{j=1}^{LMAX}E_i(t)p_{ji}^{\alpha r}\left(\vec{E},\vec{\kappa},t\right). \tag{9.16}$$

The change of the average population number $E_i(t)$ of a traffic cell i of one hour to the next hour can be calculated on the basis of the traffic flows $F_{ji}^{\alpha r}$ and $F_{ij}^{\alpha r}$ between the different traffic cells i and j. The flows in turn depend on the respective time-dependent attractiveness $u_i^{\alpha}\left(\vec{E}, \vec{\kappa}, t\right)$ of the traffic cells, the resistance functions $g^{\alpha r}\left(w_{ji}^{\alpha r}\right)$ and the scaling parameter (mobility parameter) $\varepsilon^{\alpha r}(t)$. In this way, the population stock of a traffic cell changes during the day in a similar way as the attractiveness values change. The current population stock $E_i(t)$ of a traffic cell provides the number of decision makers for the next hour group. This dynamic is required in order to count traffic demand and traffic supply in an appropriate manner. In particular, the determination and estimation of daily traffic peaks can only be carried out using a time-dependent transport model.

The individual transition rates $p_{ji}^{\alpha r}\left(\vec{E}, \vec{\kappa}, t\right)$ represent group related trip rates from one traffic cell to another, where three factor sets are essentially important:

(a) Attractiveness

The attractiveness $u_i^{\alpha}\left(\vec{E}, \vec{\kappa}, t\right)$ of a particular traffic cell depends across-the-board on e.g., the current population distribution $\vec{E}(t)$, the distribution of work places, dwellings and a set of socio-economic factors. The significance of the different socio-economic variables $\vec{\kappa}$ is determined by means of a multivariate regression procedure. Since traffic demand and supply changes during the day, the statistical weight and the significance of the different explanatory variables also changes over time. Obviously, the composition of the particular set of (significant) key variables also depends on the trip purpose α.

(b) Resistance Function

Resistances $g^{\alpha r}\left(w_{ij}^{\alpha r}\right)$ depend on trip purpose α and the transport mode r, where $w_{ji}^{\alpha r}(t)$ represents a generalised resistance or cost parameter for a trip from i to j; $w_{ji}^{\alpha r}(t) = t_{ji}^{\alpha r} + b_1^{\alpha} c_{ji}^{\alpha r} + b_2^{\alpha} K_{ji}^{\alpha r}$, where t_{ji} is the traffic density dependent travel time, c_{ji}^{r} the travel costs, $K_{ji}^{\alpha r}$ appropriate comfort parameters, and $b_1^{\alpha}, b_2^{\alpha}$ the corresponding weight factors. The following resistance function seems to be adequate to describe modal choice behaviour (second term) as well as decisions concerning the time of start of the trip (third term):

$$g^{\alpha r}\left(w_{ji}^{\alpha r}(t)\right) = c_0^{\alpha r}\left(w_{ji}^{\alpha r}(t)\right)^{c_1^{\alpha r}} \frac{\exp\left(-c_2^{\alpha r} w_{ji}^{\alpha r}(t)\right)}{\sum\limits_{s=1}^{R} \exp\left(-c_2^{\alpha s} w_{ji}^{\alpha s}(t)\right)} \frac{\exp\left(-c_3^{\alpha r} w_{ji}^{\alpha r}(t)\right)}{\sum\limits_{t'=t-1}^{t+1} \exp\left(-c_3^{\alpha r} w_{ji}^{\alpha r}(t')\right)}, \tag{9.17}$$

where the parameters $c_0^{\alpha r}, c_1^{\alpha r}, c_2^{\alpha r}$ and $c_3^{\alpha r}$ depend on trip purpose α, the travel mode r and the specific trip hour t, respectively. In a first step, only travel times t_{ij}^{r} are considered a proxy in the resistance function $w_{ij}^{\alpha r}(t)$. In accordance with Hautzinger (1982) and Steierwald and Schönharting (1993), the following simplified resistance function of the gamma type seems to be adequate:

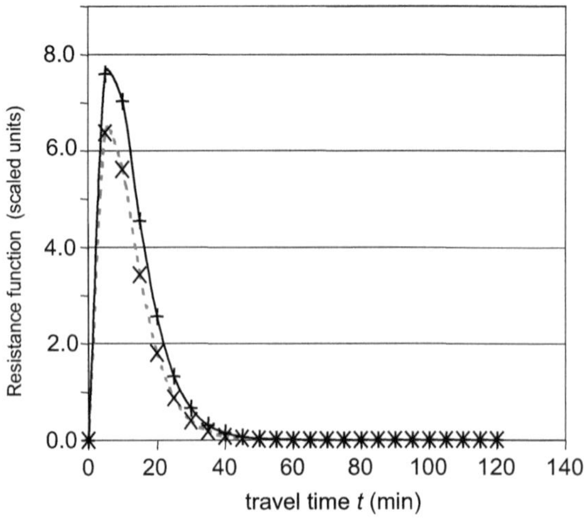

Fig. 9.1 Resistance function (9.18) for car use in the morning hours (x) and in the afternoon for the Stuttgart region (year 1997) (Source: Eurosil 1997)

$$g^{\alpha r}\left(t^r_{ji}\right) = t_{ji}^{c_1^\alpha} \exp\left(-c_2^{\alpha r} t^r_{ji}\right). \tag{9.18}$$

In Fig. 9.1, the resistance function (9.18) is shown for the Stuttgart region. The two curves describe the resistance function for using a car in the morning hours and in the afternoon.

(c) Scaling Factor

A time-dependent scaling parameter $\varepsilon^{\alpha r}(t)$ that correlates with the mean mobility behaviour of the population (e.g., mean number of trips per day) is appropriate.

For the modelling of the traffic flows, the following functional dependence proved to be effective:

$$F^{\alpha r}_{ji}(t) = E_i(t) p^{\alpha r}_{ji}\left(\vec{E}, \vec{\kappa}, t\right) = E_i(t) \varepsilon^{\alpha r}(t) f^{\alpha r} b_t g^{\alpha r}\left(w^{\alpha r r}_{ji}\right) \exp\left(u^\alpha_j\left(\vec{E}, \vec{\kappa}, t\right) - u^\alpha_i\left(\vec{E}, \vec{\kappa}, t\right)\right), \tag{9.19}$$

where the flexibility of the agents to undertake a trip is considered by the scaling parameter $\varepsilon^{\alpha r}(t)$, the width of the hour group by b_t and the parameter $f^{\alpha r}$ indicates the accessibility of traffic mode r.

The described modelling of the traffic flows is based on the general experience that agents compare the "attractiveness" of the traffic cells with respect to the trip purpose α and that the probability for a trip $i \to j$ increases with increasing difference $\left[u^\alpha_j\left(\vec{E}, \vec{\kappa}, t\right) - u^\alpha_i\left(\vec{E}, \vec{\kappa}, t\right)\right] > 0$ of attractiveness per unit of time, and simultaneously the trip probability into the opposite direction $j \to i$ decreases. In this sense, the attractiveness differences are the "driving forces" of the traffic system.

The macroscopic traffic flows are usually different even for identical attractiveness of the traffic cells since they have to be multiplied by the population numbers

E_i and E_j, respectively. On the other hand, a simultaneous increase of the attractiveness of all traffic cells by a common factor may not change the trip distribution. This is guaranteed by the normalisation of the attractiveness

$$\sum_{i=1}^{L} u_i^{\alpha}\left(\vec{E}, \vec{\kappa}, t\right) = 1. \tag{9.20}$$

The transportation system is in its (macroscopic) balance if simultaneously for all traffic cells the sum of inflows into a traffic cell equals the sum of outflows.

The internal traffic within a traffic cell is assumed to be:

$$F_{ii}^{ar} = E_i p_{ii}^{ar}\left(\vec{E}, \vec{\kappa}, t\right) = E_i \varepsilon^{ar}(t) f^{ar} b_t g^{ar}\left(w_{ii}^{ar}\right) \exp\left(\gamma^{\alpha} u_i^{\alpha}\left(\vec{E}, \vec{\kappa}, t\right)\right), \tag{9.21}$$

where the parameter γ^{α} weights the attractiveness value of the traffic cell i.

As expected, the internal traffic of a cell i depends directly on the corresponding socio-economic data of that particular cell and on the data of all other cells via the normalisation of the attractiveness.

The dependence of the attractiveness $u_i^{\alpha}\left(\vec{E}, \vec{\kappa}, t\right)$ of the traffic cells on the population distribution $\vec{E}$ and of further socio-economic variables $\vec{\kappa}$ has the consequence that the evolution equation for the population redistribution by means of traffic becomes a non-linear differential equation or difference equation. In this case, nonlinearities reflect the complexity of the involved decision processes. Depending on the initial conditions, i.e., the population distribution at a given time t and the further system parameters of the urban/regional areas, the nonlinear dynamics can lead to a complex variety of self-organising traffic flows (Nijkamp and Reggiani 1992). Consequently, the dynamics of the population distribution is described by traffic flows and feedback effects resulting from the land-use patterns (Goodwin 1994, 1995).

This transport model contains trip-chains in an aggregated form. Here, a decisive difference appears to traditional traffic models since the hourly (day time-controlled) redistribution of the agents can be considered and simulated (Binder et al. 2002). More precisely, the so called four-stage-algorithm is not able to describe trip-chains as well as time-of-day dependent effects. Because the decisions concerning the transport system depend on the current distribution of the agents, a dynamic redistribution of the population is not only plausible; it improves simultaneously the model quality. It is a further advantage of this sub-model that traffic generation and traffic redistribution occur at one model step, simultaneously.

9.1.5 Parameter Estimation of the Traffic Model

The system parameters such as mobility, attractiveness and the parameters of the resistance function can be directly linked to the trip decision processes using optimisation procedures. For this aim the transport model has to be matched to the empirical traffic flow matrices $F_{ij}^{are}(t)$ (index e), the population numbers $E_i^e(t)$ and travel time matrices t_{ij}^r, which in turn depend on the traffic volume of the hour groups, respectively. The minimisation of the functional (9.22)

$$H_t\left[\varepsilon^{ar}, c_1^{ar}, c_2^{ar}, \vec{u}^{\alpha}\right] = \sum_{i,j=1}^{L} \left[F_{ji}^{are}(t) - \varepsilon^{ar}(t) b_t E_i^e(t) g^{ar}\left(w_{ji}^{ar}\left(t_{ij}^r\right)\right) \exp\left(u_j^{\alpha}(t) - u_i^{\alpha}(t)\right)\right]^2 = \text{Min}, \tag{9.22}$$

enables to calculate the attractiveness values $u_i^{\alpha}(t) \equiv u_i^{\alpha}\left(\vec{E}, \vec{\kappa}, t\right)$ of the different hour groups t, and all further system parameters for each trip purpose α and trip mode r. Depending on the actual time, the trip purpose distribution changes within the (macroscopic) traffic flows. For example, trips between home-work and work-home may predominate in the morning hours and the late afternoon, while leisure time trips or shopping trips show another temporal distribution over the day. Therefore, the estimated parameters depend on the trip purpose and time of day. The iterative determination of the travel time via the traffic assignment procedure is shown in Fig. 9.2. Depending on the number of cars and the other modes of the transport network (network of streets) the travel time changes. Therefore, a number of iteration steps are needed to assign the travel times of the different modes and the density of those transport modes on the street network in a consistent way.

The convergence behaviour of the transport sub-model for a typical application is shown in Fig. 9.3.

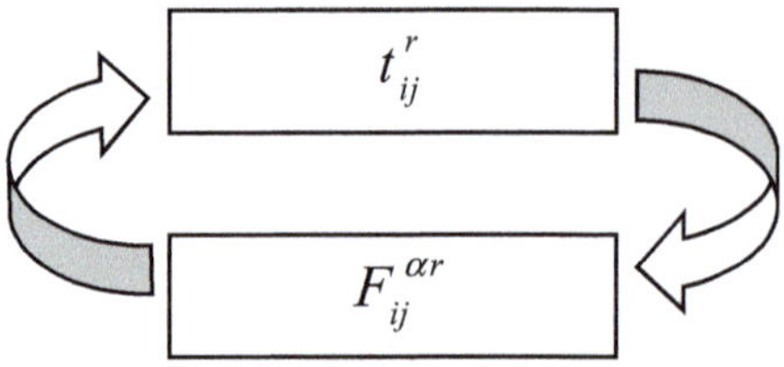

Fig. 9.2 Iterative determination of the travel time via traffic assignment procedure (Source: Eurosil 1997)

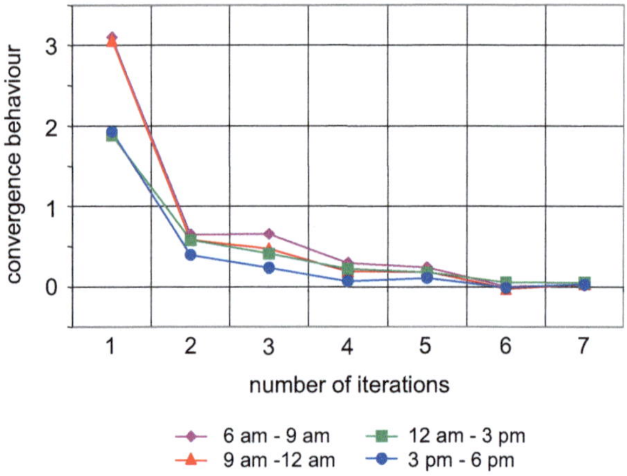

Fig. 9.3 Convergence behaviour of the transport sub-model for different time hours of the day (Source: Eurosil 1997)

The estimated attractiveness values obtained in the first step may now depend on traffic cell specific key-factors κ_i^n, $n = 1, \ldots$ gained from two different fields: on the one hand, from the class of the so-called synergy variables, describing general group effects, such as pigeons effects and on the other hand, from a sequence of potential attractiveness indicators, e.g., number of jobs available, number of vacant dwellings, regional income per capita, shop distribution and other local infrastructure depending factors. The set of key-factors (selected socio-economic variables) is determined via a multiple regression analysis in a second step:

$$u_i^\alpha(\vec{E}, \vec{\kappa}, t) = \sum_n b_n^\alpha(t)\kappa_i^n. \tag{9.23}$$

The elasticity's b_n^α, assigned to the socio-economic variables κ_i^n, are dimensionless numbers and indicate the influence of the independent variables on the dependent variable. The selection of relevant indicators is performed using appropriate statistical characteristics (T-values and other significance tests).

However, up to now, the portrayed formalism describes only short-term effects like redistribution of the population due to traffic events such as commuting and shopping trips. Long-term population redistribution effects, e.g., migratory events of the population between cells as well as changes in the land-use pattern have to be considered, too.

9.1.6 *Migration Flows and Population Development*

Due to transport processes, there is a redistribution of the population over the day within the urban system (short-time process) $E_i(t)$, described by the transport sub-model. In the long-run, by migration events and because of the demographic development, the "housing population", denoted by $n_i(\tau)$, changes as well, where τ describes the year under consideration. This slowly changing housing population $n_i(\tau)$ provides the starting conditions for the short-time population redistribution

$$E_i(t = 6\ \text{am}) = E_{i0} = n_i(\tau), \tag{9.24}$$

of traffic cell i, and $n_i(\tau) = \langle n_i \rangle_\tau$ the corresponding mean value of the residents.

The analysis and forecasting of the long-time population redistribution $n_i(\tau)$, via the Master equation approach is described in detail in Chap. 8. The migration flow between i and j is given by (8.105)

$$w_{ji}(\tau) = n_i(\tau)p_{ji}(\vec{n}, \vec{z}) = v_0(\tau)f_{ji}(d_{ij})\sqrt{n_i(\tau)n_j(\tau)}\exp(\delta_j(\vec{z}) - \delta_i(\vec{z})), \tag{9.25}$$

where $\vec{z}$ represents a set of explanatory variables z_γ, $\gamma = 1, 2, \ldots, G$, related to different significant socio-economic influences, $v_0(\tau)$ is a general mobility indicator of the population, and the distance-dependence of the migration flow is described by the deterrence function $f_{ji}(d_{ij})$. The preference of the cell (urban region) i is introduced by $\delta_i(\vec{z})$. The long-time intervals (e.g., 1 year) are indicated by τ. In order not to underestimate long-distance moves, the deterrence function (8.42) is recommended.

The long-time population redistribution reads

$$n_j(\tau+1) = n_j(\tau) + 2\nu_0 \sum_{i=1}^{L} f_{ij}(d_{ij}) \sqrt{n_i(\tau) n_j(\tau)} \sinh\left(\delta_j(\vec{z},\tau) - \delta_i(\vec{z},\tau)\right) + \rho_j(\tau), \tag{9.26}$$

where the birth- and death-processes as well as immigration and emigration events are described by $\rho_i(\tau)$, respectively.

In a second step, the estimated cell specific attractiveness values $\delta_i(\vec{z},\tau)$ are linked to particular location factors (key-factors) gained from a sequence of potential explanatory indicators, e.g., number of available jobs, number of vacant dwellings, regional income per capita, shop distribution and other local infrastructure depending factors. The set of explanatory variables and their corresponding elasticity's are determined via a multiple regression analysis

$$\delta_i^\alpha(\vec{z},\tau) = \sum_\gamma a_\gamma^\alpha(\tau)\, z_i^\gamma. \tag{9.27}$$

The elasticity's $a_\gamma^\alpha(\tau)$, assigned to the socio-economic variables, are dimensionless numbers and indicate the influence of the independent variables on the dependent variable. The selection of relevant indicators is performed using appropriate statistical characteristics (T-values, other significance tests).

The migratory preferences $\delta_i(\vec{z})$ have to be distinguished from the attractiveness parameters $u_i^\alpha(\vec{E},\vec{\kappa})$ involved in the transport model, even if some "driving forces" (key-factors) seem to be relevant for both sub-models.

The explanatory variables are related to indicators belonging to the following sub-markets:

- Labour market (number of work places, wages, unemployment rate, etc.)
- Housing market (number of apartments, dwellings, houses, price of land, rent level, etc.)
- Accessibility indicators (depending on travel time, travel costs, parking possibilities, transport infrastructure, etc.)
- Other specific indicators (availability of different services, neighbourhood, environment, recreation possibilities, etc.)

Therefore, as a consequence, the urban sub-model is linked to the traffic sub-model via accessibility measures depending on travel time, and vice versa, the traffic sub-model is linked to the urban sub-model via its coupling to the changing population stock of the traffic cells.

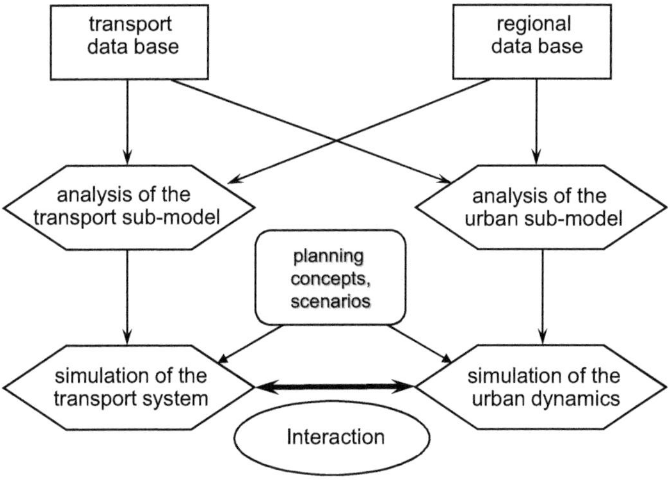

Fig. 9.4 Principle structure of the nested transport and urban model (Source: Weidlich and Haag 1999)

Moreover, the impact of commuting on the population redistribution is of importance (Fig. 9.4). Depending on the initial conditions, such as the distribution of population at a given time and the further system parameters, the non-linear dynamics lead to self-organised commuter flow pattern (Nijkamp and Reggiani 1992; Goodwin 1994).

9.1.7 The Step-vice Application Procedure

In transport modelling different scenarios have to be considered:

(a) basic scenario, describing the current traffic situation
(b) planning scenarios, describing possible realizations of modifications of the transport network or changes in the housing distribution, e.g., because of the planning of new dwellings

Especially environmental impacts of the different scenarios have to be evaluated and compared, e.g., CO_2 emissions and traffic performance indicators.

In the following eight steps, the procedure is summarized in schematic terms:

Step 1: Traffic assignment
In the first step, the different traffic flows $F_{ij}^{\alpha r}(t)$ are assigned to the corresponding traffic network for the "basic scenario". This determines the travel times t_{ij}^{r} between the single cells as well as accessibility measures of the cells for the considered modes (according to Fig. 9.2).

In the same way, the travel time matrices t_{ij}^{r} (*iteration 1*) and the corresponding accessibility measures are determined for the traffic flows $F_{ij}^{\alpha r}$ of the different planning scenarios and the corresponding network of streets.

Step 2: Estimation of the traffic sub-system
Given the traffic flows $F_{ij}^{\alpha r}(t)$ of the "basic scenario", the corresponding travel times t_{ij}^{r} and the related population numbers $E_i(t)$, the parameters of the transport sub-model are estimated by the non-linear estimation procedure (9.22):

- attractiveness "basic scenario" $u^{\alpha r}(t)$ of the cells i
- parameters $\gamma^{\alpha r}$
- mobility parameters ε^{r} for the modes r and
- parameters of the resistance function b^{r} and c^{r}.

Statistical tests (correlation, Fisher's F-value) are used to evaluate the quality of the parameter estimation.

Step 3: Multiple regression attractiveness parameters (reference scenario)
By means of a multiple regression, the dependence of the attractiveness values of the traffic cells on characteristics of the cells are determined. That means the corresponding parameters $b_n^{\alpha}(t)$ as well as statistical test measures are calculated to select the explanatory variables of the traffic sub-system.

Step 4: Analysis of the population distribution (basic scenario)
To analyse the population distribution and its dependence on the spatial preferences of the cells $\delta_i(\tau)$, the spatial preferences $\delta_i(\tau)$ and v_0, f_{ij} are calculated according the non-linear optimization procedure (8.130). This yields:

- spatial preferences "basic scenario" $\delta_i(\tau)$ of the cells i
- interaction parameters $f_{ij}(\tau)$
- mobility parameters $v_0(\tau)$

Statistical tests (correlation, Fisher's F-value) are used to evaluate the quality of the parameter estimation.

Step 5: Multiple regression spatial preferences (reference scenario)
By means of a multiple regression, the dependence of the spatial preferences $\delta_i(\tau)$ of the cells on the characteristics of the cells are determined. That means the corresponding parameters $a_\gamma^{\alpha}(\tau)$ as well as statistical test measures are calculated to select the explanatory variables of the urban sub-system. The influence of the spatial preferences of the cells for the different "planning scenarios" can be determined by a multiple regression using accessibility measures for the "planning scenarios".

Start of the Iteration Procedure
In the next steps, we will describe an iteration procedure (Steps 6–8) to forecast the evolution of the traffic sub-system and the urban sub-system for the planning scenarios.

Step 6: Simulation of the population development (planning scenario)
If it is known which explanatory measures are significant for the spatial preferences, in the next step, these preferences and therefore the changed population distribution $n_i(\tau)$ (*iteration 1*) can be calculated for the planning scenarios, using (9.26) for the planning horizon τ. If explanatory measures for the spatial preferences are not available, but we can identify and estimate a tendency of evolution for the preferences, Eq. (9.26) can also be used to estimate the temporal evolution of the spatial sub-system $n_i(\tau)$.

Step 7: Simulation of the traffic flows (planning scenario)
In this step, the (transport) attractiveness values $u_i^{\alpha r}(t)$ (*iteration 1*) of the cells for the "planning scenarios" will be calculated, using their possible dependence on population values and accessibility measures.

The traffic flows $F_{ij}^{\alpha r}$ (*iteration 1*) are determined for all modes r with these attractiveness values $u_i^{\alpha r}(t)$ (*iteration 1*), with the travel times t_{ij}^r (*iteration 1*) and the population numbers $E_i(t)$ (*iteration 1*). The mobility parameter and the parameters of the resistance function are assumed to be constant.

Step 8: Assignment of the traffic flows
The traffic flows $F_{ij}^{\alpha r}$ (*iteration 1*) are now assigned for the "planning scenarios" to the corresponding street network to get the new travel times t_{ij}^r (*iteration 2*) and the new accessibility measures. With these new travel times and new accessibility measures for each mode, the population configuration (dependent on the accessibilities) and the traffic flows change again, so that the next iteration step applies.

To accelerate this iteration procedure it is recommended to average the new travel times t_{ij}^r (*iteration 2*) and the travel times of the step before t_{ij}^r (*iteration 1*).

End of the Iteration Procedure
Steps 6–8 are repeated till the change of the flows and travel times can be ignored, in other words, when the procedure has converged. As a result, the traffic flows $F_{ij}^{\alpha r}$ for the single modes r and trip purposes α for the "planning scenarios" are obtained.

Step 9: Results
The comparison of the traffic flows $F_{ij}^{\alpha r}$ for the "planning scenarios" with the "basic scenario" shows the impact of the "planning scenarios" on modal split, traffic volume, travel time, environmental impacts and urban development.

The general framework in addition to the methodology so far is presented in Fig. 9.5.

The following example underlines the principle functioning of this integrated transport and evaluation network. This example does not aim to show and discuss all results in detail, especially regression results and their meaning. Many transport and land-use examples have been analysed, modelled and simulated at STASA during the last two decades. Hence, everybody who is interested in learning more about the application of this model to real-world tasks is recommended to read the outcomes of several European and National projects related to transport and urban

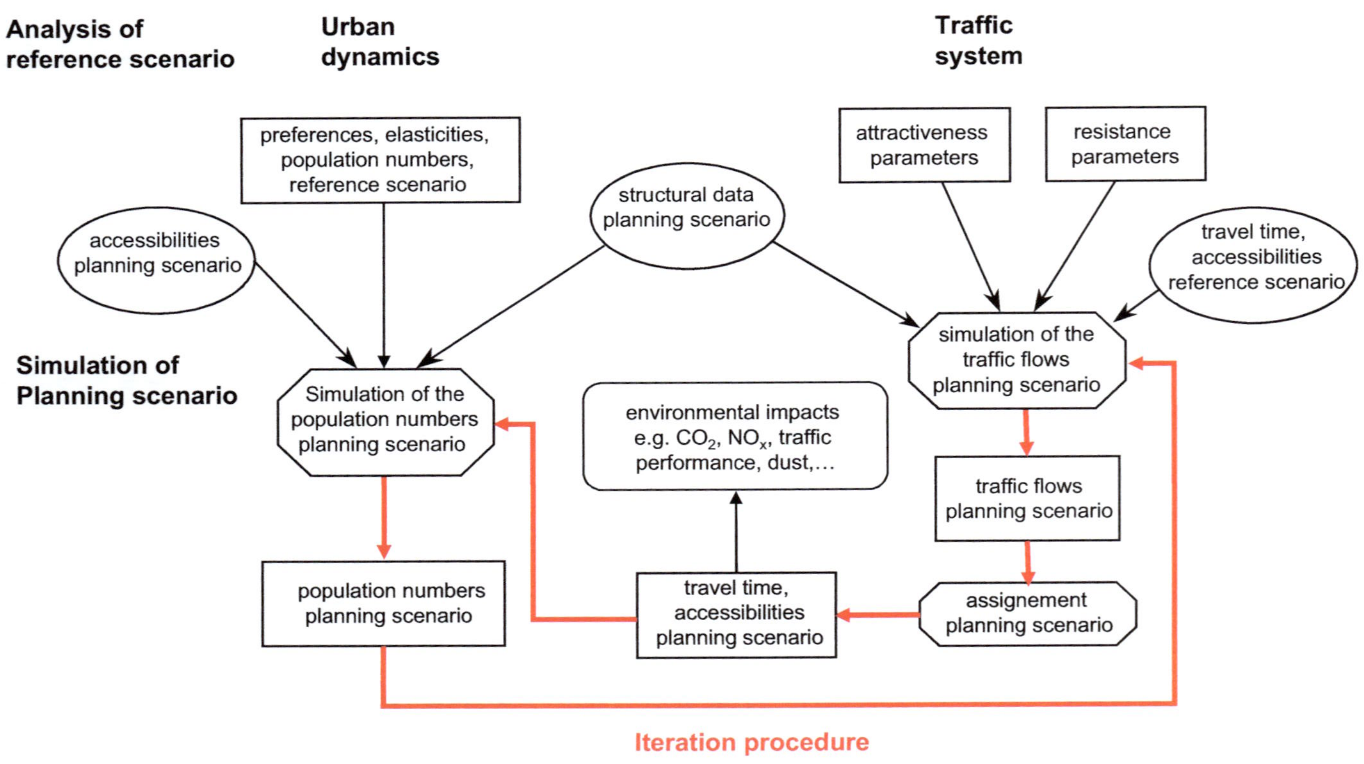

Fig. 9.5 The general framework of the nested transport and urban model for the simulation of the effects of the planning scenario (Source: Weidlich and Haag 1999)

development (e.g., Bundesminister für Verkehr 1996, 1998; Eurosil 1997; Scatter 2001).

9.1.8 The Transport Network Within the Region of Stuttgart

The Study Region

The Region of Stuttgart is situated in the south-west of Germany and covers five districts called Boeblingen, Esslingen, Goeppingen, Ludwigsburg and Rems-Murr, and the City of Stuttgart with a total of 179 communes. The overall population of the Region of Stuttgart is about 2.7 million inhabitants. Together with its state capital, the City of Stuttgart, it represents the economic and cultural centre of the state of Baden-Württemberg.

The City of Stuttgart represents an important traffic node within the Trans-European Network (TEN). With regard to the traffic infrastructure network, elements on all spatial levels are presented: in private vehicle traffic, they range from city streets to motorways: in public transport from city buses, rail rapid transit (S-Bahn), subway (U-Bahn) to high-speed trains (ICE). All modes are present: pedestrian, bicycle, bus, light rail transport, suburban rail, car, as well as intermodal transport opportunities like park-and-ride (P&R).

The traffic subsystem as well as the urban/regional subsystem form a complex intertwined system. Its dynamics take place on different time scales but are modelled making use of the same principles:

The daily flows of traffic in the region of Stuttgart are the result of very quick decision processes of the individuals to realize a trip between two traffic cells (origin–destination) with a special trip purpose. Decision processes for a certain destination, the moment of the setting out, the mode of transportation, the choice of the route, etc. take place on a very short time scale.

The development of the urban/regional subsystem, namely the development of the spatial population distribution, is a process on a long-term time scale. The population distribution changes because of migration acts and birth-death events of individuals. The equations of motion contain transition rates between the cells. The assigned flows depend among others on accessibility measures. This leads therefore to a coupling of the transport and urban/regional subsystem.

Commuting offers the possibility to choose a new home by retaining the workplace and vice versa. This is often combined with the acquisition of real estate or the improvement of working conditions by retention of the place of work without simultaneously losing the advantages of an existing residence (Steierwald and Künne 1993). Therefore, the imbalance between labour demand and labour supply on the level of communities is partially compensated by the commuter dynamics. The development of regional employment, the regional gross wage payment and parts of the community revenues are strongly dependent on the spatial distribution of commuter flows (Binder 2007; Albeverio et al. 2008).

The motorway A81 represents an important link of the Trans-European Network (TEN), especially concerning the international traffic in the Northern-Europe-Germany-Switzerland-Italy chain. Both Schemes (A81 and S1) run in parallel and concern the link of the capital to villages and new settlements far from the city centre of Stuttgart by radial elements. Due to urban and regional traffic, the transport network, especially the A81 is highly overburdened in the outskirts of Stuttgart. The consequences are overloading (congestion) leading to serious obstructions in long-distance traffic as well as an increase of the environmental pollution in the area of the Stuttgart case study.

There exists a well constructed secondary network of roads connecting the City of Stuttgart with sub-centres of the other five state districts of the Region of Stuttgart. The interactions within the traffic system include traffic diversions from the secondary road network as well as extensive diversions from parallel mayor roads.

In the Stuttgart case study of the European project Eurosil (1997), the causes for the regional traffic are related to decisions of individuals and interactions within the traffic system as well as with the settlement structure. The development of the population in the examined corridor in the nearer Stuttgart area (see Fig. 9.6) is characterised by strong growth in the periphery of the city. In addition, enormous growth in industrial and business zones of cities along the A81 corridor can be observed accompanied by an increase of workplaces and a redistribution of workplaces from the city centre of Stuttgart towards the outer urban ring.

The impacts of the following transport elements are considered:

- The A81 motorway Stuttgart-Singen (Fig. 9.6)
- The railway line S1 Stuttgart-Herrenberg

The "empirical data base" consists of traffic flows (all modes) between the traffic cells as well as population numbers and migration flows on a yearly base between all cells and data on the traffic network.

Basic Scenario: With A81, Without S1

The basic scenario describes the situation in 1990 when the motorway A81 was already open but the rail rapid transit S1 was not in use yet. Data exist for this case, containing traffic flows and travel time matrices of the individual as well as public transport, population numbers, numbers of employees and other characteristics (location factors) related to the traffic cells (communities).

Given the transport flows, there is a redistribution of the population over the day (short-time process), described by $E_i(t)$, where t is the time of the day. In the long-term, driven by population aging and migration, the housing population, denoted by E_{i0}, changes as well. This, in most of the applications slowly changing housing stock, provides the starting conditions for the short-time redistribution $E_i(t = 6\ \text{am}) = E_{i0} = n_i(\tau)$.

These data base is used to estimate the system parameters of the transport and urban/regional subsystems, e.g., "transport attractiveness" and "resistance"

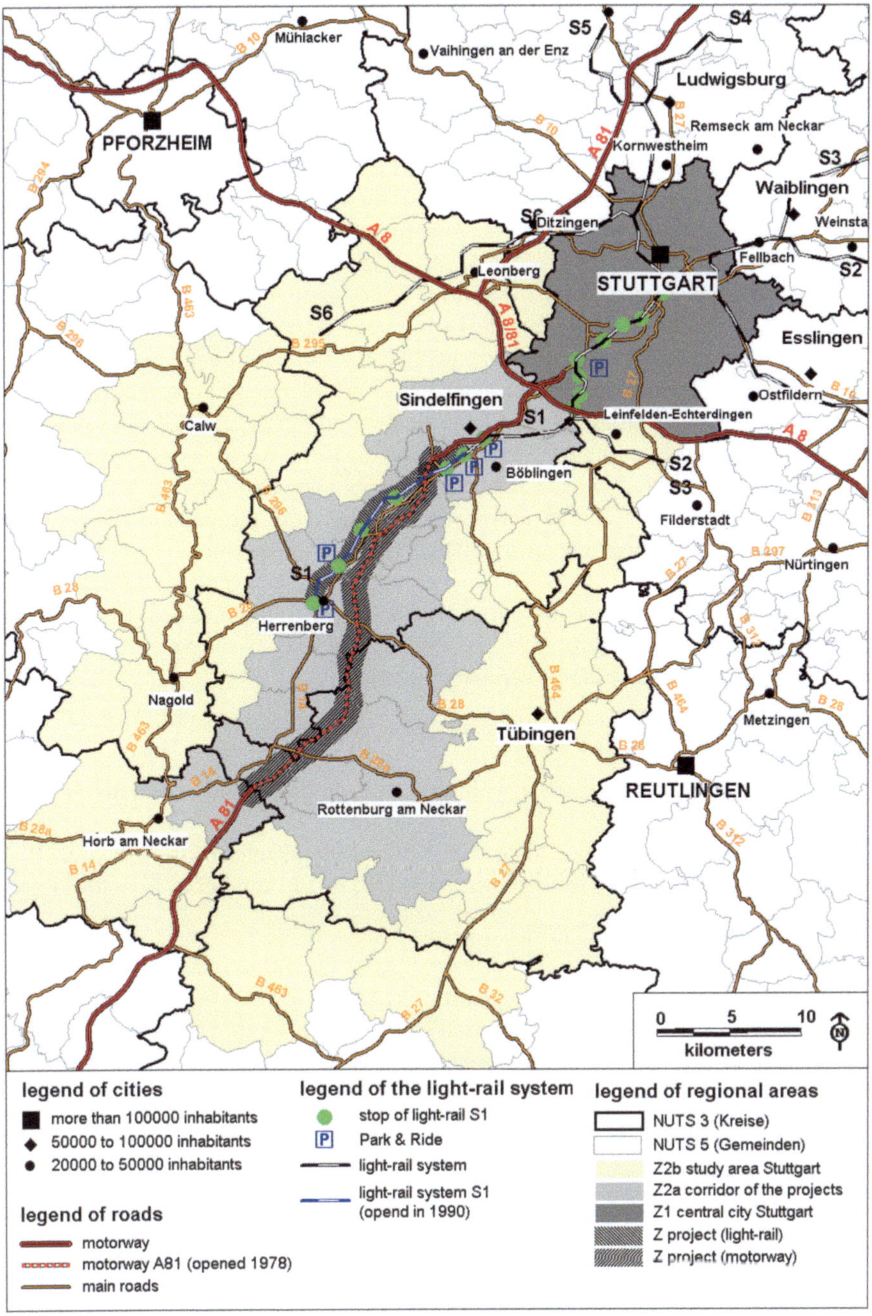

Fig. 9.6 Area of the Stuttgart case study (Source: Eurosil 1997)

parameters for the transport subsystem and "regional preferences" for the urban/regional subsystem as already described.

A comparison of the empirical flow matrix with the modelled flow matrix, calculated with the STASA transport model underlines the quality of the estimation. The result of the estimation for the basic scenario shows a high significance with a correlation coefficient of $R^2 = 0.87$, *Fischer's F* $= 703.4$ (Eurosil 1997).

The estimated resistance function (9.18) for public transport exhibits higher travel times combined with a smaller overall value compared with the resistance function for private transport. This leads to fewer trips being made by public transport compared with private transport. The transport attractiveness of the traffic cells in the region of Sindelfingen/Boeblingen, where many work places are located, is above the average. This means that the probabilities for a trip in those cells are favoured. This is plausible, since in the considered time interval from 6 am to 10 am, the trip purpose home-work dominates and many people drive from their place of home in the south of the study area towards the work places located in the northern part of the study area Sindelfingen/Boeblingen and the City of Stuttgart.

In a further step, these "attractiveness" values and "preferences" have been connected with appropriate macro-variables (key-variables) making use of a multiple regression.

The attractiveness values of the traffic cells for the traffic model are related to a number of indicators by multiple regressions. On the basis of the 58 traffic cells, the following (almost independent) variables are used:

- Population size
- Number of employees
- Number of persons accessible within 30 min by individual transport (IT) and public transport (PT)
- Number of employees accessible within 30 min by IT and PT
- Indicator for travel time per distance, weighted by the number of trips for IT and PT

These variables were normalised according to the attractiveness values of the cells. The variable showing the highest significance is the number of employees, as expected, for the considered time interval from 6 am to 10 am. The population size of the cells has only a low positive significant impact as well as the accessibility measures. These variables have been used for the simulation of the impacts of the different planning scenarios.

Planning Scenarios

In comparison to the basic scenario

- **Basic scenario:** with A81, without S1

  the following planning scenarios are considered:

- **Planning scenario A:** without A81, without S1
- **Planning scenario B:** with A81, with S1
- **Planning scenario C:** without A81, with S1

For these planning scenarios, changes in the transport system and population distribution have been calculated, following the already described step-vice application procedure of Sect. 9.1.7.

Assessment of the Impacts

The assessment of the impacts of the three planning scenarios on traffic related indicators is shortly presented next.

Population distribution

For the cases without A81, the population density in the area Boeblingen/ Sindelfingen, where the A81 ended before 1978, is higher than in the case with A81. For the area of Herrenberg, the population density without A81 is not as high as in the case with A81. The effect of the motorway on the population distribution is therefore very strong, and a corresponding strong shift of population in the direction of the community of Herrenberg, caused by the extension of the A81, can be found. This result confirms that the extension of radial transport lines, accompanied with a drastic reduction of travel times to the city centre, support urban sprawl.

For the scenario with A81 and with S1, the population density increases a little compared to the reference scenario in the area along the light rail system S1, i.e., in areas in vicinity to the train stops of the S1. The effects caused by the S1 on population distribution are not as strong as those caused by the motorway. In other words, the radial extension of the public transport system S1 has a rather moderate effect on population redistribution compared with the extension of the motorway A81. This indicates that the extension of the radial transport axis and the improvement of accessibility indicators in the hinterland accelerate the effect of urban sprawl in the Stuttgart region.

Transport system

If the attractiveness of a community (traffic cell) is high compared with other communities, the probability for a trip into this community increases. For the reference scenario, the attractiveness of the communities in the areas of Stuttgart and Boeblingen/Sindelfingen is high. For the planning scenarios without A81, the attractiveness of Boeblingen/Sindelfingen increases by at least 5–10%, whereas the attractiveness of the area of Herrenberg decreases compared to the reference scenario. This leads to corresponding shifts in the traffic flows for the time-group 6 am to 9 am, respectively.

Due to the increase in the traffic volume and the overall average travel distance, the average travel time increases in the study area as well. As public transport increases, private vehicle traffic also increases. The modal split shifts towards private transport (car) due to the road construction measure A81.

The construction of the line S1 increases public transport use and thus shifts the modal split towards public transport. Through diverting public transport trips from bus transport in the Herrenberg area to the rail, the extension of the S1 towards Herrenberg leads to a clear decrease in public transport mileage as well as of average travel distances in public transport. A slight decrease in the mileage of

Table 9.1 Traffic effects in the study area

		Basic scenario: (with A81, without S1)	Planning scenario A: (without A81, without S1) (%)	Planning scenario B: (with A81, with S1) (%)	Planning scenario C: (without A81, with S1) (%)
		Change in relation to the basic scenario			
Passanger trips	Private vehicle traffic	276,000	−14.8	+0.2	−14.6
[trips/24 h]	Public transport	85,000	−4.3	+1.9	+3.5
Modal-split [%]		23.5%	+9.2	+1.3	+15.4
Milage [km/24 h]	Private vehicle traffic	2,906,000	−23.0	−0.4	−21.7
	Public transport	775,000	−7.6	−27.1	−24.5
Average travel distance [km]	Private vehicle traffic	11	−9.6	−0.7	−8.3
	Public transport	9	−3.5	−28.5	−27.1
Average travel time [min]	Private vehicle traffic	10	−1.9	−0.4	−2.0
	Public transport	26	−3.0	−3.8	−2.7

private vehicle traffic can also be registered, which is caused by the construction of the S-Bahn S1.

Assignment Results

A few assignment results for the different planning scenarios for the Stuttgart case study are shown in Table 9.1.

Conclusions

The STASA transport model is based on the framework of the Master equation approach. It is behaviourally oriented, highly nonlinear and containing several links between the traffic subsystem and the urban/regional subsystems. Contrary to the classical gravitational approach and/or four-stage model, where trip generation, distribution, mode-choice and assignment are sequentially treated, the STASA model proved to be able to estimate not only directly induced traffic components but also indirect effects. Attractiveness indicators are assigned for each trip purpose to all traffic cells, representing their socio-economic structure, e.g., number of work places, inhabitants, leisure time possibilities, to mention a few. This leads to a strong coupling between the traffic level and the urban/regional subsystem. In this way, the time-of-the-day dependent redistribution of the population between the traffic cells can be calculated, leading to changing population numbers over the day and therefore to a changing population stock for the next following traffic decision of the agents. The long-time evolution of the place of home population, which represents the starting population in the morning hours for the daily traffic redistribution, is also influenced by changing urban/regional attractiveness indicators as well as by migration and birth-death events. This completes the STASA model, which, therefore, corresponds to an aggregated trip-chain model.

9.2 Spatial Interaction Models and Their Micro-Foundation

In this chapter, the model building framework of the Master equation is once more used for the construction of spatial interaction models. It will be outlined how some of the usual limitations in model building can be circumvented. Our broader overall framework with full probabilistic dynamic underpinning also facilitates extensions in a number of directions. For instance, it becomes obvious how conventional spatial-interaction models can be derived and generalized, especially how fundamental dynamic aspects can be introduced (Rabino and Occelli 1999; Pumain et al. 1989). The decision based modelling offers a new approach to parameter estimation, and to see and understand how nonlinearities enter the final macro-equations of motion.

The early urban models of Lowry (1964), Hill (1965), and Forrester (1968) have often been criticized and essentially ignored by economists. The main point of criticism stressed that they were not based on economic theory (Mills 1972; Nelson and Winter 1982). Practical questions such as doing cost–benefit analysis or predicting the response of real estate prices to public investment could not be answered. However, these models initiated a great amount of creative work in this field. The pioneering work in modelling dynamics of Forrester (1971) and his group and their cooperation with *The Club of Rome* underlined in their first book "The Limits to Growth" (Meadows et al. 1974) the basic premises and the dynamic characteristics of exponential growth and the physical limits imposed by a finite planet. A creative rethinking in policy and economics partly started and led to an enormous effort in modelling of more economically adapted huge complex systems (Meadows and Meadows 1973; Dosi et al. 1994).

Allen and Sanglier (Allen and Sanglier 1978a, b; 1979) introduced a spatial-interaction model where jobs and populations of different kinds follow a logistic path in each region and compete with each other via corresponding capacities. The model includes concepts like capacities, distance effects and some economic aspects. Of course, such a logistic model structure could be derived from a stochastic background theory, as the Master equation, assuming a birth/death structure of the transition rates. But there are important processes in spatial dynamics that cannot be fitted into this restrictive birth/death frame, since flow processes such as migration or other transport phenomena are important as well.

The spatial-interaction model of Wilson (1970, 1971) and Harris and Wilson (1978) is also of a logistic type. The imbalance between revenue attracted to a region and the cost of supply to maintain a certain level of the facility stock leads to changes in the level of the facility stock. Contrary to the model of Allen and Sanglier, however, Harris and Wilson take into account expenditure flows, but neglect some important economic factors which are included in the first model. The expenditure flows are directly linked to the facility stock assuming a distance decreasing interaction term.

These important papers raised the interest of geographers and economists to apply these models and ideas to real-world topics. Three main problems became obvious:

- difficulties in calibration, since the number of trajectories which can be compared with empirical data is smaller than the number of fitting parameters
- the models focus on the supply side. There are no dynamic equations for the demand-side describing reactions of the expenditure flows to changes in the facility stock
- the lack of a micro-economic foundation.

Our aim is to illustrate the main construction principle of spatial-interaction models, using the Master equation framework. Therefore, we start with a simple example; however, the approach can be easily extended in a straightforward manner.

9.2.1 A Service System as Basis of a Spatial-Interaction Model

Our starting point is the dynamic decision process of individuals who decide to use facilities of different regions as well as of the entrepreneurs, retailers and landowners who decide on changing the facility stock, the process of goods and services and the land rents (see Table 9.1). We shall not use any equilibrium assumption in deriving the fundamental equations of motion for the relevant macro-variables of the system (Haag 1986, 1989a, b; Bertuglia et al. 1990).

The different decision processes are modelled using the Master equation approach. Uncertainties in the decision processes or fluctuations are introduced via the Master equation. As a result, we obtain the time-dependent probability that a certain decision configuration is realized. The framework also provides the link between micro-economic considerations and the macro-economic equations for aggregate mean values. These mean values are obtained from the Master equation as described in the Chaps. 3 and 4.

Table 9.2 The different agents of the service sector model

Agent	Decision choice	Controlled variable	
Consumer	Destination choice	Expenditure flow	$T_{ij}(t)$
Developers	Facility size	Facility stock	$Z_j(t)$
Retailers	Prices of goods and services	Price level	$p_j(t)$
Land owners	Rent of land	Rent level	$r_j(t)$

We consider four types of agents (see Table 9.2), whose decisions control destination choice (consumers), facility size (developers), prices of goods or services (retailers) and land rent (land owners). For the purpose of this chapter, we will assume residential location to be fixed, though this assumption can easily be relaxed, as shown in Chap. 8.

Let us consider a discrete urban system consisting of L zones connected with a given transport network. The transportation costs between zone i and zone j are denoted by c_{ij}. The expenditure configuration is denoted by $\underline{T}(t) = \left(T_{ij}(t)\right)$ at time t. The $T_{ij}(t)$ describes expenditure flows from a zone i (of consumer's residence) to zone j (where the consumer is using facilities) at time t. We assume that this array is determined by utility functions $v_{ij}(t)$ through a mechanism to be described later. Moreover, the utility functions $v_{ij}(t)$ have two subscripts since they relate to both, housing and service sector.

The total revenue $D_j(t)$ attracted to each zone j is obtained by

$$D_j(t) = \sum_{i=1}^{L} T_{ij}(t) \tag{9.28}$$

and the total expenditure $E_i(t)$ of consumers living in zone i is

$$E_i(t) = \sum_{j=1}^{L} T_{ij}(t). \tag{9.29}$$

The utility functions $v_{ij}(t)$ are assumed to depend on supply side variables. Let $Z_j(t)$ be the scale of provision of facilities in zone j at time t, let $p_j(t)$ be the unit price of the service (assuming one type of good or services for simplicity) and let $r_j(t)$ be the land rent to be paid per unit of facilities provided at zone j.

Therefore, the modelling task is to determine the expenditure flows $T_{ij}(t)$ at time t, the spatial and temporal pattern of provision $Z_j(t)$, prices $p_j(t)$ and rents $r_j(t)$ for given initial conditions $T_{ij}(0)$, $Z_j(0)$, $p_j(0)$ and $r_j(0)$.

9.2.2 The Master Equation of the Service Sector Model

Decision processes are stochastic processes since we are interested in the probability $P(\vec{c},t)$ that a configuration $\vec{c}=\left(\underline{T},\vec{Z},\vec{p},\vec{r}\right)$ is realized at time t. The corresponding Master equation reads:

$$\frac{dP(\vec{c},t)}{dt}=\sum_{\vec{k}}w\left(\vec{c},\vec{c}+\vec{k}\right)P\left(\vec{c}+\vec{k},t\right)-\sum_{\vec{k}}w\left(\vec{c}+\vec{k},\vec{c}\right)P(\vec{c},t) \qquad (9.30)$$

with the normalization condition

$$\sum_{\vec{c}}P(\vec{c},t)=1. \qquad (9.31)$$

The Master equation shows how this distribution changes over time. In this special case, we make the justifiable assumption that the probability distribution can be factorized as follow:

$$P\left(\underline{T},\vec{Z},\vec{p},\vec{r},t\right)=P\left(\underline{T},t\right)P\left(\vec{Z},t\right)P(\vec{p},t)P(\vec{r},t). \qquad (9.32)$$

However, the distribution function does not factorize in general, it is an assumption. The total transition rate $w\left(\vec{c}+\vec{k},\vec{c}\right)$ is obtained as the sum over contributions of different socio-economic processes:

$$\begin{aligned}w\left(\vec{c}+\vec{k},\vec{c}\right)=&\sum\nolimits_{i,j,k}^{L}w_{ik,ij}^{(1)}\left(\underline{T}+\vec{k}^{(1)},\vec{Z},\vec{p},\vec{r};\underline{T},\vec{Z},\vec{p},\vec{r}\right)\\&+\sum\nolimits_{j}^{L}\left(w_{j+}^{(2)}\left(\underline{T},\vec{Z}+\vec{k}^{(2)},\vec{p},\vec{r};\underline{T},\vec{Z},\vec{p},\vec{r}\right)+w_{j-}^{(2)}\left(\underline{T},\vec{Z}+\vec{k}^{(2)},\vec{p},\vec{r};\underline{T},\vec{Z},\vec{p},\vec{r}\right)\right)\\&+\sum\nolimits_{j}^{L}\left(w_{j+}^{(3)}\left(\underline{T},\vec{Z},\vec{p}+\vec{k}^{(3)},\vec{r};\underline{T},\vec{Z},\vec{p},\vec{r}\right)+w_{j-}^{(3)}\left(\underline{T},\vec{Z},\vec{p}+\vec{k}^{(3)},\vec{r};\underline{T},\vec{Z},\vec{p},\vec{r}\right)\right)\\&+\sum\nolimits_{j}^{L}\left(w_{j+}^{(4)}\left(\underline{T},\vec{Z},\vec{p},\vec{r}+\vec{k}^{(4)};\underline{T},\vec{Z},\vec{p},\vec{r}\right)+w_{j-}^{(4)}\left(\underline{T},\vec{Z},\vec{p},\vec{r}+\vec{k}^{(4)};\underline{T},\vec{Z},\vec{p},\vec{r}\right)\right)\end{aligned} \qquad (9.33)$$

The terms $w_{ik,ij}^{(1)}$ refer to changes in the expenditure flow configuration $\underline{T}$, due to decisions of consumers to change from a state having residence in i and using facilitates in j to a state of still having residence in i but now using facilitates in k. Changes of the residential location are not taken into account, since it is reasonable to assume that the housing mobility will be considerably slower than the shopping mobility. Indeed decisions to buy in another zone can be made much more easily than moving from one housing location to a housing location in another zone. This means it is assumed that the spatial distribution of households can be given exogenously. Since households make their location decisions not only with respect to their place of work, but also in relation to public and private facilities, the problem cannot be so readily decoupled. Feedback effects should be allowed. However, the assumption of a fixed residential location can easily be relaxed.

The terms $w_{j+}^{(l)}$ and $w_{j-}^{(l)}$ describe decisions of entrepreneurs ($l = 2$), retailers ($l = 3$) and land owners ($l = 4$) to make available or to remove one unit of the facility stock Z_j to increase or to decrease the prices p_j and the rents r_j, respectively.

In (9.33) we have neglected contributions of simultaneous transitions to the total transition rate. This means that a transition to a neighbouring state in the configuration space is assumed as a sequential process of single steps, instead of a simultaneous transition of all of them.

Consumer Dynamics

Let $n_{ij}(t)$ be the number of consumers having residence in i and using facilities in j. Then

$$n_i(t) = \sum_{j=1}^{L} n_{ij}(t) \tag{9.34}$$

is the number of consumers living in i and the total number of consumers is given by

$$N(t) = \sum_{i=1}^{L} n_i(t). \tag{9.35}$$

The individual transition rate $p_{ik,ij}(\vec{c})$ for a consumer living in i for changing his shopping area $j \rightarrow k$ is assumed to depend on the expected utility gain $(v_{ik} - v_{ij})$

$$p_{ik,ij}(\vec{c}) = \varepsilon_1(t) \exp\left(v_{ik}(t) - v_{ij}(t)\right), \tag{9.36}$$

where $\varepsilon_1(t)$ is a time scaling parameter. Then the total number of changes of consumer trips $(i,j) \rightarrow (i,k)$ is given by

$$\widetilde{w}_{ik,ij}^{(1)} = n_{ij} p_{ik,ij}(\vec{c}). \tag{9.37}$$

Often, it is reasonable to assume that the expenditure flows $T_{ij}(t)$ can be matched to the consumer numbers $n_{ij}(t)$ via

$$T_{ij}(t) = g(t) n_{ij}(t), \tag{9.38}$$

where $g(t)$ can be seen as the average of individual needs. By a comparison of the total stocks, $g(t)$ can be determined in an easy way. Thus, we are able to derive from the consumer dynamics (9.37) transition rates per unit of time for the expenditure flows

$$w_{ik,ij}^{(1)}\left(\vec{T}+\vec{k}^{(1)},\vec{Z},\vec{p},\vec{r}\right)=T_{ij}\varepsilon_1(t)\exp\left(v_{ik}(t)-v_{ij}(t)\right)\ \text{for}\ \vec{k}^{(1)}$$
$$=\left(0,\ldots,1_{ik},\ldots,(-1)_{ij},\ldots,0\right). \tag{9.39}$$

The transition rates (9.39) have to be inserted into the Master equation, to find a certain expenditure flow pattern. However, the facility stock, prices and rents interact with these flow patterns. Therefore, the other decision processes of the other agents have to be considered as well.

Developers, Retailers and Land Owners

The dynamics of the provision of facilities $\vec{Z}(t)$, prices $\vec{p}(t)$, and rents $\vec{r}(t)$ will be considered together since the same kind of methods can be used. Let $\vec{X}^{(l)}=\left(X_1^{(l)},\ldots,X_L^{(l)}\right)$ be the facility stock configuration $\left(l=2,\vec{X}^{(2)}=\vec{Z}\right)$, the price configuration $\left(l=3,\vec{X}^{(3)}=\vec{p}\right)$ or the rent configuration $\left(l=4,\vec{X}^{(4)}=\vec{r}\right)$, respectively. By definition of the corresponding agents, the configuration $\vec{X}^{(l)}=\left(X_1^{(l)},\ldots,X_L^{(l)}\right)$ will change in the course of time. A rather general formulation for the birth- and death rates reads (Haag 1990; Haag and Wilson 1986)

$$w_{j\pm}^{(l)}\left(\vec{X}^{(l)}+\vec{k}^{(l)},\vec{X}^{(l)}\right)=\frac{1}{2}f_j^{(l)}\left(\vec{X}\right)\exp\left(\pm\Phi_j^{(l)}(\vec{c})\right)\ \text{for}\ \vec{k}^{(l)}=\left(0,\ldots,(\pm1)_j,\ldots0\right) \tag{9.40}$$

for $l=2,3,4$ and for rather general functions $f_j^{(l)}\left(\vec{X}\right)>0$ and $\Phi_j^{(l)}(\vec{c})$. The functions can be modified to include saturation effects and immigration effects as well. The factor $f_j^{(l)}\left(\vec{X}\right)>0$ describes the speed of adjustment. Here, it is assumed that the speed of adjustment is proportional to the realised scale of the stock $\vec{X}^{(l)}$

$$f_j^{(l)}\left(\vec{X}\right)=\varepsilon_l(t)X_j^{(l)}(t). \tag{9.41}$$

The function $\Phi_j^{(l)}(\vec{c})$ takes into account the imbalance between costs of supply and the revenues attracted to facilities in the zone under consideration. If $\Phi_j^{(l)}(\vec{c})>0$ or $\Phi_j^{(l)}(\vec{c})<0$, it is likely that the facility stock expands or decreases, the prices and the rents increase or decrease, respectively. Reasonable assumptions are

$$\Phi_j^{(l)}(\vec{c})=\lambda_l(t)\left(D_j(t)-C_j(t)\right) \tag{9.42}$$

or alternatively

$$\Phi_j^{(l)}(\vec{c})=\lambda_l(t)\left(D_j(t)-C_j(t)\right)/Z_j(t). \tag{9.43}$$

The scaling parameter $\lambda_l(t) > 0$ describes the intensity of response of an agent. The total revenue attracted to j is $D_j(t)$, and the total costs are denoted by $C_j(t)$. If there is an economic surplus $D_j(t) > C_j(t)$ the probability of an extension of the facility stock exceeds the probability of a reduction due to the profit expectations of the agents.

We further assume, that the total costs of supply of facilities in zone j are proportional to the number of units of the facility stock

$$C_j(t) = k_j + r_j(t)Z_j(t), \tag{9.44}$$

where k_j is a fixed constant and $r_j(t)$ is a measure for the rent of one unit of the facilities in zone j. Of course, other cost function can also easily be introduced.

9.2.3 *Deterministic Equations of the Dynamic Service Sector Model*

By inserting the transition rates (9.39) and (9.40) into the Master equation (9.30), approximate equations of motion for the mean values of the expenditure flows $\langle T_{ij} \rangle$, the facility stock $\langle Z_j \rangle$, prices $\langle p_j \rangle$ and land rent $\langle r_j \rangle$ can be derived according to Chap. 4. Therefore, we skip the derivation and proceed immediately to the approximate mean value equations

$$\frac{d\langle T_{ij} \rangle}{dt} = \varepsilon_1(t)\left(\sum_{k=1}^{L} \langle T_{ik} \rangle \exp\left(v_{ij}(t) - v_{ik}(t)\right) - \sum_{k=1}^{L} \langle T_{ij} \rangle \exp\left(v_{ik}(t) - v_{ij}(t)\right)\right) \tag{9.45}$$

$$\frac{d\langle Z_j \rangle}{dt} = \varepsilon_2(t)\langle Z_j \rangle \sinh \lambda_2 \left(\sum_{i=1}^{L} \langle T_{ij} \rangle - \left(k_j + \langle r_j \rangle \langle Z_j \rangle\right)\right) \tag{9.46}$$

$$\frac{d\langle p_j \rangle}{dt} = \varepsilon_3(t)\langle p_j \rangle \sinh \lambda_3 \left(\sum_{i=1}^{L} \langle T_{ij} \rangle - \left(k_j + \langle r_j \rangle \langle Z_j \rangle\right)\right) \tag{9.47}$$

$$\frac{d\langle r_j \rangle}{dt} = \varepsilon_4(t)\langle r_j \rangle \sinh \lambda_4 \left(\sum_{i=1}^{L} \langle T_{ij} \rangle - \left(k_j + \langle r_j \rangle \langle Z_j \rangle\right)\right) \tag{9.48}$$

for $i, j = 1, \ldots, L$.

Equations (9.45)–(9.48) constitute the dynamic service sector model if the dynamic utility functions are specified. Given the initial conditions of the service sector model $\langle T_{ij} \rangle_0$, $\langle Z_j \rangle_0$, $\langle p_j \rangle_0$, and $\langle r_j \rangle_0$, the trajectories of the system for $t > 0$ can be computed. The hyperbolic sine function leads to an amplification of the reaction of the agents on economic disequilibrium.

Near equilibrium (9.46) yields the hypothesis of Harris and Wilson (1978), if we further assume constant prices and rents and situate consumers in the equilibrium state.

By summing up (9.45) over j it can easily be seen that the total expenditure stock $\langle D_i \rangle$ in all zones remains constant:

$$\frac{d\langle D_i \rangle}{dt} = 0 \tag{9.49}$$

for $i = 1, 2, \ldots, L$, and therefore, satisfies a conservation law. If this statement should be relaxed, migration effects of consumers have to be taken into account (see Chap. 8).

The dynamic service sector model comprises very different time scales. For example, the time scale for altering a shopping area $\varepsilon_1(t)$ is certainly quite different from that of changing the facility stock $\varepsilon_2(t)$, the price dynamics $\varepsilon_3(t)$ and the rental dynamics $\varepsilon_4(t)$. Customers have in general no long commitments with the shops and can easily make another choice every day. Therefore, it is a reasonable assumption in certain applications to replace (9.45) by its stationary solution (indicated by a hat):

$$\widehat{T}_{ij} = \widehat{D}_i \frac{\exp(2\widehat{v}_{ij})}{\sum_{k=1}^{L} \exp(2\widehat{v}_{ik})} \tag{9.50}$$

for $i, j = 1, \ldots, L$. Equation (9.50) can easily be verified by inspection. If we insert (9.50) in (9.46)–(9.48), this leads to analytical simplifications and is called adiabatic elimination procedure (Haken 1977, 1983). Using this adiabatic elimination procedure, the $\langle T_{ij} \rangle$ are in equilibrium with the momentary values of $\langle Z_j \rangle_t$, $\langle p_j \rangle_t$ and $\langle r_j \rangle_t$.

The Eqs. (9.46)–(9.48) are in their equilibrium state if

$$\widehat{D}_j = \widehat{C}_j \tag{9.51}$$

for all $j = 1, 2, \ldots, L$ is fulfilled, simultaneously. Therefore, there exists a coupling between all variables.

9.2.4 *Service Sector Model of Harris and Wilson*

By comparison of (9.50) with the special form of the flow pattern, assumed by Harris and Wilson (1978), Clark and Wilson (1983) and Birkin and Wilson (1985), the corresponding utility function $v_{ij}^{(W)}$ belonging to this flow pattern can be derived

$$v_{ij}^{(w)} = \frac{1}{2}\left(\alpha \ln \langle Z_j \rangle - \gamma \ln \langle p_j \rangle - \beta c_{ij}\right). \tag{9.52}$$

Near equilibrium, or in other words, if $\left|\Phi_j(\vec{c})\right| << 1$ and assuming (9.50), (9.52), the Harris and Wilson service sector model is obtained (prices and rents are not considered)

$$T_{ij} = \frac{D_i(t) Z_j^{\alpha}(t) \exp(-\beta c_{ij})}{\sum\limits_{k=1}^{L} Z_k^{\alpha}(t) \exp(-\beta c_{ik})} \tag{9.53}$$

and

$$\frac{d\langle Z_j \rangle}{dt} = \varepsilon_2(t) \langle Z_j \rangle \left(\sum_{i=1}^{L} \langle T_{ij} \rangle - k_j \langle Z_j \rangle \right). \tag{9.54}$$

Therefore, the application of the Master equation framework yields an alternative and systematic derivation of the Harris and Wilson spatial interaction model, now based on microeconomic dynamic decision making and statistical averaging procedures (Haag and Wilson 1986). The Master equation framework has the advantage of introducing explicitly dynamic utility functions involved in the decision process of the consumers.

From an economic point of view, however, it is not satisfactory that prices of goods $\langle p_j \rangle$ and transportation costs c_{ij} are treated in a quite different way (compare the terms $\ln \langle p_j \rangle$ with c_{ij}) in the utility function (9.52). Therefore, an alternative assumption concerning the utility function is proposed

$$v_{ij}^{(H)} = \frac{1}{2}\left(\alpha \ln \langle Z_j \rangle - \beta(\langle p_j \rangle + c_{ij}\right). \tag{9.55}$$

The utility function $v_{ij}^{(H)}$ of consumers having residence in zone i and using facilities in zone j thus depends on the perceived prices of goods $(\langle p_j \rangle + c_{ij})$. It seems to be very plausible to assume that the shopping attitude of a consumer living in zone i will primarily depend on the offer of facilities in a certain retailing zone j and his travelling costs c_{ij} from i to j.

9.2.5 Expenditure Flows and Transportation Costs

The structure of the stationary expenditure flow pattern of (9.50) is highly influenced by the deterrence parameter β, or in other words by the transportation costs. For two important limiting cases $\beta \to 0$, and $\beta \to \infty$, the structure of the flow pattern can be obtained analytically:

(a) Transportation Costs Can Be Neglected
If transportation costs can be neglected compared with other costs $\beta \to 0$, the flow pattern reads

$$\lim_{\beta \to 0} \widehat{T}_{ij} = \frac{\widehat{D}_i \widehat{Z}_j^{\alpha}}{\sum_{k=1}^{L} \widehat{Z}_k^{\alpha}}. \tag{9.56}$$

Obviously, $\beta \to 0$ will lead to high competition between the different zones. We will have the same effect if transportation costs are independent of the distance, since then transportation costs play no rule in the decision process of consumers. Costumers will not hesitate to go shopping in zones far away from their residence. An improvement of the transportation network, especially the radial lines out of cities towards shopping centres located in suburbs intensifies the competition between shopping centres. Thus, if transportation costs can be neglected or are the same for different shopping areas, a situation of fierce competition between shopping centres exists, which may lead to decreasing prices calling for investors activity.

(b) High Transportation Costs
In case of very high transportation costs, $\beta \to \infty$, the flow pattern collapses

$$\lim_{\beta \to \infty} \widehat{T}_{ij} = \widehat{D}_i \delta_{ij}. \tag{9.57}$$

In this limiting case a quite different flow pattern can be observed. For high values of $\beta, \beta \to \infty$ the flows within each zone dominates. On the one hand, this will give all shopping centres a good chance to survive since customers would not travel to other zones because of the high travel costs. On the other hand, because of a lack of competition between the shopping centres, the price dynamic favours increasing prices and decreasing quality measures.

More detailed simulations of this service sector model are given in Haag and Frankhauser (1989).

9.2.6 Concluding Comments

Starting from the Master equation framework, a dynamic service sector model of the decision behaviour of consumers, developers and retailers has been derived. Therefore, a microeconomic foundation has been provided for the macroeconomic equations of motion for expenditure flows, facility stock, prices and rent dynamics. The link thus provides the possibility to estimate the model parameters in analogy to the migration dynamics in Chap. 8. This is an important aspect if the model should be used for real-world applications.

The dynamic service sector model of Harris and Wilson can be related to the decision behaviour of consumers and developers, when it is assumed that consumers are in a stationary state (adiabatic elimination procedure). However, in the Master equation framework, the utility and profit functions can easily be modified and adapted to the data base to provide a more realistic dynamic.

Although it is true that the final stationary state of the system is independent of initial changes in the scale of provision, the long relaxation time of spatial urban systems in general prevents the system from approaching its stationary state. Variations and changes of some system parameters may lead to delayed effects. Therefore, it is reasonable to simulate the full dynamic service sector model (9.45)–(9.48) in real-world applications.

References

Albeverio S, Andrey D, Giordano P, Vancheri A (eds) (2008) The dynamics of complex urban systems. Physica-Verlag, Heidelberg

Allen P, Sanglier M (1978a) Dynamic models and urban growth. J Soc Biol Struct 1:265–280

Allen P, Sanglier M (1978b) Dynamic models and urban growth. J Soc Biol Struct 2:269–278

Allen P, Sanglier M (1979) A dynamic model of growth in a central place system. Geogr Anal 11:256–272

Aoyama Y (1989) A historical review of transport and land-use models in Japan. Transp Res 23A:53–61

Batten DF, Casti J, Thord R (eds) (1995) Networks in action. Springer, Berlin

Batty M (1994) A chronicle of scientific planning: The Anglo-American modelling experience. J Am Plan Assoc 60:7–16

Bertuglia CS, Leonardi G, Wilson AG (1990) Urban dynamics. Designing an integrated model. Routledge, London

Bertuglia CS, Clark GP, Wilson AG (1994) Modelling the city, performance, policy and planning. Routledge, London

Binder J (2007) Modellierung von Pendlerströmen und Pendlerstromänderungen über ein nichtlineares Entscheidungsmodell basierend auf dem Mastergleichungsansatz. Sierke Verlag, Göttingen

Binder J, Haag G, Koller M (2002) Modelling and simulation of migration, regional employment development and regional gross wage payment in Germany: the bottom-up approach. In: Regional economies in transition, Research report 1, University of Trollhättan, Uddevalla

Birkin M, Wilson AG (1985) Some properties of spatial-structural-economic-dynamic models. Working paper 440, School of Geography, University of Leeds

Boyce DE (1988) Renaissance of large-scale models. Pap Reg Sci Assoc 65:1–10

Bundesminister für Verkehr (1998) Modellierung agglomerationsbedingter Einflüsse auf wegebauinduzierte Beförderungsprozesse für die Bundesverkehrswegeplanung (BVWP), Bundesministerium für Verkehr, Bau und Stadtentwicklung (FE-Nr. 96546/98)

Bundesminister Verkehr (1996) Qualifizierung, Quantifizierung und Evaluierung wegebauinduzierter Beförderungsprozesse. FE-Nr 90436/95

Clark M, Wilson AG (1983) The dynamics of urban spatial structure. Progress and problems. J Reg Sci 23:1–18

Courgeau D (1985) Migrants and migrations. Population selected papers. INED, Paris

Courgeau D (1995) Migration theories and behavioural models. Int J Popul Geogr 1:19–27

Dosi G, Freeman C, Fabiani S (1994) The process of economic development: Introducing some stylized facts and theories on technologies, firms and institutions. Ind Corp Chang 3:1–45
Eurosil (1997) European strategic intermodal links. European Commission, FP4-Transport, ST-96-SC1131
Forrester JW (1968) Urban dynamics. MIT Press, Cambridge
Forrester JW (1971) World dynamics. Wright-Allen Press, Cambridge
Fotheringham AS, O'Kelly ME (1989) Spatial interaction models: formulations and applications. Kluwer Academic Publishers, Dordrecht
Gayda S, Boon F, Batty M, Besussi E, Chin N, Haag G, Binder J, Lautso K, Martino A, Noël C, Dormois R (2003) The SCATTER project – sprawling cities and transport: from evaluation to recommendations. Paper presented at the European Transport Conference, Strasbourg
Goodwin PB (1994) Traffic growth and the dynamics of sustainable transport policies. ESRC Transport Studies unit, University of Oxford, Oxford
Goodwin PB (1995) Empirical evidence on induced traffic: a review and synthesis. ESRC Transport Studies unit, University of Oxford, Oxford
Haag G (1986) A stochastic theory for residential and labour mobility including travel networks. In: Nijkamp P (ed) Technological change, employment and spatial dynamics, Lecture notes in economics and mathematical systems, vol 270. Springer, Berlin
Haag G (1989a) Dynamic decision theory: Application to urban and regional topics. Kluwer Academic Publishers, Dordrecht
Haag G (1989b) Spatial interaction models and their micro-foundations. In: Andersson A, Batten B, Johansson B, Nijkamp P (eds) Advances in spatial theory and dynamics. North Holland, Amsterdam
Haag G (1990) Services 2 – a master equation approach. In: Bertuglia CS, Leonardi G, Wilson AG (eds) Urban systems. Design for an integrated dynamic model. Routledge, London
Haag G (2002) Sprawling cities in Germany. FrancoAngeli, Milano
Haag G, Frankhauser P (1989) A stochastic model of intraurban supply and demand structures. In: Hauer J et al (eds) Urban dynamics and spatial choice behaviour. Kluwer Academic, Dordrecht, pp 219–246
Haag G, Gruetzmann K (2001) Transport and urban development. In: Clark G, Madden M (eds) Regional science in business. Springer, Berlin
Haag G, Weidlich W (1984) Model construction in dynamic spatial theory. Proceedings of the task force meeting in dynamic analysis of spatial development. IIASA, Laxenburg
Haag G, Wilson AG (1986) a dynamic service sector model – a master equation's approach with prices and land rents. Working paper 447, School of Geography, University of Leeds
Haken H (1977) Synergetics: an introduction. Springer, Heidelberg
Haken H (1983) Advanced synergetics. Springer, Heidelberg
Harris B (1985) Urban simulation models in regional science. J Reg Sci 25:545–567
Harris B, Wilson AG (1978) Equilibrium values and dynamics of attractiveness terms in production-constrained spatial-interaction models. Environ Plan A 10:371–388
Hautzinger H (1982) Reiseweite- und Reisezeiteffekte von Geschwindigkeitszuwächsen im Personenverkehr. Int Verkehrswesen 34:227–249
Haynes KE, Fotheringham AS (1984) Gravity and spatial interaction models. Sage, Beverly Hills
Hill DM (1965) A growth allocation model for the Boston region. J Am Inst Plann 31:111–120
Hutchinson B, Nijkamp P, Batty M (eds) (1985) Optimization and discrete choice in urban systems. Springer, Berlin
Isard W (1954) Location theory and trade theory: Short-run analysis. Q J Econ 68:305–322
Kutter E (1972) Demographische Determinanten des städtischen Personenverkehrs, Thesis, University of Brunswick
La Barra D (1990) Integrated land use and transport modelling. Cambridge University Press, Cambridge
Lowry IS (1964) A model of metropolies. Rand Corporation, Santa Monica

Meadows DL, Meadows DH (eds) (1973) Towards global equilibrium. Collected papers. Wright-Allen Press, Cambridge
Meadows DL, Behrens WW III, Meadows DH, Naill RF, Randers J, Zahn EK (1974) Dynamics of growth in a finite world. Wright-Allen Press, Cambridge
Meier E (1990) Neuverkehr infolge Ausbau und Veränderung des Verkehrssystems, Schriftenreihe des Instituts für Verkehrsplanung, Transporttechnik Straßen- und Eisenbahnbau der ETH Zürich 81
Mills ES (1972) Studies in the structure of urban economics. John Hopkins University Press, Baltimore
Nelson RR, Winter SG (1982) An evolutionary theory of economic change. Harvard University Press, Cambridge
Nijkamp P, Reggiani A (1992) Interaction, evolution and chaos in space. Springer, New York
Pumain D (1982) La dynamique des Villes. Economica, Paris
Pumain D, Haag G (1991) Urban and regional dynamics-towards an integrated approach. Environ Plan A 23:1301–1313
Pumain D, Sanders L, Saint-Julien T (1989) Villes et Auto-Organisation. Economica, Paris
Rabino GA, Occelli S (1999) Understanding spatial structure from network data: theoretical considerations and applications. Cybergeo, 29
Rosser JB (2011) Complex evolutionary dynamics in urban-regional and ecologic-economic systems: From catastrophe to chaos and beyond. Springer, New York
Sactra (1994a) Trunk roads and the generation of traffic. The Standing Advisory Committee on Trunk Road Assessment. The Department of Traffic, London
Sactra (1994b) Trunk roads and the generation of traffic. Response by the Department of Traffic to the Standing Advisory Committee on Trunk Road Assessment. The Department of Traffic, London
Sactra (1998) Transport investment, transport intensity and economic growth. The Standing Advisory Committee on Trunk Road Assessment. The Department of Traffic, London
Scatter (2001) Sprawling cities and transport: from evaluation to recommendations, European Commission, Key Action 4: City of Tomorrow and Cultural Heritage, EVK4-CT-2001-00063 RS
Stasa (1998) Modellierung agglomerationsbedingter Einflüsse auf wegebauinduzierte Beförderungsprozesse für die Bundesverkehrswegeplanung (BVWP), Bundesministerium für Verkehr, Bau und Stadtentwicklung (FE-Nr. 96546/98)
Stasa (2002) Entwicklung eines Bewertungsmoduls für den induzierten Verkehr und Abschätzung seiner Auswirkungen für den Bundesverkehrswegeplan (BVWP 2003), Bundesministerium für Verkehr, Bau und Stadtentwicklung 2002
Steierwald G, Künne HD (eds) (1993) Stadtverkehrsplanung. Springer, Berlin
Steierwald G, Schönharting J (1993) Personen-Kfz-Verkehr 1990. Nachbarschaftsverband, Stuttgart
Transecon (2000) Urban transport and local socio-economic development. European Commission, GMA1–2000-27049
Webster FV, Bly PH, Paulley NJ (eds) (1988) Urban land-use and transport interaction. Policies and models. Report of the International Study Group on Land-Use/Transport interaction (ISGLUTI) Aldershot, Avebury
Wegener M (1986) Integrated forecasting models of urban and regional systems. In: Batey PW, Madden M (eds) Integrated analysis of regional systems, London papers in regional science, vol 15. Pion, London, pp 9–24
Wegener M (1987) Transport and location in integrated spatial models. In: Nijkamp P, Reichman S (eds) Transportation planning in a changing world. Gower, Aldershot
Wegener M (1994) Operational urban models: state of the art. J Am Plan Assoc 60:17–29
Wegener M (1998) Applied models of urban land use, transport and environment: state of the art and future developments. In: Batten D, Kim TJ, Lundquist L, Mattsson LG (eds) Network infrastructure and the urban environment. Springer, Berlin

Weidlich W, Haag G (1983) Concepts and models of a quantitative sociology: the dynamics of interacting populations, Springer-series synergetics, vol 14. Springer, Berlin
Weidlich W, Haag G (eds) (1999) An integrated model of transport and urban evolution with application to a metropole of an emerging nation. Springer, Berlin
Wilson AG (1970) Entropy in urban and regional modelling. Pion Books, London
Wilson AG (1971) A family of spatial interaction models and associated developments. Environ Plan 3:1–32
Wilson AG (1997) Land use/transport interaction models – past and future. JTEP 32:1

MIX
Papier aus verantwortungsvollen Quellen
Paper from responsible sources
FSC® C105338
FSC
www.fsc.org

If you have any concerns about our products,
you can contact us on
ProductSafety@springernature.com

In case Publisher is established outside the EU,
the EU authorized representative is:
Springer Nature Customer Service Center GmbH
Europaplatz 3, 69115 Heidelberg, Germany

Printed by Libri Plureos GmbH
in Hamburg, Germany